# Turtles, Tortoises and Terrapins

SURVIVORS IN ARMOR

# Turtles, Tortoises

## SURVIVORS IN ARMOR

and Terrapins

*Ronald Orenstein*

FIREFLY BOOKS

*To my parents*
*Charles and Mary Orenstein,*
*with endless thanks and love*

**A FIREFLY BOOK**

Published by Firefly Books (U.S.) Inc. 2001



First Printing

---

U.S. CATALOGING IN PUBLICATION DATA
(Library of Congress Standards)

Orenstein, Ronald.
Turtle, tortoises and terrapins: survivors in armor / Ronald Orenstein. -1st ed.
[304] p. : col. photos. : maps ; cm.
Includes bibliographic references and index.
Summary: Turtle anatomy, habitat, and life cycles through the ages.
ISBN 1-55209-605-X
1. Turtles. I. Title.
597.92 21 2001

---

Published in the United States in 2001 by
Firefly Books (U.S.) Inc.
P.O. Box 1338, Ellicott Station
Buffalo, New York, USA
14205

Published in Canada in 2001 by Key Porter Books Limited.

CONSULTANT EDITORS:

Dr. Jean Mortimer, Chair, Hawksbill Task Force, IUCN Marine Turtle Specialist Group

Dr. Peter C. H. Pritchard, Director. Chelonian Research Institute

Dr. George Zug, Curator Of Herpetology, National Museum of History

Electronic formatting: Jean Peters
Design: Peter Maher

Permission to quote from *The Aye-Aye and I* was kindly granted by the Estate of Gerald Durrell.

Printed and bound in Canada

# Contents

# Acknowledgments

Writing this book has been, among other things, a great learning experience. Many of the people who helped me along the way deserve my thanks not only as advisers but as teachers. My only regret is that the book is not—indeed, cannot—be long enough to include everything they taught me.

My first thanks must go to my three consultant editors, Jeanne Mortimer, Peter Pritchard, and George Zug. All three read the manuscript in its entirety, offering much advice and information and making many valuable suggestions. My visits to Peter's home and George's office were both productive and memorable, and I only wish I could have joined Jeanne in the Seychelles and Aldabra! Nonetheless, Jeanne and I worked together at the CITES Meeting in Nairobi in 2000. I count all three of my editors as valued friends. Of course, any mistakes in this book are my responsibility, not theirs.

I must also thank Anders Rhodin and Marydele Donnelly for their encouragement and help in the early stages, and Anders deserves a vote of thanks from all turtle enthusiasts for spearheading the invaluable journal *Chelonian Conservation and Biology*, which he co-edits with John Behler and Peter Pritchard.

I received help, encouragement, and advice on specific portions of the text from a number of turtle biologists and other experts. Once again, the responsibility for any errors is mine, but my thanks must go to Randall Arauz, Brian Bagatto, John Behler, Peter Bennett, Kurt Buhlmann, John Cann, Joanna Durbin, Lee Durrell, Jack Frazier, Terry Graham, Karen Holloway-Adkins, Donald Jackson, Rod Kennett, Ursula Keuper-Bennett, Faith Kostel-Hughes, Colin Limpus, Ron Nussbaum, Gary Packard, Frank Paladino, Charles Peterson, Grahame Webb, and particularly to Robert Reisz, who not only read over the entire chapter on turtle evolution, but also took the time to invite me to his office and untangle, slowly and carefully, the almost baroque controversy over turtle origins.

My thanks, also, to the many friends and colleagues (and a few total strangers) who answered my questions, provided reprints, chased down obscure bits of information, or simply cheered me up when the slogging got heavy. They include Sandra Altherr, Solomon Aquilera, Sali Bache, George Balazs, April Beer, Allen Bolten, Marny Bonner, Carol Britson, Juan Carlos Cantu, Kathryn Craven, Osha Gray Davidson, Michael deBraga, Claudia Delgado, Kaaren Dickson, Karen Eckert, Scott Eckert, Daniela Freyer, Eugene Gaffney, Hedelvy Guada, Marinus Hoogmoed, Sue Hudson, Douglas Hykle, Frank Ippolito,

R.C. "Hank" Jenkins, Shirley McGreal, Barry Kent Mackay, John Levell, Peter Lutz, Peter Meylan, Ann Michels, Russell Mittermeier, Nicholas Mrosovsky, Jose Truda Palazzo, Doug Perrine, Nicholas Pilcher, Pamela Plotkin, Damaris Rotich, Anne Roy, Allen Salzberg, Chloe Schauble, Jeanette Schnars, Diane Scott, Dionysius S.K. Sharma, Jessica Speart, James Spotila, Teresa Telecky, Wayne Van Devender, John Wehr, Blair Witherington, Frank Bambang Yuwono, and many posters to the Internet discussion list CTURTLE. I'm sure there are others I have forgotten to mention, and to them I offer both my thanks and my apologies.

In particular, I would like to thank the organizers, speakers, poster presenters, and the many other people who attended the 21st Annual Symposium on Sea Turtle Biology and Conservation held in Philalephia in February 2001. For education, enthusiasm, and inspiration, the symposium was hard to beat. This book would be much poorer had I not been able to be there, and my thanks to Dan Morast and the International Wildlife Coalition for making it possible.

To Key Porter Books, and in particular to Susan Renouf and to Janice Zawerbny, my long-suffering editor, my thanks for their patience, their encouragement, and their occasional threats as this seemingly endless project wound down to the wire. Special thanks also to my copy editor Liba Berry.

Finally my thanks to my parents, Charles and Mary Orenstein, without whose support I would have accomplished very little, and to my children, Randy and Jenny, for their (more or less) reasonable tolerance as their father disappeared from view, for days on end, into a world of turtles.

# *A Word About Words*

*"When we were little," the Mock Turtle went on at last, more calmly, though still sobbing a little now and then, "we went to school in the sea. The master was an old Turtle—we used to call him Tortoise—"*

*"Why did you call him Tortoise, if he wasn't one?" Alice asked.*

*"We called him Tortoise because he taught us," said the Mock Turtle angrily: "really you are very dull!"*

—Lewis Carroll (Charles Lutwidge Dodgson), Alice's Adventures in Wonderland

The first problem that faces someone trying to write a popular book about the members of the Order Testudines is what to call them. Different corners of the English-speaking world use the names "turtle," "tortoise," and "terrapin" in contradictory ways. Originally, the word "turtle" meant sea turtles only. Freshwater turtles were "terrapins," and their heavy-footed land-based relatives were "tortoises." That is still the usage in England (where, of course, there are no turtles, except for the occasional sea turtle). South Africans have tended to follow British usage, calling their freshwater turtles "terrapins" even though most of them belong to a quite different branch of the Turtle Order from the terrapins of the northern hemisphere. In North America, the name "terrapin" became restricted to a single species, the diamondback terrapin (*Malaclemys terrapin*). All other freshwater turtles, unless they were specifically called something else, like cooter (an African-derived word) or slider, became simply "turtles."

Meanwhile, in Australia, where "true" tortoises do not exist, everything that was not a sea turtle, including some animals that almost never leave the water, was called a "tortoise." Today, the tendency in Australia is to call them all "turtles" instead.

I can find no way to make sense out of this Babel of usages. Furthermore, I do not want to fall into the pedantic circumlocution of calling all these animals "chelonians" after one of the older technical names for their order, Chelonia. I have therefore decided, for the purposes of this book, to be parochial and adopt the usage of my home continent, North America. In this book, then, when I say "turtles" I mean the whole lot, tortoises and terrapins thrown in. For the names of individual species, I use "terrapin" only for the diamondback terrapin (*Malaclemys terrapin*) of North America, the African mud terrapins of the family Pelomedusidae (following South African usage), and the Asian terrapins of the genera *Batagur* and *Callagur*. I restrict "tortoise" to the "true" tortoises of the family Testudinidae. I use the South

African name "padloper" (which means "trail-walker" in Afrikaans) for the pygmy tortoises of the genus *Homopus*, for no better reason than that I like the way it sounds. My consultant editor, Dr. George Zug, is trying to convince the world to call marine turtles "seaturtles," written as a single word. With the greatest deference and gratitude to him, I will follow a growing growing modern usage and call them sea turtles, written as two words.

There is, unfortunately, no universally recognized list of English names for turtles. I am therefore free, I think, to be somewhat arbitrary. My starting point has been Ernst and Barbour's Turtles of the World, but I have frequently departed from it. For example, for Australian species I follow John Cann's *Australian Freshwater Turtles*, South African names largely follow R. Boycott and O. Bourquin's *The Southern African Tortoise Book*. In general, I do not use seperate English names for turtle subspecies.

I also use scientific names, but to avoid overburdening the text with them I usually only use a scientific name the first time a species is mentioned in any given chapter. Turtle biologists, equally unfortunately, do not always agree on scientific names either, and there is no standard list. To any of them who are reading this, my choice of names here is not intended to be an expression of scientific opinion. Thus, for example, I use *Geochelone* for tortoises that many scientists assigned to *Chelonoidis*, *Dipsochelys* and other genera, because *Geochelone* is still in wide usage, not because I really believe that the South American tortoises belong in the same genus as the leopard tortoise of Africa, the Aldabra tortoise or the Indian star tortoise. On the other hand, I follow the split of the softshell genus *Trionyx* into nine genera, because that change seems to have won broad acceptance in the past ten years.

PREFACE

# *Why Turtles Matter*

"Turtles," writes Anders Rhodin of the Chelonian Research Foundation, "are in terrible trouble."

Few herpetologists—the scientists who study reptiles and amphibians—would disagree with him. There is hardly a place left on earth, on land or sea where turtles are safe. In some places, most particularly in southern Asia where forests and rivers are being swept clean of turtles to supply growing and voracious markets for food and pets, their situation is little short of desperate. Some species have probably already disappeared; more will almost certainly do so, despite efforts we make to save them. Pollution, habitat destruction, overhunting, climate change, and disease strike at species after species. Populations of the largest turtle in the world, the leatherback sea turtle (*Dermochelys coriacea*), collapsed throughout the Pacific Ocean during the last five years of the 20th century. Poachers are stealing the beautiful radiated tortoise (*Geochelone radiata*), even from national parks. The unique Central American river turtle (*Dermatemys mawii*), the only living representative of its family, is being eaten out of existence.

There is more to turtles than most of us know. We think of them as the quintessence of slowness. When Camille Saint-Saëns assigned music to the tortoise in *Carnival of the Animals*, it was Jacques Offenbach's famous cancan—played at a glacial pace. But anyone who lets a careless hand get too close to an angry snapper or softshell will learn just how rapidly a turtle can move. The big-headed turtle (*Platysternon megacephalum*) of Southeast Asia can scale a slippery boulder or even climb into a tree. The pig-nose turtle (*Carettochelys insculpta*) of Australasia can dart away at four times the speed of a swimming human, and a sea turtle can fly through the water with balletic grace.

Imagine that turtles had vanished long ago, with the dinosaurs, and we knew them only from fossils. Surely we would be amazed that such bizarre creatures, sealed in bone, ribs welded to their shells, had existed; had ranged successfully almost throughout the world, in desert, river, and forest, and far out into the open sea; had dug burrows that became homes for other creatures; had a role to play in the habitats where they lived. We would regret

that we had missed the opportunity to see them plodding their way through ancient forests, beneath the feet of monsters.

But turtles, unlike so many other reptiles of past ages, did survive, and for many of us they are a commonplace. Some of us think of them with amusement, as comic-strip characters, plush toys for children, or dancing, top-hatted figures on a box of candy. For others, turtles are a source of food and income, whether from selling a tortoise as a pet or showing tourists a sea turtle laboriously digging its nest in the sand. For some, turtles are even an object of veneration, to be protected and fed on the grounds of a temple. Humankind sees turtles as anything but what they really are: highly evolved, remarkable creatures, necessary components of their shrinking and ever more degraded ecosystems. We in the West have ceased to be amazed by them.

I have written this book because turtles do amaze me. I am not a herpetologist but an ornithologist, a student of birds, and turtles were always on the periphery of my attention. I could not help, though, collecting bits and pieces of information about them, and the more I learned the more astonished I became at the sheer range of adaptation in such superficially humble creatures.

As I have gone on from ornithology to a career in wildlife conservation, and a lobbyist's role in dealing with the excesses of the international wildlife trade, turtles have come more and more into the center of my vision. In recent years, I have found myself supporting the ban on international trade in tortoiseshell, the beautiful scutes of the hawksbill sea turtle (*Eretmochelys imbricata*), and trying to fathom the almost uncontrolled turtle markets of eastern Asia. I have tried to become not just an admirer of turtles but one of their advocates.

If you are not one already, I hope that this book will make you, too, their admirer and their advocate. Turtles should fill you with a sense of wonder, and our treatment of them should fill you with a sense of concern. I know it is entirely unscientific to ascribe human qualities to the processes of evolution, but it is hard not to admire turtles for their sheer doggedness in having made it this far, and this successfully. That they are here for us to wonder at means that we should wonder at them, and make sure that our children and grandchildren have the chance to do the same.

In 1953, the authors of *Reptiles and Amphibians: A Guide to Familiar American Species* wrote that turtles (and lizards and snakes) "are interesting and unusual, although of minor importance. If they should all disappear, it would not make much difference one way or the other." Although we know better today, it is our generation that is presiding over their disappearance.

It is up to us to get turtles out of the terrible trouble we have placed them in. Turtles matter because of what they are, because of the path they have taken, because of their role in the natural world, because of their impact, over the centuries, on our society, culture, and even our religions. They matter because it would be shameful if their long tread through 200 million years of evolutionary history should end through our negligence, our greed, and our failure to act.

CHAPTER ONE

# *The Essential Turtle*

The carapace and plastron of this long-necked turtle (*Chelodina* sp.) from northern Australia are connected by a bony bridge.

I ONCE ASKED a professor of mine, a noted authority on the anatomy and biology of turtles, why he had devoted his life to their study.

"I have great respect," he replied solemnly, "for any animal that can get its shoulder girdle inside its rib cage."

My professor (whose name was Thomas S. Parsons) was only half joking. He had, in fact, put his finger on one of the things that makes turtles so puzzling and fascinating: the fact that almost everything about them, from their inside out, represents a tremendous evolutionary distortion of what we expect from the body of a land vertebrate. That distortion—including the weird relationship between ribs and shoulder girdle that so fascinated Dr. Parsons—is the product of a marvelous series of evolutionary adjustments that

A green sea turtle swims over the reefs of Sipadan Island, Malaysia, beneath a school of bigeye jacks (*Caranx sexfasciatus*).

have refitted turtles for life encased in their single most notable and distinctive feature: the shell. The shell does not merely encase a turtle; to a great extent, it defines what a turtle is. It has meant redesigning much of a turtle's internal anatomy simply to allow it to carry out normal, everyday functions—functions as basic as breathing.

How does a turtle, locked inside its shell, take a breath? Other armored vertebrates, like armadillos, have soft bellies that can move as they inhale or, like the trunkfish puttering around coral reefs, do not have lungs at all. Only turtles face the challenge of filling their lungs while sealed, above and below, in a case of armor. They cannot expand their ribs to breathe, as we do, because their ribs are fused to the shell itself.

As we shall see at the end of this chapter, turtles have found not just one way to breathe, but several. As specialized, constrained, and uniform as turtles seem to be, they have shown, in this and other ways, a remarkable ability to adapt and change. This ability has carried them through 200 million years of evolution, around the world and into almost every conceivable habitat except the polar wastes and the deep sea. In all of these areas, turtles have found ways to thrive. Their shell, far from being a restriction, has been a key to their survival, and to their adaptability.

Therefore, to understand the essential nature of turtles, we must begin with the shell.

### Scutes and Plates

The turtle shell is a composite, built of elements from the surface strata of the skin, others from deeper skin layers, and still others from deep within the turtle's body. It is made up of two main sections: one covering the upper side of the body, called the *carapace*, and a second, the *plastron*, protecting its underside. In most turtles, the carapace and plastron are locked together on each side by a *bridge*. The bridge is more than simply a link; it is a brace. The bridges and plastron form a solid arch of bone across the bottom of the carapace, and make the whole shell considerably harder to crush.

Turtle carapaces may be quite flat or steeply domed, almost circular or elongate (or even vaguely rectangular), smooth and polished or rough, sculptured, saw-backed or even spiny. The plastron may be an extensive,

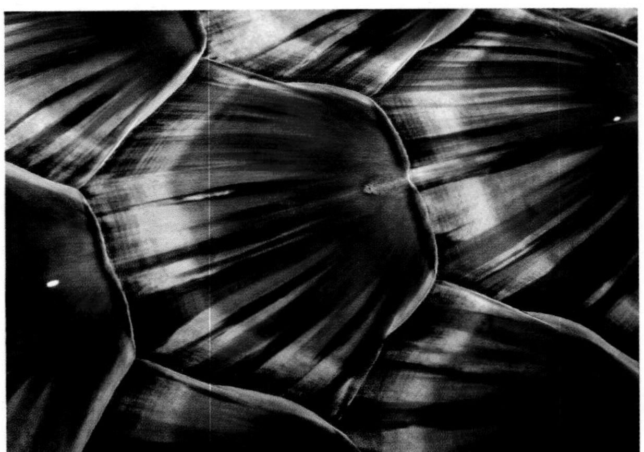

The keratin plates, or scutes, covering a turtle's carapace may be smooth and glossy (green sea turtle *Chelonia mydas*, left) or roughened and ornamented (spurred tortoise *Geochelone sulcata*, right).

The carapace of softshell turtles, like this spiny softshell (*Apalone spinifera*), is covered with leathery skin.

thickened breastplate or a cross-shaped remnant. It is usually flat, though in many male turtles—tortoises, which mate on land, in particular—the middle or rear portion of the plastron is concave—a useful shape for an animal that has to mount a female with an unyielding, rounded carapace.

The outer surface of the shell usually consists of a series of epidermal plates, called *scutes*, arranged in distinctive patterns. A typical turtle has 54 scutes, 38 on the carapace and 16 on the plastron. The scutes around the margin of the carapace are called *marginals*, while the row running down the center of a turtle's back, over its vertebral column, are the *vertebrals* (except for the very first one, which is called the *cervical* or *nuchal* because it overlies the neck). Running down the sides of the carapace between the vertebrals and marginals are a row of four *pleurals*. The scutes of the plastron have names, too (see the diagram on page XXX).

There are variations on this arrangement in some turtles. The number of scutes may differ, even among individuals. Though most turtles have 12 pairs of marginals, mud turtles and musk turtles (Kinosterninae) have only 11 (or 9, according to authorities who consider one pair, the *humerals*, to be single split scutes rather than two separate ones). Their Central American relatives in the subfamily Staurotypinal have only 8 pairs. The alligator snapping turtle (*Macrochelys temminckii*), with some fossil turtles, has an extra row of small scutes above the marginals, called *supramarginals*. Softshell turtles (Trionychidae), the pig-nose turtle (*Carettochelys insculpta*) and the leatherback sea turtle (*Dermochelys coriacea*) have no scutes at all, though there are vestiges of them in very young pig-nose turtles and leatherbacks. Instead, their shells are covered by leathery skin.

Scutes normally abut one another, interlocking like tiles in a mosaic. In some turtles, the scutes on the carapace may overlap. The carapacial scutes of a young hawksbill sea turtle (*Eretmochelys imbricata*) are drawn out, as it grows, into backwards-facing points lying over one other like shingles.

Scutes are really giant scales. They are not made of bone, and for good reason. Bone is highly porous, full of passages for blood ves-

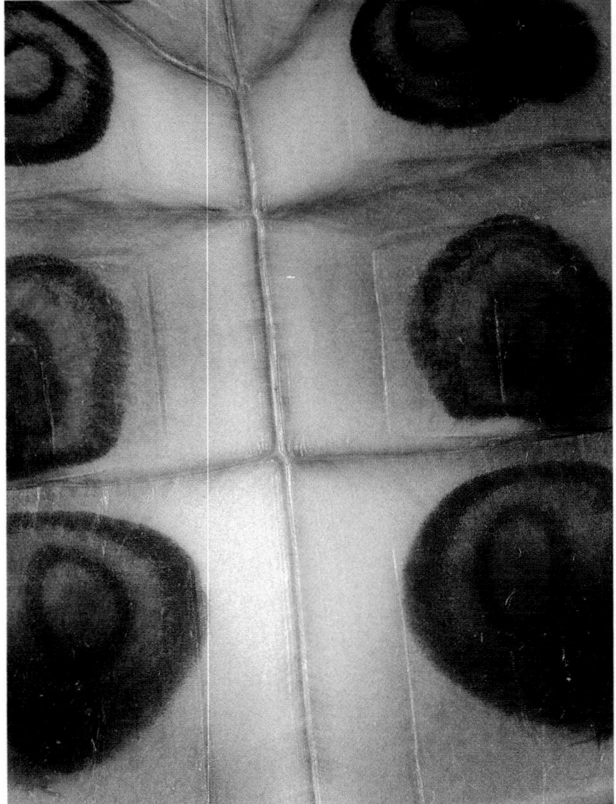

Colored rings form as the plastral scutes of this red-eared slider (*Trachemys scripta elegans*) grow outward from a central nucleus.

Bright colors and patterns adorn the plastra of turtles in many families, including the common toad-headed sideneck (*Phrynops nasutus*, Chelidae, above) and the western painted turtle (*Chrysemys picta bellii*, Emydidae, below).

sels, and weathers rapidly when exposed to the open air. Scutes are made, instead, of a much more durable substance: the protein *keratin*, the same material that makes up your hair and fingernails, the feathers of birds, or the horn of a rhinoceros.

Like other scales, scutes form in the outer layer of skin, the *epidermis*. Turtle epidermis produces two types of keratin: a flexible material called α-keratin found in almost all vertebrates, including ourselves, and β-keratin, a hard and brittle compound found only in reptiles (including birds). The carapace of softshell turtles, pig-nose turtles and leatherbacks is covered by a leathery skin with a surface made entirely of α-keratin. β-keratin, though, is the only kind of keratin in the shell of most turtles. Tortoiseshell—the scutes of the hawksbill sea turtle—is β-keratin. Plates of it can even be welded together.

As each scute grows outward from its central nucleus, or *areola*, deposits of color may form rings or eyespots, as on the plastra of young sliders (*Trachemys scripta*), or rays streaming out from the areolae to form the sunburst patterns of tent tortoises (*Psammobates* spp.) or the Indian star tortoise (*Geochelone elegans*). Each scute develops its

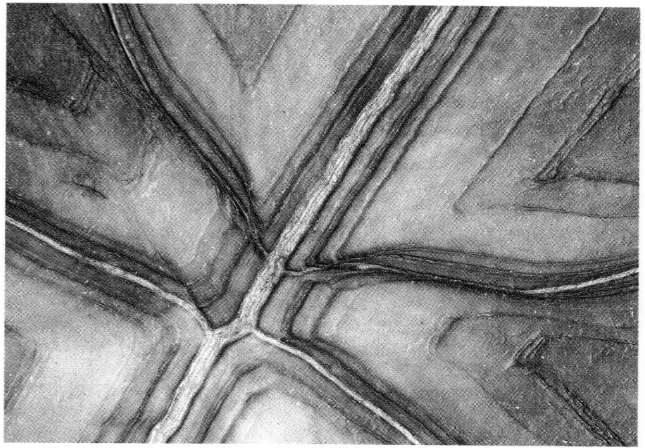

Growth rings, or *annuli*, on the plastron of a helmeted terrapin (*Pelomedusa subrufa*, left) and a Blanding's turtle (*Emydoidea blandingii*, right). The wider the annuli, the more rapidly the turtle has grown.

own pattern. Only rarely do the patterns cross from scute to scute; it is as though each scute is a separately designed tile, independently painted but forming part of a greater mosaic.

Each year, a new scute forms beneath the previous year's. As it is larger than its predecessor, its outer edge shows around the perimeter of the older scute in the form of a growth ring or band (often called an *annulus*). The bands are arranged in concentric polygons around the areola. In strongly seasonal climates, turtles alternate periods of activity and rapid growth with months of inactivity and slow growth. The bands trace this annual cycle of feast and famine, and are pronounced in turtles that shed the outer layer of their scutes at more or less regular intervals. Like the growth rings of a tree, the Annuli can give some idea of the age of their bearer; how accurate this is may vary from species to species, from area to area, from youth to old age, or even from individual to individual. Multiple rings may occasionally form in a single year.

Wide bands show good growth, while narrow bands imply poor growth. A fair-sized Bell's hinged tortoise (*Kinixys belliana*) from Africa may have only a few annuli on the scutes of its plastron, suggesting that it has grown very rapidly over a short time. By contrast, a spider tortoise (*Pyxis arachnoides*) from Madagascar, a much smaller animal, may have numerous bands closely packed together, a sign that it has taken many years to reach even its diminutive proportions.

## The (Very) Old Shell Game

If you could peel the scutes away, you would uncover another layer of plates, this time made of bone. There are usually 59 of them, though the exact count may vary: 50 in the carapace, and 9 in the plastron. They are arranged in a pattern similar to the arrangement of the scutes, but not, as you might expect, identical to it. Every scute is not exactly matched to a bone below. This mismatch adds to the overall strength of the shell, like the different layers in a sheet of plywood.

The bones of the shell have their own names. Those around the margin of the carapace are not called marginals but *peripherals*. The bones underlying the vertebral scutes are called *neurals* because they correspond to the neural spines of the vertebrae, the arches

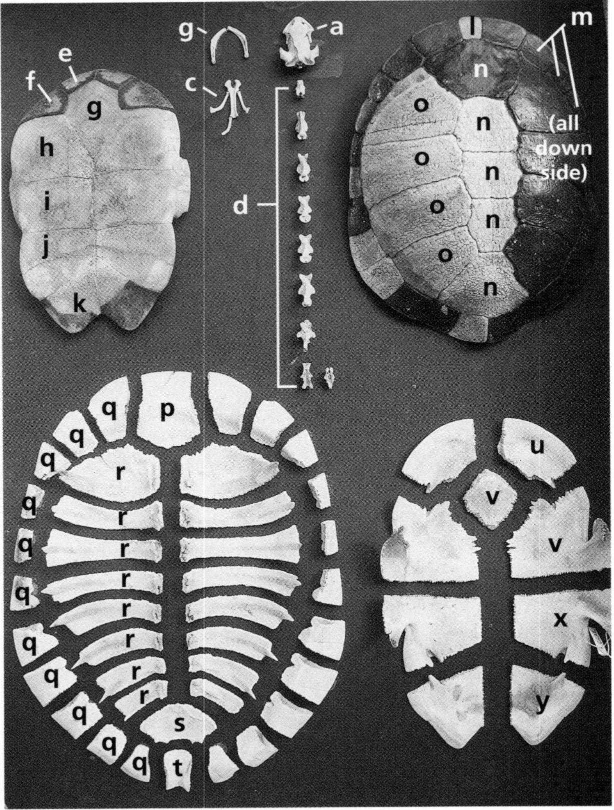


The shell and some skeletal elements of an Australian snake-necked turtle (*Chelodina longicollis*). **a.** *skull*; **b.** *mandible*; **c.** *hyoid apparatus*; **d.** *vertebrae.* Scutes of the plastron [upper left]: **e.** *gular*; **f.** *humeral*; **g.** *intergular* [not present on all turtles]; **h.** *pectoral*; **i.** *abdominal*; **j.** *femoral*; **k.** *anal.* Scutes of the carapace [upper right]: **l.** *cervical*; **m.** *marginals*; **n.** *vertebrals*; **o.** *pleurals.* Bones of the carapace [lower left]: **p.** *nuchal*; **q.** *peripherals*; **r.** costals; **s.** *suprapygial*; **t.** *pygial* (NB: the neural bones are either not visible or absent in chelid turtles, including *Chelodina*). Bones of the plastron [lower right]: **u.** *epiplastron*; **v.** *entoplastron*; **w.** *hyoplastron*; **x.** *hypoplastron*; **y.** *xiphiplastron.*

through which the spinal cord passes. Between the neurals and the vertebrals lie the *costals*, so-called because they overlie the ribs, or *costae.* A few of the other bones of the carapace have individual names, and the bones of the plastron all have names ending, quite reasonably, in "plastron."

Bone is an invention of the vertebrates. There is nothing particularly unusual about a vertebrate covered with bony armor. There are armored fishes, armored lizards, armored crocodiles, and armored mammals. The armor of turtles is unusual because of the kind of bone that goes into its makeup.

We soft-skinned humans tend to think of bone as something that forms deep within our bodies. The very earliest bone to evolve, though, actually formed within the skin. More specifically, it formed within the underlayer of the skin called the *dermis*; and so it is known as *dermal bone.* The bones of our face are largely dermal, and so is the only part of our skeleton we can actually see, the enamel of our teeth.

The first fishes, 450 million years ago, had a skeleton of dermal bone around the outside of their bodies, but no bone at all on the inside. Internal bone did not come along for millions of years, and when it appeared it formed in a completely different way. The earliest internal skeletons—like the skeletons of sharks today—were formed not of bone but of cartilage. The first internal bone either formed around this cartilage or within it. Bone that forms within cartilage, like the bones of our vertebral column, shoulder and hip girdles, and arms and legs, is called *endochondral bone.*

All of the armored vertebrates we know of, except turtles, have their armor composed entirely of dermal bone. The turtle shell, alone among all the types of armor that vertebrates have developed since the first fishes swam, combines dermal bone and endochondral bone. As it has evolved, portions of the turtle's internal skeleton have migrated towards the surface.

You can see this most easily on the inside of an empty turtle carapace. Here, its vertebral column and ribs stand out like roof beams, bracing the shell. The ribs are fused to the dermal plates. So are most of the 10 trunk vertebrae; only the first and last are merely attached to the carapace rather than fused with it. This is the source of Thomas Parsons's paradox: in order for the ribs to become the supporting buttresses of the carapace, they have to lie not inside the shoulder and hip girdles, but above and outside them.

How do they get there? As an embryo turtle develops within its shell, the scutes form before the bony plates beneath them. Only late in development do the ribs, still formed of cartilage, begin to grow from the embryonic vertebral column. As they reach outward, they actually pierce the dermis—an invasion by endochondral bone into the realm where dermal bone is formed. Still later, plates of bone, including the future peripherals, begin to form a ring around the edges of the carapace. At this stage, the combination of the vertebral column, ribs, and encircling plates resembles a carriage wheel. The enclosed spaces between the ribs and the shell margin—between the spokes of the wheel—are called *fontanelles.* In a few species the fontanelles remain more or less open, but in most turtles they eventually fill in with dermal bone, spreading from rib to rib until the carapace is complete. Hatchling turtles usually have another space, or fontanelle, in the center of the plastron. In some turtles it, too, never closes completely.

There are endochondral bones in the plastron, as well, though their identity is much less obvious. Two of the plates at the front end of the plastron may be the transformed remnants of the clavicles (the bones that form our collarbone) and the coracoids, bones attached to the shoulder girdle in other reptiles. The rest of the plastron appears to be purely dermal bone, an entirely new structure with no equivalent in the skeletons of other reptiles. Just possibly, though, they may be partly derived from abdominal ribs, or *gastralia*, that float freely in the underside of some reptiles where we have our breastbone.

**What Are Turtle Shells Good For?**

If there is any question about turtles that might not seem to need answering, surely it is this one. The turtle shell is, as almost anyone would tell you, it's "house"—the place the turtle retreats to for protection from its enemies. Turtle shells certainly do protect their owners against attack by predators. This is probably true even for sea turtles, which are incapable of withdrawing their head and limbs. This obvious answer, though, is not the only one.

A turtle or tortoise on land may spend, or its ancestors may once have spent, much of their lives under the feet of herds of bison or troops of wildebeest. They may have had less to fear from being eaten than from being stepped on. The ornate box turtle (*Terrapene ornata*) of central North America has a taste for the insects in dung, and risks being trampled as it roots for them on cattle trails. Its high, domed carapace may be harder to crush than the flattened shells of its more aquatic relatives. In a tortoise, the bony bridge and the plastron combine to lock the sides of the carapace together, making it even more difficult to crush from above. Some tortoises, like the leopard tortoise (*Geochelone pardalis*) of Africa, further protect themselves by growing rapidly to a size too great for even the

The domed shape of the carapace of this ornate box turtle (*Terrapene ornata*) makes it harder to crush if the animal is stepped on.

most negligent ungulate to miss.

Box turtles, though, have lost the bony bridge, and with it some of the structural rigidity of a completely braced shell. Possibly to compensate, they have fused the bones of their carapace into a single unit, a process called *ankylosis*. This fusion may make the shell less likely to spread apart and split open if something steps on it. It may, however, also make it easier to shatter. If a crack starts, it may continue right across the carapace instead of stopping at a suture between separate plates of bone.

Ankylosis comes with a cost. The bones of the shell grow at their edges, so without those edges, a turtle can no longer grow. Once the process of ankylosis is completed, an individual box turtle is as big as it will ever get. Box turtles, even when full-grown, are fairly small animals; their shoulder girdles, which must rock backwards when they close their shells, cannot support much weight anyway. The trade-off in increased shell strength may make up for their inability to grow large enough for a bison to notice.

Oddly enough, the Central American river turtle (*Dermatemys mawii*), which almost never comes out of the water, also has a fully ankylosed shell at maturity. So do two highly aquatic Asian turtles, the river terrapin (*Batagur baska*) and the painted terrapin (*Callagur borneoensis*). Why, we don't know. The only tortoises to have full ankylosis belonged to an extinct genus, *Cylindraspis*, from the Mascarene Islands in the Indian Ocean. *Cylindraspis* tortoises were medium to large animals, sharing the islands with nothing larger than turkey-size flightless birds like the dodo. It is hard to imagine what they did with the extra strength, assuming they

The broken carapace of this Florida red-bellied turtle (*Pseudemys nelsoni*) has received some human first aid.

had it at all; the shells of *Cylindraspis* tortoises were quite thin.

In tortoises, the shell may be as much for offense as for defense. Many male tortoises use the thickened leading edge of the plastron as a butting and ramming weapon, either to drive off a rival or to batter a female into submission. In a few species, including the angulate tortoise (*Chersina angulata*) of South Africa and, especially, the rare ploughshare tortoise or angonoka (*Geochelone yniphora*) of Madagascar, the front of the male's plastron juts forward to form a prominent and highly effective hook. This hook can be a lever to flip a rival over, or even a goring weapon that can cause serious injury. The more inoffensive Asian brown tortoise (*Manouria emys*) uses its thickened gulars to help it burrow in leaf litter, and the Texas tortoise (*Gopherus berlandieri*) uses them to help scoop out its resting place, or *pallet*, in the soil (though this does not explain why they are larger in males).

The shell can provide protection against the elements. Most tortoises live in open country, where desiccation in the hot sun is a real threat. By greatly reducing the area of skin that it exposes to the sun, its shell may keep a tortoise from drying out. Moreover, the shells of many tortoises and box turtles may help them to survive the fires that can sweep their habitats. In northern Greece, fires caused a 40 percent mortality rate in a population of Hermann's tortoise (*Testudo hermanni*), but large adults were able to survive even if all their scutes burned away. Healing and reconstruction, even in such tough animals as these, has its limits. If an adult box turtle is damaged by fire, when its shell heals it may no longer have separate, identifiable scutes.

The shell of this Gulf Coast box turtle (*Terrapene carolina major*) has been damaged by fire, but the turtle has survived.

Even sea turtles, which surely cannot encounter fire very often, may be able to recover from burning. Some indigenous people used to remove hawksbill sea turtle scutes by roasting the animal alive over a fire, and then return the victim to the sea in the belief that the animal would heal and grow a new set of scutes. How many turtles survived this, though, is unknown.

There are more subtle uses for a turtle's shell. The shell of the North American painted turtle (*Chrysemys picta*) may be crucial to its ability—remarkable for an air-breathing animal—to spend months every winter beneath the ice of a frozen pond. During those long weeks the turtle functions entirely without breathing air, and to a very great extent without oxygen at all (see Chapter 5). The lactic acid that builds up in its tissues as a result of this oxygen deprivation would surely kill it, or drive it to the surface, were it not for the vast storehouse of minerals in the bones of its shell. These minerals act as buffering agents, neutralizing the acid before it can build to harmful levels.

The critical importance of shell minerals to the survival of the painted turtle is a fairly recent discovery, although the painted turtle has been the most intensively studied of all turtles. Perhaps equally vital reasons for turtles to have shells, especially for the many turtles that we know less well, remain to be discovered.

### Variations on a Theme

Turtle shells have changed little in their basic architecture since the earliest days of the dinosaurs. That does not mean that all turtle shells are alike. Variations on the basic shell theme have had much to do with the adaptability of the turtle body plan. Most land-dwelling turtles have higher, more domed

This juvenile spiny turtle (*Heosemys spinosa*) from Southeast Asia bears a spiked carapace and a beautifully-patterned plastron.

One of the "sawbacks," the black-knobbed map turtle (*Graptemys nigrinoda*) of rivers in Alabama and Mississippi.

shells than their water-living relatives. Turtles that spend much of their lives swimming tend to have flat, smooth shells, the better to slice through the water. The round, flattened shells of many softshell turtles cut through water like a discus through air. They are not, though, of a shape that is stable in a crosscurrent, and while some softshells enter the ocean, they generally avoid the open sea. Sea turtle shells, on the other hand, are elongate teardrops, a highly efficient shape for reducing drag and ensuring stability. The sea turtle with the most elongate shell, the leatherback, is also the one that ranges most widely through the open ocean (though the green sea turtle is a better swimmer than the hawksbill, which has the more elongate shell).

Some turtles have highly sculptured shells. The spiny turtle of Southeast Asia (*Heosemys spinosa*) has the scutes of its carapace produced into prickly spikes that may not only make it an unpleasant mouthful for a predator, but a difficult animal to detect in the leaf litter of the forest floor. Among the North American map turtles (*Graptemys*), the so-called "sawbacks" have a saw-toothed keel running down the midline of the carapace. The scutes of a wood turtle (*Clemmys insculpta*) carapace are heaped up into rough pyramids. This does not involve the underlying bone; but in tortoises the bone itself may be raised into humps. Leopard tortoises in the Serengeti frequently have humped shells, perhaps because they get more protein than usual by scavenging the carcasses of large mammals or eating the feces of Serengeti carnivores. However, some tortoises, including the Indian star tortoise and the tent tortoise (*Psammobates tentorius*) of South Africa, develop humps as a matter of course as they age. These rough growths may save the tortoise's life. A tortoise on its back may die from heatstroke if it cannot get out of the sun, and of course is vulnerable to any predator that happens by. If it becomes overturned accidentally, the humps may tilt its body to one side, making it easier for the tortoise to flip itself back over and continue on its way.

Some Asian river turtles (Bataguridae), including the closely related river terrapin and painted terrapin, have paired, internal sheets of bone running like buttresses from carapace to plastron, one pair at the front of the shell and one near the rear. These may allow the shell to withstand the crushing bites of crocodiles. Female river and painted terrapins have solid bony shells, but the smaller males retain throughout life a series of fontanelles along each side, like a row of portholes. In a living terrapin these openings are covered by the scutes; the porthole effect is only visible in a skeleton. What these openings are for—assuming they have any function at all, and are not simply a reflection of the fact that males mature at a smaller size than females—no one knows, though they may, somehow, allow a male terrapin to sense vibrations in the water. Why the male should need to do this, and not the female, is hard to imagine, unless it somehow helps him find a mate.

The Central American river turtle has such thin scutes that they can hardly be called armor at all. They are very easily damaged or abraded, and once injured heal very poorly. The leathery skin that replaces the scutes of a softshell turtle may increase its ability to function beneath the surface. Many aquatic turtles can take up oxygen through their skin when they are underwater (see Chapter 5), and the greater the area of skin, the more oxygen a softshell may be able to absorb. The shells of softshell turtles, by the way, are not really soft, except at the edges; most of their bony plates are quite thick and solid. They are rubbery around the rim, though, because softshells, except for the Indian and Burmese flapshells (*Lissemys* spp.), have lost the peripheral bones around the edge of the carapace (see Chapter 3).

For the ultimate in bone loss, though, we must turn to the leatherback. This largest of all turtles has almost no identifiable bones in its shell. Even its ribs only lie up against the underside of its carapace, rather than being.embedded in the carapace, as in all other turtles. Instead, its shell is made of thick, tough connective tissue covered by leathery skin, lying over a layer of thousands of tiny, separate bony plates, some as little as 3 mm (.12 in) thick, arranged like the tiles in a mosaic. The largest of these bones lie along the seven ridges running down the length of the carapace. When a leatherback dies and rots, its shell simply disintegrates.

**Flexible Shells**

Hatchling turtles have quite flexible shells; after all, they have to curl enough to fit inside an egg. Normally, however, as a turtle grows, the sutures where the bones of its shell meet become rigid and immovable. Their edges weave back and forth in a complex series of interlocking fingers, like dozens of tiny mortise and tenon joints. The soft tissue between the bones diminishes or, in a few species, even disappears.

In many turtles, though, this seemingly rigid suit of armor is not entirely inflexible, even in adults. Snapping turtles and the bigheaded turtle (*Platysternon megacephalum*) of Southeast Asia can flex the plastron slightly along the midline, at the bridges. A sea turtle—especially a leatherback, whose bony plastron is reduced to a ring of splints—can make its whole plastron bulge. If you turn one on its back (something that, of course, you should not do), you can see its chest heave as it breathes.

African hinged tortoises, like this Bell's hinged tortoise (*Kinixys belliana*), are the only turtles with a hinge across the carapace.

A Malayan box turtle (*Cuora amboinensis*) seals itself in its shell by raising its plastral hinges.

Turtles from five different families have developed hinges, with associated muscles, that allow portions of their shells to move up or down. These hinges are almost always somewhere on the plastron, except in the hinged tortoises of the African genus *Kinixys*, which have theirs across the rear portion of the carapace, between the 7th and 8th marginal scutes. Turtles that sport hinges do not usually have them when they hatch (the spider tortoise of Madagascar is an exception), but develop them as they grow. Before a hinge can develop, the scutes and underlying bones must line up in the same spot. In hinged tortoises, the scutes and the bones along the hinge lines do not finish lining up until later in their development, and some individuals never develop a hinge at all.

Hinges have evolved over and over again, even within a single family like the Asian river turtles (Bataguridae). In the spider tortoise, the two southern subspecies have a hinge, while the third, which lives to the north, has none. In some tortoises of the genus *Testudo*, including the spur-thighed tortoise (*Testudo graeca*), females have hinges but males do not.

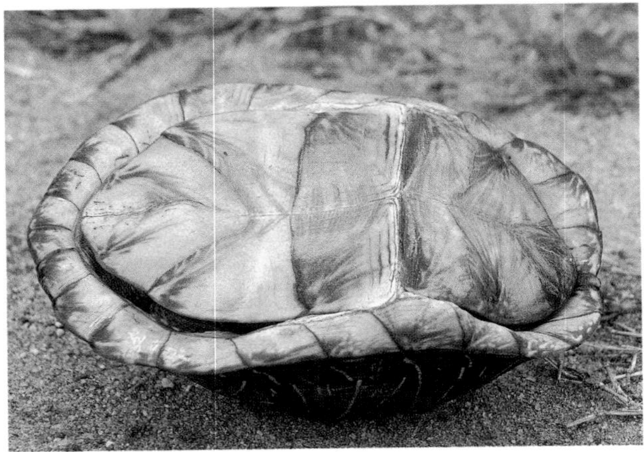

Box turtles have a single hinge across the plastron (Florida box turtle *Terrapene carolina baurii*, above); mud turtles have two, one on either side of the bridge (eastern mud turtle *Kinosternon subrubrum*, below).

There is usually only one hinge on the plastron, on one side of the bridge or the other. Of course, that means that only one lobe of the plastron can normally move, because the bridge locks the other one in place. However, a number of American pond turtles (Emydidae) and Asian river turtles have solved this problem by replacing the bony bridge with a flexible, fibrous ligament. That permits animals like the emydid American box turtles (*Terrapene* spp.) and the batagurid Asian box turtles (*Cuora* spp.) to close both lobes of their plastra around a single hinge. Mud turtles (*Kinosternon* spp.) keep the bony bridge, but have two hinges, one in front of the bridge and one behind it.

A hinge seems an obvious antipredator device, the equivalent of a portcullis or a drawbridge. This, though, may not be its only function. The hinge on a spider tortoise, for example, is well developed, but the muscles that control it, and would be needed to raise the "drawbridge", seem quite weak. In Central America, the white-lipped mud turtle (*Kinosternon leucostomum*) is partly terrestrial, while its larger relative (and frequent predator), the Mexican or northern giant musk turtle (*Staurotypus triporcatus*), is a full-time water-dweller. The white-lipped mud turtle has a much more extensive plastron than the giant musk, with two hinges, one on either side of the bridge, that allow its shell to close completely. This may be an adaptation to avoid drying out during its time on land.

Some turtles may need a hinge, or a flexible shell, to pull the head in at one end or to pass eggs out at the other. Though snapping turtles lack a hinge, their slightly flexible plastral joints make it easier for them to retract their very large heads and fold up their long necks, and their reduced, cross-shaped plastra leave "bulging room" around the bases of the limbs. Giant musk turtles have large heads and a powerful bite. When they retract their heads into their shells, they open their mouths widely as a threat to any attacker. If they could not lower the front of the plastron around the hinge, there would not be enough room to fit their gaping heads and enormous necks within the shell. Their relative, the narrow-bridged musk turtle (*Claudius angustatus*), does manage this without a hinge; it pivots the anterior part of the plastron downwards instead.

The need to lay a fairly large egg may explain why female spur-thighed tortoises, but not males, have a hinge on the rear lobe of the plastron, though their close relative, the Egyptian tortoise (*Testudo kleinmanni*), has a hinge in both sexes. Some small turtles in the family Chelidae, such as the twist-necked turtle (*Platemys platycephala*) of South America, lay large eggs for their size but do not have a hinge. Like box turtles, their bridges are fibrous, allowing the space between the carapace and the plastron to stretch a bit during egg-laying. The neotropical wood turtles of the genus *Rhinoclemmys* lay enormous, hard-shelled eggs; a Guyana wood turtle (*Rhinoclemmys punctularia*), with a carapace 20 cm (8 in) long, may lay an egg nearly 7 cm (2.8 in) long and 4 cm (1.6 in) wide. They can flex the sutures between the bones at the posterior end of the carapace and plastron, possibly to make room for the egg.

The shell of the Sonoran wood turtle (*Rhinoclemmys pulcherrima*) may have to flex to allow its enormous egg to be laid.

Turtles have other ways of getting around the problem of how to lay a large egg. The yellow-footed tortoise (*Geochelone denticulata*) of South America has a semicircular notch at the rear edge of its plastron, just large enough to allow an egg to fit through. A number of turtles and, particularly, tortoises, have simply changed the shape of their eggs from spherical to long and narrow, making it easier for a large egg to fit through what may be a very small space between the carapace and the plastron.

Though it lacks a hinge, the Indian flapshell (*Lissemys punctata*), one of the soft-shelled turtles, can close its shell openings by flexing the rear part of its carapace downward, and the front part of its plastron upward, with the help of a pair of fleshy flaps that it pulls up to conceal its hindlimbs. The ultimate in shell flexibility, though, is reached by two

The Indian flapshell (*Lissemys punctata*) has fleshy flaps on its plastron that can be drawn up to conceal its hindlimbs.

The shell of the Malayan softshell (*Dogania subplana*) must bulge to make room for its enormous head and neck.

The flexible shell of the pancake tortoise (*Malacochersus tornieri*) locks the animal in place in its rocky crevice.

remarkable and quite different species, the Malayan softshell (*Dogania subplana*) in Asia and the pancake tortoise (*Malacochersus tornieri*) in Africa.

Though softshell turtles have a fair bit of flexibility between the bones of their plastra, their bony carapaces are usually quite rigid. The exception is the Malayan softshell. Its individual shell bones are rigid enough, but every bone in a Malayan softshell's carapace is free to move. No other turtle in the world has a shell that can do this—not even a leatherback, whose shell is bound together by a heavy layer of fibrous tissue into a tough and fairly inflexible unit. The Malayan softshell may need such a flexible shell because it has a relatively enormous head and long neck. When it draws them back into its body, its whole shell must bulge to let them in. This extreme flexibility may also help it slip into a crevice, or under a boulder in the bed of a forest stream.

The ability to hide in a crevice—though on dry land rather than in the water—is the raison d'être for one of the most peculiar of all turtle shells, that of the pancake tortoise of Kenya and Tanzania. Pancake tortoises spend much of their lives in deep crevices on rocky outcrops. If disturbed, they wedge themselves so tightly in place that it is difficult to remove them. The tortoise does this by bracing its hind legs, forcing the rear of its carapace against the ceiling of the crevice. If the opening is narrow enough, it pulls in its legs as tightly as it can, forcing its carapace and plastron to bulge outward and lock the tortoise in place.

This locking mechanism requires extreme flexibility, but the pancake tortoise does not achieve this by a system of hinges or by loosening the joints between the bones of its shell. Instead, it has reduced the bones themselves to such a remarkable degree that they have become flexible. The peripheral bones that ring its carapace are as solid and rigidly locked together as in any other tortoise, but the neural and pleural bones that form its center are reduced to paper thinness and penetrated by a series of fontanelles. The fontanelle in the middle of its plastron is so large that, when the turtle braces itself, its body bulges through it like a balloon. Far from being a rigid, armored creature, a pancake tortoise—though it looks solid enough from the outside—is so soft and pliable that when you hold one in your hand, you can feel its heart beating through its shell.

### Turtle Heads, How to Hide Them, and Other Matters

Though the shell is certainly the most distinctive and unique feature turtles have, it is not the only thing that separates them from other living reptiles. Turtles have a much more solid and rigid skull than, for example, lizards and snakes. The evolution of snakes has led to a skull in which every bone but the braincase can move freely, an adaptation to swallowing prey that is much wider than the snake itself. Turtle evolution has gone in the other direction. In turtle skulls, the bones are sutured together into a rigid, immovable whole. Its form can be remarkably varied; though most turtle skulls would be immediately recognizable even to a non-expert, a few might give a trained zoologist pause.

The elongate skulls of softshell turtles reach an extreme in the narrow-headed softshells (*Chitra* spp.) of southern Asia. Their skulls are ridiculously slender, and carry the eye sockets, or *orbits*, so far forward that they can be easily mistaken for oversize nostrils. This peculiar placement means that the turtle can bury its enormous body in the mud and still watch for prey, with only a tiny snout tip projecting into the open to reveal the presence of the more than 100 kg (200 lb) of predator concealed beneath.

Even more peculiar-looking is the skull of the matamata (*Chelus fimbriatus*) of northern South America. It looks as though the turtle's nose had been run over by a steamroller. Its front half is so thoroughly flattened that it is less than a centimeter thick, even in an adult

The skull of a common snapping turtle (*Chelydra serpentina*) has deep emarginations on its posterior edge.

weighing some 10 kg (22 lb). The back half, by contrast, sweeps upward at a steep angle, and sports on each side a huge projecting flange like the decorations on a particularly extravagant murex snail. These flanges bear the turtle's ears, spacing them widely apart, a positioning that may help the matamata home in on its prey in murky river waters. You might be forgiven for overlooking the tiny space, in the midst of this riot of bone, reserved for the animal's brain.

In most living turtles, deep embayments, or *emarginations*, cut into the back of the skull on either side to make room for the jaw muscles (in the Chelidae, the family of turtles that includes the matamata, the embayment sweeps up instead from the side of the skull in front of the ear). However, in some turtles, including sea turtles and the big-headed turtle (*Platysternon megacephalum*), the embayments are greatly reduced or even absent, making for a skull with a complete, solid bony roof. It is probably not a coincidence that these are turtles that cannot withdraw their heads into their shells, either because (in the case of the big-

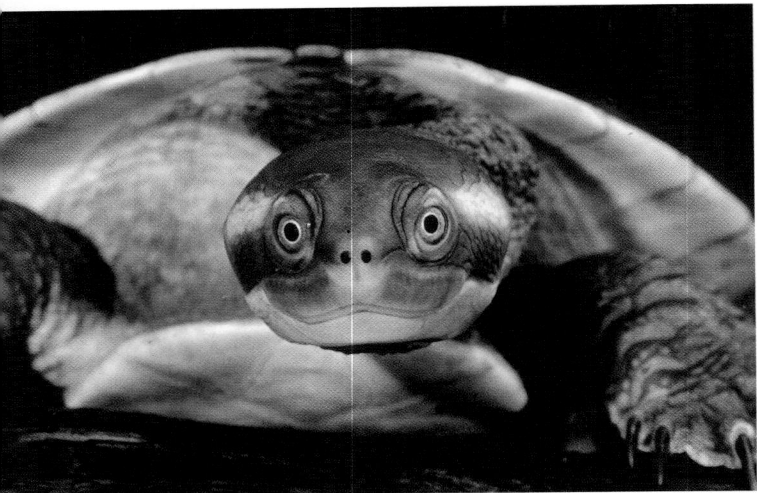

Turtle sight is quite keen. This is Krefft's turtle (*Emydura krefftii*) from northeastern Australia.

headed turtle) the head is simply too large, or (in sea turtles) because the requirements of streamlining have meant that the neckline flows smoothly into the shell, leaving no place to tuck the head out of the way.

No living turtle has teeth. Some of the very earliest fossil turtles had teeth in the center of the palate, but no known turtle, living or fossil, has teeth in what we think of as the normal place, along the edges of the jaws. Instead of teeth, the jaws of the turtle are lined with sheaths of keratin, either sharpened into cutting edges or broadened into crushing plates, depending on what the turtle does with them. A baby turtle in the egg has another hard, sharp bit of keratin, the *egg-tooth* or *egg-carbuncle*, on its snout. It uses it to slice its way out of the eggshell; a few days later, its job done, the egg-carbuncle falls off.

Turtle sight is quite keen, at least at close range, both above and below the water. We know from experiments that they are capable of color vision, particularly at the red end of the spectrum.

Turtles also have a good sense of smell. Like other reptiles, they have specialized structure called a Jacobson's organ that allows them to detect tiny chemical particles; in turtles, Jacobson's organ is located in the same chamber as the animal's nasal membranes.

How well they hear is difficult to judge. Turtles lack an external ear opening. The ear of modern turtles is surrounded by the *otic capsule*, a bony box found in no other reptile. The eardrum, or *tympanum*, is covered with scales. Turtles are, though, acutely aware of low-frequency vibrations in the water and, to a lesser extent, on land. It is probably this awareness, rather than hearing as we experience it, that sends a turtle slipping into the water long before you can approach it.

Turtles have extremely flexible necks. This flexibility perhaps makes up for the rigidity of their shells. Turtle necks range from very short, as in sea turtles, to remarkably long and snakelike, as in the Australian snake-necked turtle (*Chelodina longicollis*). Long or short, all turtle necks have the same number of vertebrae—eight. This kind of consistency is unusual in reptiles, but has a parallel in mammals: all mammals, except sloths, have only seven cervical vertebrae, whether you are talking about a giraffe or a whale.

Necks, or the way turtles use them, are a key to their relationships. Living turtles fall into two fundamental groups, based on the way they tuck their heads beneath their shells. By far the larger group accomplishes this by bending the neck into a vertical s-curve, pulling the head straight back into a slot in the body cavity. Since this action tucks the neck completely out of sight, the group is known as the *Cryptodira* or hidden-necks—though some cryptodires, including sea turtles, cannot tuck their heads in at all. All the turtles of North America and Europe, all but one of the turtles of Asia, all sea turtles, and

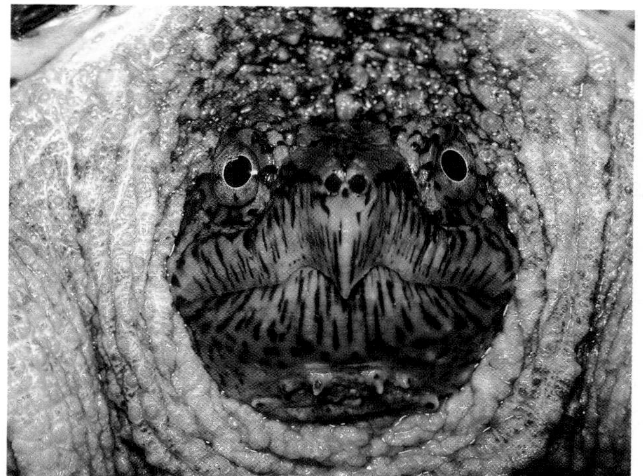

Sidenecked turtles or pleurodires (helmeted terrapin, *left*), tuck their heads to one side; hidden-necked turtles or cryptodires (common snapping turtle, *right*) bend their necks in a vertical s-curve, pulling the head into the shell.

all "true" tortoises are cryptodires.

The second group is much less well known to us Northerners. It includes an extensive array of turtles from South America, Madagascar, and Africa, and all but one of the freshwater turtles of Australia. The members of this group cannot pull their necks into an s-curve. Instead, they bend them around to the side, tucking their heads beneath an overhanging lip of the shell. They are the *Pleurodira*, or side-necked turtles.

There are other, more subtle distinctions between cryptodires and pleurodires. In pleurodires, for example, the pelvic girdle is sutured to the carapace and fused with the plastron. It is free in cryptodires. There are also important differences in the jaw mechanism of the two groups.

**Skin, Muscles, Limbs, and Tails**

Most turtles have rather smooth skins, but some sport a variety of tubercles, flaps, and bumps, especially on the head and neck. The alligator snapper, the Fitzroy River turtle (*Rheodytes leukops*) of Australia, and especially the matamata, carry such processes in considerable number and variety. The twist-necked turtle, a pleurodire, has a series of spiny tubercles on its neck. When it tucks its head away beneath its carapace, the turtle presents a barrier of spines to any predator. In the matamata, flaps of skin along the side of its neck are well supplied with nerves, and apparently help the turtle find prey in the murky waters where it lives (see Chapter 6).

Many freshwater turtles, particularly snapping turtles (Chelydridae), mud turtles and musk turtles (Kinosternidae), and pleurodires, have fingers of skin, called *barbels*, dangling under their chins. Most pleurodires have only a single pair; the big-headed Amazon River turtle (*Peltocephalus dumerilianus*) and the yellow-spotted river turtle (*Podocnemis unifilis*) of South America, as the latter's scientific name implies, only have one. Snapping turtles have four pairs, and the scorpion mud turtle (*Kinosternon scorpioides*) of Central and South America may have two pairs, three pairs, four pairs, or some irregular number haphazardly arranged. We do not know

The spot-bellied sideneck (*Phrynops hilarii*), a South American pleurodire, bears sensitive barbels under its chin.

exactly what these barbels are for, but they appear to be sensitive. Some Australian sidenecks touch each other's barbels as part of their courtship. I have watched yellow-spotted river turtles in the Frankfurt Zoo resting and stroking their barbels on the carapaces of other turtles as they swam.

Like other reptiles, turtles lack the facial muscles we mammals have, other than the muscles that open and close their eyelids, swivel, focus and adjust their eyes, or operate their jaws. A turtle's limbs, neck, and tail are well muscled, and turtles with hinged shells may have special muscles to operate them. Within the shell, though, there is little need for the trunk musculature that other vertebrates use to flex their spines or expand their ribs. These muscles, accordingly, have been lost.

Turtle limbs vary from the columnar, heavy-footed legs of tortoises to the elongate, bladelike flippers of sea turtles. Most turtles have five toes, often webbed, on each limb (a number that, some humans may be disappointed to learn, is the primitive condition in all but the earliest land vertebrates). Some tortoises have lost one of the toes on the front foot, while some American box turtles have only three toes on each limb. The most modified limbs of any turtle, the swimming paddles of sea turtles, retain only one or two claws, and in the leatherback even these are lost.

Sea turtle forelimbs have been modified into flippers, their individual digits drawn out to great lengths and fused together to form a flat paddle like a penguin's wing. A sea turtle has the normal complement of separate toe bones, though they are far longer on the forelimbs than in typical turtles, but these bones are bound together into a single stiffened unit by layers of fibrous connective tissue and an overlying scaly skin. The hindlimbs of sea turtles are much smaller than the front limbs, and are stabilizing and steering rudders (and nest-digging tools) rather than propelling paddles. Sea turtle limbs, both fore and aft, have other functions, too: males use their forelimbs to grip the shell of the female during mating; females have to be able to haul themselves back up onto the beach with their forelimbs, dig out a nest burrow using both forelimbs and hindlimbs, and then cover it up again; and hatchlings of both sexes use them to dig their way into the air and scramble down to the sea.

Sea turtles swim in a manner quite different from that of any other turtle (except for the pig-nose turtle, which also has transformed its limbs into flippers and rudders). In a typical pond turtle, the hindlimbs are larger than the forelimbs; they provide the swimming stroke, thrusting outward alternately, one after the other. Sea turtles, though, have transferred the main propulsive role to the long, flipperlike forelimbs. Instead of pushing outward on one side and then the other, they usually sweep them up and down simultaneously, weaving a figure-eight pattern. This pattern is basically the same as the wing strokes of a flying bird, or of a swimming penguin—the reason both penguins and sea turtles are said to fly underwater.

Finally, turtle tails vary considerably in length and thickness. They are particularly long in snapping turtles and in the big-headed turtle, which can actually use its tail as a brace to help it climb. The largest Australian side-neck, the Mary River turtle (*Elusor macrurus*), even takes its name from its oversize tail. Male turtles often have longer and heavier tails than females, and for a reason. The *cloacal* opening—the common opening of the digestive and reproductive tract in turtles—is partway down the underside of the tail, so a male with a longer tail has an easier job arranging himself in an appropriate position for mating. Further, the male's penis lies inside the cloacal opening in the base of the tail, and the tail has to be large enough to accommodate it. In some turtles, including mud turtles (Kinosternidae), the tip of the tail sports a hard, horny spine in males and, in some cases, in females. The spine may assist the male in bracing himself in position while he accomplishes the acrobatically difficult task of mounting his mate.

The forelimbs of these hatchling leatherbacks (*Dermochelys coriacea*) have been modified into swimming paddles.

## How Do Turtles Breathe?

Turtles have evolved modifications, alterations, and workarounds to carry on life in a variety of niches while still remaining essentially turtles. Some of the fundamental changes they have undergone must have happened at the very beginning of their evolution (see Chapter 2). As the flexibility of their rib cage was sacrificed to the protective architecture of the shell, turtles must have simultaneously developed ways to draw breath into their lungs without it. Their solutions—and there are more than one—to their breathing problems must be very old indeed.

We come back, then, to the challenge turtles faced early in their history. How do turtles breathe? It has taken turtle biologists a long time to find out, and we still do not know the answer for more than a few species. Two of them, the common snapping turtle (*Chelydra serpentina*) and the spur-thighed tortoise, do not fill their lungs in exactly the same way. Surely other turtles have tricks of their own. A number of turtles, including the Indian flapshell, have a special sheet of muscle that envelops the lungs. When this muscle contracts it probably helps the turtle exhale, but we still do not understand precisely how.

Turtles have perfectly good, multichambered lungs. They lie just under the carapace, with the rest of the internal organs—the viscera—lying below them. Their upper surface is actually attached to the carapace, while their lower surface is tied to the viscera by a sheet of connective tissue. If the viscera move downward, away from the carapace, they will drag the lungs open. That creates a negative pressure, and the air flows into the lungs. When they move upward and inward, they push against the lungs and squeeze the air out again. But how can a turtle move its viscera?

Part of the answer is, by using both special muscles and the movements of its limbs. A common snapping turtle (*Chelydra serpentina*) relies on four sets of muscles. Two lie near the front of the shell. Two more, at the rear, attach to a sort of membranous abdominal sling, somewhat equivalent to the human diaphragm. When the snapper contracts one set at each end, particularly the abdominal set which pulls the sling forward, it forces the viscera inward and upward; humans can produce the same effect with the abdominal muscles, as every opera singer knows. The other sets have the reverse effect, including pulling the abdominal sling backward. But the turtle can also force its viscera inward by drawing in its limbs, or let them slump back by extending them again. A freshwater turtle floating at the water's surface is free to use its legs as a sort of piston pump. Even the simple act of walking, which rocks its shoulder girdles back and forth within its shell, may help a turtle pump air in and out of its lungs.

Part of the answer, too, is to let outside forces do some of the work. When a turtle walks on land, gravity alone pulls its viscera downward. For a snapping turtle, with its slightly flexible plastron, that means their sheer weight will hold the lungs open above them, making inhalation a fairly effortless process. The trick is getting the air out again, and for that the snapping turtle must make a physical effort. A snapping turtle floating in the water faces a different problem. Here, it must deal not with gravity, but with water pressure. Instead of pulling the plastron and viscera outward, water pressure will tend to push them in. The deeper the turtle holds its body while it stretches its neck upward to take a breath of air, the greater that pressure will be. If it is strong enough, the turtle will have no problem exhaling, because the pressure will do the work. This time, though, it will have to use effort to inhale. If its body rises closer to the surface, it may reach a point where gravity and water pressure cancel each other out, and the turtle must fill and empty its lungs without any outside help. It is one of the remarkable things about turtles that they seem able to adjust to these changes without difficulty, so that they can breathe with the least amount of effort no matter what their circumstances.

The spur-thighed tortoise works matters a bit differently. Its plastron is too rigid to allow gravity to help it inhale. Gravity probably helps

When a Galápagos tortoise (*Geochelone nigra*) walks, the movement of its shoulder girdles pumps air in and out of its lungs.

its viscera return to their normal position after the tortoise exhales, but it cannot drag the viscera down enough to create the negative pressure the tortoise needs to fill its lungs. For that, and to exhale, the tortoise must resort to its muscles and, particularly, to its limbs. It does not have the anterior muscle, the *diaphragmaticus*, that a snapping turtle uses to exhale. Instead, it rocks its shoulder girdle inward and upward, or outward and downward, accomplishing the same thing by different means.

That may take care of normal breathing, but what happens when a turtle pulls in its head and limbs? They have to go somewhere, particularly in cryptodire turtles that draw the whole head and neck back into the body cavity. With the head, the limbs, and the viscera all tucked within its shell, a turtle has no room for air in its lungs. As long as it stays that way, it cannot breathe. Nor can it breathe, or at least breathe air, during a deep dive, or, for the turtles that do so, while it overwinters at the bottom of a pond. Turtles have, therefore, not only developed their own unique ways of breathing air; they have evolved methods for getting along without air, and even, in some circumstances, without oxygen. How they do this we shall see in Chapter 5.

Our story so far, though, should be enough to show that turtles are far more variable and adaptable than their humble appearance suggests. In the next chapter, we will reach back over 200 million years to see, as best we can, how these remarkable creatures came to be.

CHAPTER TWO

# *Turtles in Time*

If we were to take a journey back in time, passing, as we went, generation after generation of turtles evolving steadily in the opposite direction, we would probably see disappointingly little in the way of changes from modern turtles back to their distant ancestors. We could traverse the whole of the Cenozoic period, watching, in reverse, the passage of the Age of Mammals; continue through the great cataclysm that ended the reign of the dinosaurs 65 million years ago; cross the Cretaceous and the Jurassic periods, while dinosaur empires rose and fell and continents split and drifted across the globe; and all through that span of over 200 million years, turtles would still, basically, be turtles.

We would, of course, pass species and even families of turtles that have not survived to our own time. We would pass such oddities as the meiolaniids, great tortoise-like creatures sporting steerlike horns. We would cross the trail of giants, creatures paleontologists have named, with suitable awe, *Colossochelys* and *Stupendemys.* But all these animals would still, very obviously, be turtles, and their individual peculiarities would lead us no closer to the creatures from which, in the distant past, turtles first arose.

Our journey back would lead us, finally, to the late Triassic, somewhere between 220 and 225 million years into the past, and an animal called *Proganochelys. Proganochelys*, despite a few anatomical peculiarities, was—no question about it—a turtle. If one crossed the road in front of you today, you probably wouldn't give it a second glance. You might be surprised to find, were you to open its mouth, that unlike any turtle living today, *Proganochelys* had teeth, but you would still be highly unlikely to mistake it for any other

A restoration of *Proganochelys*, the earliest-known turtle. *Proganochelys* lived during the Triassic, 225 million years ago.

*Eunotosaurus*, despite paddle-like ribs that look like a shell in the making, is not now considered to be a turtle ancestor.

## CHART 1: Geologic Time Scale

| Eon | Era | Period | | Epoch | Dates | Age of | |
|---|---|---|---|---|---|---|---|
| Phanerozoic | Cenozoic | Quaternary | | Holocene | 0-2 | Mammals | Humans |
| | | | | Pleistocene | | | |
| | | Tertiary | Neogene | Pliocene | 2-5 | | |
| | | | | Miocene | 5-24 | | |
| | | | Paleogene | .Oligocene | 24-37 | | |
| | | | | Eocene | 37-58 | | |
| | | | | Paleocene | 58-66 | | Extinction of dinosaurs |
| | Mesozoic | Cretaceous | | | 66-144 | Reptiles | Flowering plants |
| | | Jurassic | | | 144-208 | | 1st birds/mammals |
| | | Triassic | | | 208-245 | | First Dinosaurs |
| | Paleozoic | Permian | | | 245-286 | Amphibians | End of trilobites |
| | | Carboniferous | Pennsylvanian | | 286-320 | | First reptiles |
| | | | Mississippian | | 320-360 | | Large primitive trees |
| | | Devonian | | | 360-408 | Fishes | First amphibians |
| | | Silurian | | | 408-438 | | First land plant fossils |
| | | Ordovician | | | 438-505 | Invertebrates | First Fish |
| | | Cambrian | | | 505-570 | | 1st shells, trilobites dominant |
| Proterozoic | Also known as Precambrian | | | | 570-2,500 | | 1st Multicelled organisms |
| Archean | | | | | 2,500-3,800 | | 1st one-celled organisms |
| Hadean | | | | | 3,800-4,600 | | Approx age of oldest rocks 3,800 |

sort of animal. *Proganochelys*, the earliest turtle we have found, has very little to tell us about the changes that must have taken place to turn an ordinary reptile into the highly specialized, almost distorted armored creature that turtles have become.

### The Case of the Missing Ancestor

With *Proganochelys*, our trail into the past runs cold. We do not know from whence it came. We may be not much closer to knowing today than we were more than a century ago, in the 1880s, when *Proganochelys* was first discovered.

Turtles are so different from any other reptile that their peculiarities are practically useless as a guide for distinguishing among potential ancestors, and the origin of turtles remains one of the great unanswered questions of evolutionary biology. By comparison, the more famous question of the origin of birds is a comparatively minor disagreement. All the candidates for the ancestral bird—dinosaurs, crocodile relatives, and others—are to be found within one subgroup of advanced reptiles, the archosaurs. By contrast, the possible choices for the original turtle span almost the entire range of reptiles, living and extinct.

They include a menagerie of peculiar creatures that died out either long before the rise of the dinosaurs, or just as those most famous

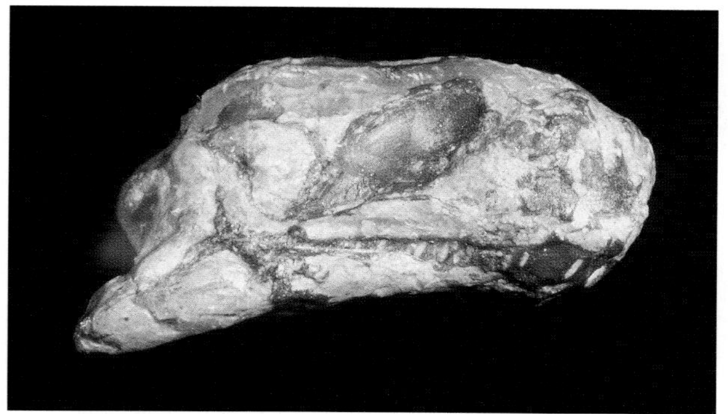

The skull of the Permian reptile *Eunotosaurus* is anapsid, with no temporal fenestrae, but is otherwise not turtlelike.

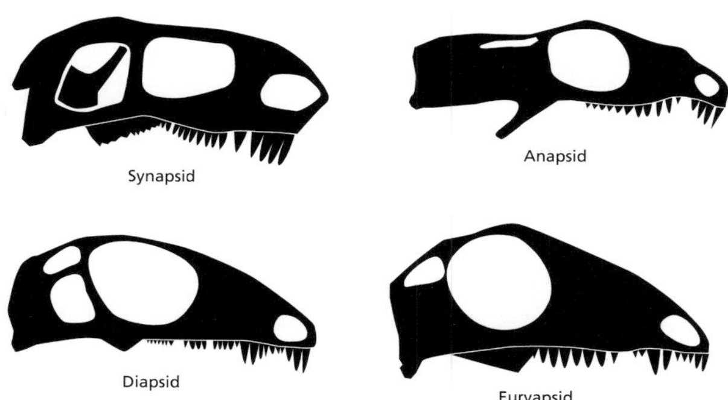


The four traditional categories of amniote skull based on the number and position of the temporal fenestrae.

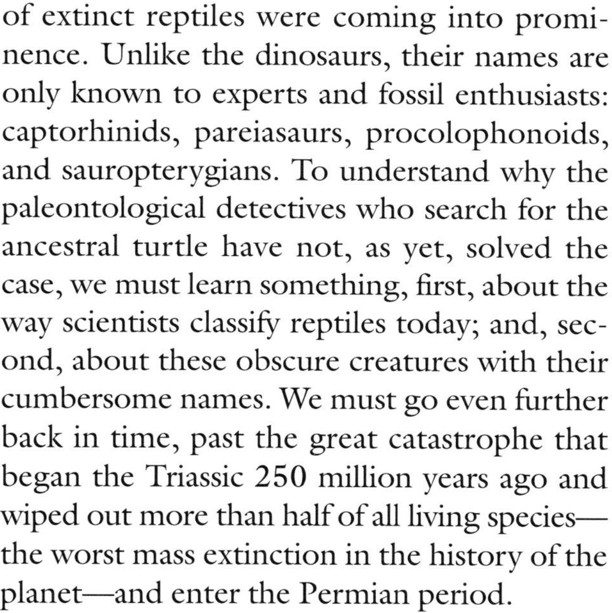

of extinct reptiles were coming into prominence. Unlike the dinosaurs, their names are only known to experts and fossil enthusiasts: captorhinids, pareiasaurs, procolophonoids, and sauropterygians. To understand why the paleontological detectives who search for the ancestral turtle have not, as yet, solved the case, we must learn something, first, about the way scientists classify reptiles today; and, second, about these obscure creatures with their cumbersome names. We must go even further back in time, past the great catastrophe that began the Triassic 250 million years ago and wiped out more than half of all living species—the worst mass extinction in the history of the planet—and enter the Permian period.

For the sake of completeness, we should first take a brief look at what was once thought to be the best candidate for an ancestral turtle. This was a little reptile, called *Eunotosaurus*, that lived in South Africa during the Permian. *Eunotosaurus* had a very peculiar rib cage. Instead of being a thin bow of bone, each rib was a broad, flattened paddle. It did not take much imagination to visualize the paddles growing even broader, fusing together, and forming the turtle shell. Unfortunately for this idea, that is not the way turtle shells are constructed. Turtle ribs are not expanded; they are fused, instead, with plates of dermal bone (see Chapter 1). Furthermore, a more careful study of *Eunotosaurus* showed that in almost every feature it was nothing like a turtle. *Eunotosaurus* has accordingly been dropped from the list of suspects.

So, it appears, has a group of extinct reptiles, the captorhinids, that were on the candidate list in the 1980s. Captorhinids were the commonest reptiles during the Permian period. Given the way the debate has seesawed back and forth they may yet reemerge as possibilities, but as they seem to be out of the turtle picture for now, we can (thankfully) ignore them.

## Holes in the Head

At one time, the most important thing to know about turtles for anyone seeking their ancestors was the absence of the right kind of holes in their skulls. The earliest reptiles had a solid, massive skull roof that covered the top of their heads, with major openings only for eyes, ears, and nostrils. The skull roof formed separately from the braincase inside it. Early reptiles were, in fact, rather like Darth Vader in his helmet, the helmet being the equivalent of the skull roof.

With time, however, openings, called *temporal fenestrae*, began to develop in the skull roof between the eye and the ear. The temporal fenestrae lightened the skull and gave the jaw muscles, previously sandwiched between the skull roof and the braincase, somewhere to go when they bulged. Some later reptiles had a single opening high on the side of the skull. In others, the opening was lower down, while still others had two openings, one above and one below.

Accordingly, students of fossil reptiles once classified their subjects into four main categories: Reptiles with solid skull roofs were called *anapsids*, meaning "without openings." Those with a single high opening were called *euryapsids*, meaning, as you might expect, "high opening." Those with the lower opening were called *synapsids*, and those with both openings were called *diapsids*. All living reptiles except turtles, as well as dinosaurs, (and their descendants, the birds), pterosaurs, and some other, less well known animals, are diapsids. The euryapsids included extinct marine reptiles like the plesiosaurs, which we will be discussing later because they have a possible role in the turtle story. Today, though, euryapsids are considered to be diapsids that lost their lower opening, and so the "euryapsids" as a group have disappeared from the scientific literature.

The synapsids, before dying out in the Triassic, gave rise to the mammals, including us. That may surprise you, because you may not think that you have a hole in the side of your head. In fact, you do. Your temporal fenestrae, however, have become so large that almost all of the bone that once framed them is gone. There is practically nothing of your original skull roof left.

Turtles have no temporal fenestrae. Instead, in most modern turtles the jaw muscles bulge out of the emarginations in the skull (see Chapter 1). Under the system of reptile classification that held until very recently, therefore, turtles are anapsids, and the ancestral turtle had to be sought somewhere among the cluster of primitive reptiles lacking temporal fenestrae.

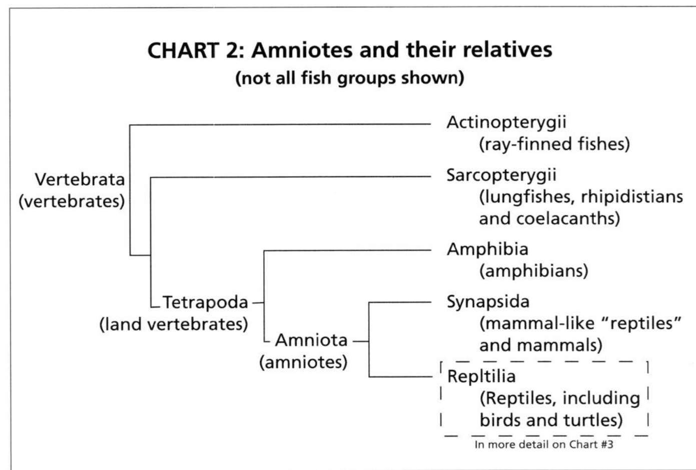


## The New Reptiles

In recent years, though, the very idea of what constitutes a reptile has changed. Modern biologists no longer define animal groups on the basis of how similar their members look. Instead, they try—usually using computers to analyze their data—to set up categories that reflect the actual branching patterns of evolution. A modern taxonomic category may include animals that seem very different from one another, but are grouped together because they are all on the same branch of their family tree. This approach is known as *cladistics*. Under a cladistic approach, for example, birds are reptiles because they sit on the same branch of the family tree as lizards, snakes, crocodiles, and dinosaurs. In fact, they share a sub-branch of that tree with crocodiles and dinosaurs, but not with lizards and snakes, so that if lizards, snakes, and crocodiles are all reptiles then, to be consistent, birds must be reptiles too. By the same logic, all mammals, including humans, are synapsids.

All land vertebrates, except amphibians (the group that includes the living frogs and salamanders), evolved from animals that laid a *cleidoic* egg; that is, an egg sealed within a shell that allowed it to survive out of water. The embryo that grows within this type of egg is surrounded by a membrane called the *amnion*, so the descendants of those first ancestors are called *amniotes.* The amniotes form a natural group—a single branch of the vertebrate family tree that includes modern reptiles, birds (which are, as we have explained, reptiles), and mammals.

Zoologists used to use the term "reptile" to cover all of the early amniotes. However, when we use cladistics to examine how the amniote branch evolved, we find that the synapsids (including the mammals) are not on the same sub-branch as the one leading to living reptiles (including the birds). In other words, if we define the term "reptile" to include all of the animals, descended from the common ancestor of reptiles, that are alive today, the synapsids are amniotes, but not reptiles—not even such well-known synapsid "reptiles" as the sail-backed *Dimetrodon* and the whole range of creatures we have been calling "mammal-like reptiles" for years. Using cladistics, the synapsid "mammal-like reptiles" are not reptiles.

Turtles, though, are unquestionably reptiles, even in the modern sense of the term. We can therefore ignore the synapsids (barring an occasional glance in the mirror) for the rest of this story, and concentrate on the occupants of the reptile family tree. Aside from a few early side branches, the reptile tree apparently split into two main limbs over 300 million years ago, near the beginning of the Permian.

One limb contains the lizards, snakes,

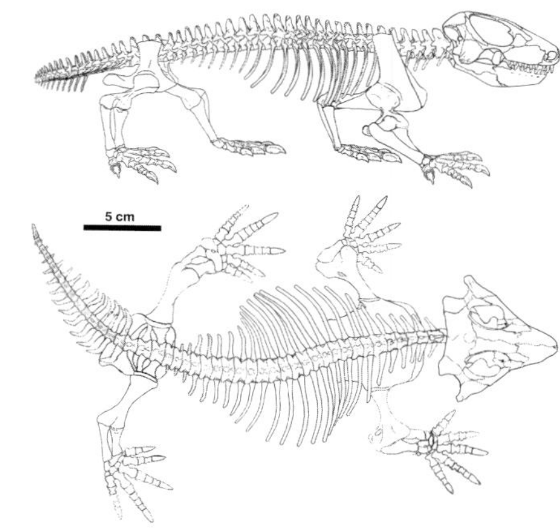


Skeletal reconstruction of *Procolophon*, a 225-million-year-old procolophonid from the early Triassic of South Africa.

*Scutosaurus*, a 2.5 m (8 ft), 250-million-year-old pareiasaur (and possible turtle ancestor) from the Permian of Russia.

crocodiles, dinosaurs, and birds. Except for the turtles, it contains every reptile that survived the Triassic. Because this branch contains most of the animals we traditionally think of as reptilian, its members are called *eureptiles.* This means true or good reptiles. They are called a number of other things as well; scientists have not yet agreed, to put it mildly, on the various choices. We'll continue to call them eureptiles here.

The second limb contains most of the animals with anapsid skulls. These are the *parareptiles*. As paleontologist Robert Reisz of the University of Toronto, an expert on the parareptiles, points out on his web site, "Just as paramedics are not exactly doctors, and paramilitary forces are not exactly military forces, parareptiles are not exactly true reptiles"—that is, they are not eureptiles, but, unlike the synapsids, they are perfectly good reptiles.

### The Usual Suspects

Do turtles belong on the parareptile branch, or on the eureptile branch? Up until about 10 years ago, the answer would have been obvious. Turtles, with their solid skull roofs, were clearly anapsid. Therefore, they were parareptiles—the only parareptiles to survive to this day. Which parareptile group, though, is the best candidate for a turtle ancestor?

Some paleontologists champion the pareiasaurs, great lumbering armored Permian behemoths, something like a cross between a buffalo and a toad. Others prefer the procolophonoids, much less unusual-looking animals by our standards, rather like large, clumsy lizards. The procolophonoids actually survived into the Triassic, the era of *Proganochelys*. They are the only parareptile group to have done so, something that may make them a bit more likely to have evolved into turtles than the pareiasaurs.

Supporters of the procolophonoids point to skeletal similarities between turtles and such animals as *Owenetta*, the earliest known procolophonoid. *Owenetta* is a vaguely lizard-like little animal, less than 30 cm (1 ft) long, from the Upper Permian of the Karoo in South Africa. We have very good fossils of *Owenetta*. Most importantly, we have well-preserved skulls.

Fossil skulls are rare, but they provide particularly valuable clues for turtle detectives. Most of a turtle's skeleton has changed so radically, as the shell has evolved, that it provides very few clues as to its origin. Though turtle skulls are highly variable in shape and proportion, their fundamental structure has changed far less from the typical reptilian condition. That makes it easier to compare the skulls of potential ancestors, bone for bone, with modern turtles, providing, of course, that we are lucky enough to find them. We were lucky with *Owenetta*, and its skull seems to share a number of similarities with, and only with, turtles. If those similarities hold up, and are truly found in no other animal groups, an early procolophonoid may be a very good candidate for the ancestral turtle.

However, Michael Lee of the University of Queensland, Australia, has made an extended and detailed argument that turtles belong with the pareiasaurs. According to Lee, turtles are not just related to pareiasaurs, they *are* pareiasaurs, descendants of some of the smaller members of that lineage. He points to a number of features that he says turtles and pareiasaurs share. Like turtles, many of the pareiasaurs carried bony dermal armor. Pareiasaurs also share a projection of the shoulder blade called the *acromion process*, a structure found elsewhere only in turtles. Lee has even proposed a series of evolutionary steps by which small pareiasaurs could have developed into turtles.

Lee, though, has his critics. The last known pareiasaur died out 40 million years before *Proganochelys*, the first known turtle. That is a tremendous length of time. Pareiasaur opponents point out that if these large, armored animals really did survive to become the ancestors of turtles, we should have found at least some fossil evidence that they were there. That may seem to be a weak argument; after all, we might still find the missing fossils, or perhaps the animals that would have left them lived somewhere like modern Antarctica, where their fossils are out of reach. But there are other reasons to question Lee's conclusions.

Robert Reisz points out that bony armor alone does not make an animal a good candidate for the ancestral turtle. Armor made from dermal bone is not just found in pareiasaurs and turtles, but in a long list of other vertebrates, living and extinct (including, for example, armadillos). The special thing about turtle armor, as we discussed in the first chapter, is that it combines dermal with endochondral bone—and pareiasaurs don't do that. In fact, nothing else does. If the armor of any fossil animal is to be evidence that we are dealing with a genuine turtle ancestor, it will have to share some of the features that make a turtle shell unique.

Furthermore, the acromion process in pareiasaurs may be an entirely different structure from the one in turtles. Michael deBraga of the University of Toronto and Olivier Rieppell of the Field Museum, Chicago, have argued that the "acromion" in turtles is actually a modified version of a bone that turtles are otherwise missing, the anterior coracoid. If that is true (and Lee certainly argues that it is not), then it cannot be the same bone as the one in pareiasaurs, because pareiasaurs have both the acromion and the anterior coracoid. Until this argument is settled, the case for the pareiasaurs is questionable.

### Surprising Developments

That would seem to bring us back to the procolophonoids. However, a number of recent studies have forced paleontologists to consider the almost disturbing notion that turtles may belong with the eureptiles.

How could turtles, with their solid skull roofs, be eureptiles? The fact that turtles are, technically, anapsids may be misleading us about what their true ancestors were. Some eureptiles—members of an extinct group called araeoscelids—may have been losing their temporal fenestrae. Dwarf caimans (*Palaeosuchus* spp.), living relatives of alligators from South America, have actually lost their upper fenestrae. Some parareptiles, conversely, had fenestrae of their own. The solid skull roof of turtles may mean that they lost temporal fenestrae that their ancestors possessed, not that, like other anapsids, they never had them at all. Instead of being the last survivors of an ancient anapsid lineage, turtles may be highly specialized relatives of lizards, or perhaps even of crocodiles.

What could have led to such a surprising conclusion? In a number of studies, including one by Rafael Zardoya and Axel Meyer, and another by S. Blair Hedges and Laura Polling, the evidence was molecular—a computer-generated analysis of similarities among molecules of mitochondrial DNA and other genetic material from a variety of living animals. Both studies put turtles among the diapsids. The Hedges and Polling study even placed them closer to crocodiles than to lizards (the Zardoya/Meyer study was equivocal on this point).

Unfortunately, it did the same for a peculiar reptile from New Zealand, the tuatara (*Sphenodon punctatus*). On the basis of just about every other kind of evidence available, the tuatara, though not a lizard, is closer to lizards than to crocodiles. If the molecular evidence is misleading about the tuatara, then it may be wrong about turtles too. Further, it is a bit difficult to use DNA evidence as a guide to whether turtles are eureptiles or parareptiles. The parareptiles, of course, are all extinct, and despite *Jurassic Park*, we are in no position to analyze the DNA of animals that vanished 180 million years ago.

Molecular evidence is not the only thing pointing to the eureptiles. Hedges and Polling argued that some crocodiles, and an extinct

*Placodus*, a placodont from the early to middle Triassic of Europe, grubs for shellfish in the mud of a sea bottom.

Not a turtle but the cyamodontoid placodont *Placochelys* from the middle to late Triassic of Europe; 90 cm (3 ft) long.

group of vaguely crocodile-like animals called aetiosaurs, had bony armor, but as we have seen with the pareiasaurs, that alone doesn't mean much. DeBraga and Rieppell, and later Rieppell and Reisz, subjected old-fashioned anatomical characters—bumps, processes, and shards of bone from animals both living and extinct—to modern cladistic analyses, and to their surprise found the turtles ensconced within the eureptiles. Their closest eureptilian relatives, however, appeared to be neither lizards nor crocodiles, but another group of peculiar, extinct creatures: the sauropterygians. The sauropterygians are a group of reptiles that had returned to the sea to live.

**Pseudo-turtles**

The best-known sauropterygians are the plesiosaurs, creatures like the long-necked *Elasmosaurus*, once aptly described as looking like a snake threaded through the body of a turtle. Their short-necked, crocodile-headed cousins, the pliosaurs, included immense creatures that must have been the largest and most fearsome predators of the entire Age of Reptiles, *Tyrannosaurus rex* not excepted.

For our purposes, though, the most interesting sauropterygians were the placodonts, a group that never made it out of the Triassic. Placodonts may have lived like reptilian walruses. Most of them had broad, flat, powerful teeth that were almost certainly used to crush mollusks or crustaceans. In appearance, a typical placodont would probably have reminded you of a cross between a manatee and a marine iguana. The last group of placodonts to survive, though, the cyamodontoids, were far stranger.

Some cyamodontoids had extremely flattened bodies and long whiplike tails. They may have lived something like modern stingrays. Others were the closest any animal group has come to producing the equivalent of a turtle. One, *Placochelys*, was even portrayed as a fossil turtle on a 1969 Hungarian postage stamp. Another, the meter-long *Henodus*, had a broad turtlelike shell splayed out at the sides like the head of a double-bladed ax, complete with carapace and plastron. Like the earliest turtles, *Henodus*, which lived in saltwater lagoons in Europe, had lost most of its teeth. *Henodus* had even shifted its shoulder girdle inside its rib cage, the only animal other than turtles ever to achieve this.

Turtles are not placodonts, though, and *Henodus* was certainly not a turtle. Its shell is actually quite unlike a turtle's carapace. Its ribs are not involved, and its armor is made up of a series of interlocking hexagonal plates of dermal bone laid out like the tile on a bathroom floor. The turtlelike characteristics of *Henodus*, which certainly evolved independently from those of real turtles, tell us less about turtle relationships than it does about the sort of conditions that might have led to the evolution of both groups.

One of the chief difficulties in imagining how turtles evolved has been the breathing problem. As turtles became sealed in their bony shells, they had to switch from a breathing system that involved expanding and collapsing the rib cage to one that involved using the legs as a sort of piston pump. Since land animals need their legs for support, it is hard to see how this transition could have occurred. What would the intermediate evolutionary steps have been like? A land animal whose legs were only partly adapted for breathing, but had lost in the process some of their ability to hold up its body, might seem an unlikely candidate for evolutionary success. Tortoises use their legs in breathing, of course, and seem quite able to walk and breathe at the same time; however, as they almost certainly evolved from river turtles, their breathing system would have been quite advanced by the time they took to a terrestrial life.

Rieppell and Reisz pointed out that *Henodus*, which presumably faced the same problem, certainly evolved its shell in the water. An aquatic animal, whatever else it may use its legs for, does not need them to hold itself up. If the ancestor of turtles also lived in the water instead of on land, the breathing problem disappears. Rieppell and Reisz came up with a number of other arguments for an aquatic origin for turtles, including the suggestion that armoring one's underside—as both turtles and cyamodontoid placodonts do—only makes sense if you are likely to be attacked from below. This is much more likely to happen to a swimming animal than to one walking about on dry land.

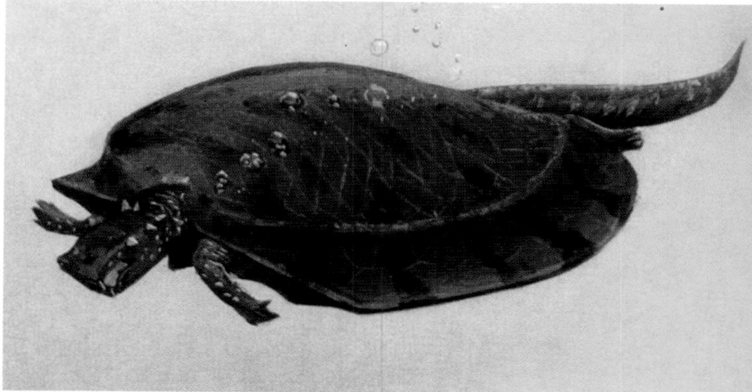

*Henodus*, strangest of the placodont "pseudo-turtles," from the late Triassic of Germany; about 1 m (3.3 ft) long.

There are some flaws in this argument. The plastron could have evolved, not as armor, but as a brace locking the sides of the carapace together (see Chapter 1). Other armored land vertebrates, like armadillos, may have no belly armor, but unlike turtles they either can roll into a ball to protect their undersides, as armadillos do, or are so huge that they could have protected their bellies by simply squatting on the ground, as the extinct glyptodonts, giant armored relatives of armadillos, may have done.

If Rieppell and Reisz are right, though, that leaves us with a dilemma. If turtles evolved in the water, it is unlikely that their ancestors were the pareiasaurs. All the pareiasaurs we have discovered were certainly land animals. However, if both turtles and sauropterygians started out as swimming animals, perhaps the similarities between them have less to do with relationship than with the results of similar evolutionary pressures. This phenomenon, in which unrelated animals can come to resemble one another because of similarities in the way they live, is called *convergence.* It is the bane of anyone trying to figure out where animals fit on their family tree.

Is convergence misleading us about the relationships between turtles and sauropterygians? If so, where do turtles really belong? Rieppell and Reisz tried to find out by running their analysis again, leaving out the sauropterygians. To their surprise, the turtles, instead of winding up among the eureptiles, landed back among the parareptiles. In other words, the reason that the turtles had come out among the eureptiles in the first place was because the sauropterygians were there. The characters turtles and sauropterygians share had pulled the turtles into the eureptile camp. If those characters are shared because of convergence instead of evolutionary relationship, then Rieppell and Reisz's study does not prove that turtles are eureptiles after all. The key word, of course, is "if."

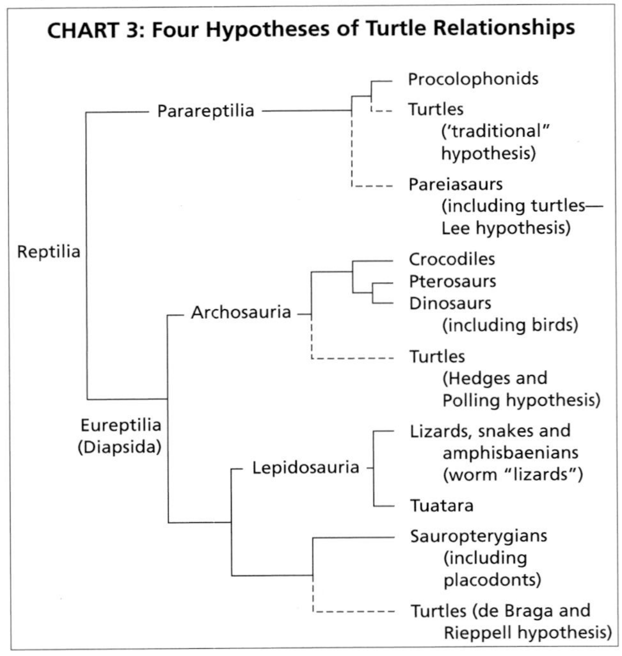


## An Unsolved Mystery

Where does all this leave us? Are turtles, despite all the new evidence, really parareptiles, as scientists have thought for years? Are procolophonoids, which seem to share so many characters with turtles, the best candidates after all?

There are reasons to think so. Part of a turtle's unique cheek structure is a bone, the *quadratojugal*, which parareptiles have but eureptiles do not—assuming, of course, that we have identified it correctly in parareptile fossils. On the other hand, turtles in the family Chelidae have lost the quadratojugal, and some others have reduced or lost it too, so its absence may not tell us much. Or should we believe what the computers tell us, and accept that turtles belong among the eureptiles, even if this seems, at the very least, counterintuitive? There are some features of the ankle joint, as well as a unique jaw ligament, that turtles seem to share only with eureptiles. If turtles really evolved from a swimming ancestor, should we look for that ancestor near the sauropterygians? If not, what sort of swimming creature was it? Or is Lee right, and is the ancestral turtle to be found among the land-based, lumbering pareiasaurs? Or somewhere else altogether?

At this stage, in the year 2001, we simply don't know. Further analysis of the fossils we have may bring us no closer to a solution. What we really need is a fossil we have yet to discover, something transitional between a genuine turtle and whatever its ancestor might have been. Unfortunately, the most likely time for such an animal to have existed would be the early Triassic, a period that has left us comparatively poor fossil remains. Since the center of parareptile evolution was probably in the former southern hemisphere supercontinent of Gondwanaland, and since one of the biggest chunks of Gondwanaland, Antarctica,

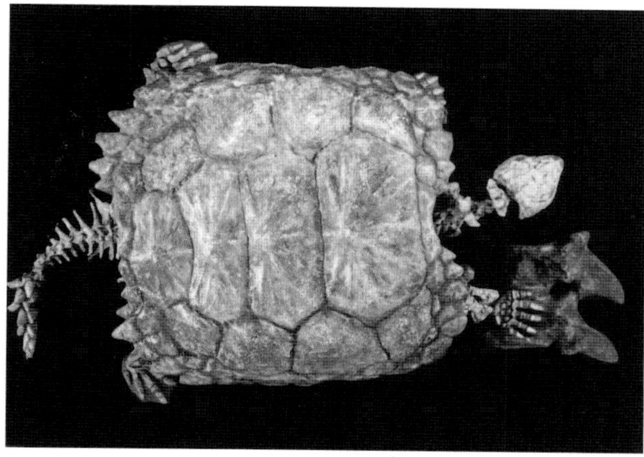

Thanks to some excellent fossil specimens (this is a cast), we know a great deal about the skeleton of *Proganochelys.*

is buried beneath tons of ice, the fossil we need may be far beyond our reach. The ancestral turtle, whatever it was, may keep its mystery.

**The Oldest Turtle**

It is, I confess, with some relief that I lead my readers a short distance forward in time, away from the welter of potential proto-turtles, pseudo-turtles, quasi-turtles, and definitely-not-turtles that littered the Permian and the early Triassic, to the first genuine, unquestionable turtle that we have. It is the animal we met at the beginning of this chapter: *Proganochelys.*

In contrast to its missing ancestors, we know quite a bit about *Proganochelys.* This is largely thanks to Dr. Eugene Gaffney of the American Museum of Natural History, New York, who has published an extremely detailed, bone-by-bone description of a number of excellent, near-complete specimens. There are fragments of *Proganochelys*-like turtles known from Greenland and Thailand, but the best specimens, the ones Dr. Gaffney examined, come from Germany. *Proganochelys* looked rather like a modern common snapping turtle (*Chelydra serpentina*), and, perhaps, lived like one. It probably spent much of its time walking along the bottom of shallow swamps, with occasional forays onto dry land. There, at least in Germany, it would probably have encountered *Plateosaurus*, one of the few popular Triassic dinosaurs.

*Proganochelys* could not withdraw its head beneath its shell, but perhaps made up for this with a row of spines protecting the back of its neck. The scutes of its carapace were rough and peaked. Its last few tail vertebrae were fused into a club, though it is not all clear if *Proganochelys* could have used them as one. And, as we have already mentioned, *Proganochelys* had teeth, not on the edges of its jaws, where it apparently had a horny beak like modern turtles, but on the roof of its mouth.

*Proganochelys* shows that turtles did not acquire all their special features at once. Though Dr. Gaffney identified six characteristics that *Proganochelys* shares only with other turtles, he found six more that it shares only with other early reptiles. Most of these involve the skull.

*Proganochelys* has a fairly advanced shell, its most obvious turtle feature. Its skull, though, is in many ways still the open, unfused, flexible structure of a typical early reptile. *Proganochelys* still has two bones, the *lacrimal* and the *supratemporal*, found in early reptile skulls but missing in more advanced turtles. It has two *vomers*, bones that lie side by side in the roof of the mouth. In later turtles these have been reduced to one or, in some South American river turtles (Genus *Podocemis*), lost.

Most importantly, perhaps, while modern turtles have a solid box of bone hemming in their middle ear, *Proganochelys* has the same open, unossified ear region as other early reptiles. We have only learned recently how the change from the condition in *Proganochelys* to that in modern turtles may have taken place.

The skull of *Australochelys africanus* from the early Jurassic of South Africa, the oldest African turtle ever found.

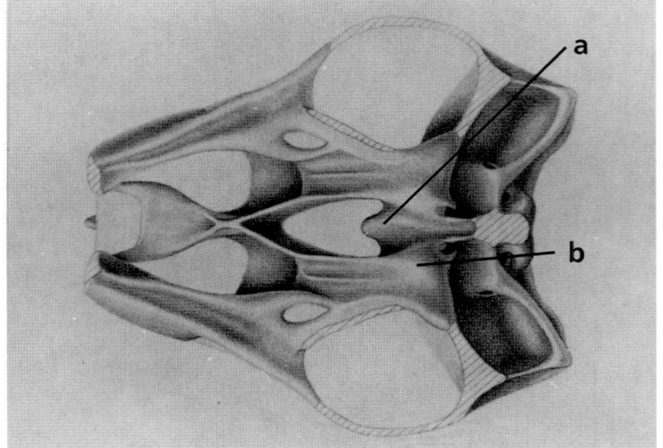


*Australochelys* skull from below: a. braincase (basisphenoid bone); b. patlate (pterygoid bone).

## Early Advances

The clue to this major step in turtle evolution came with the description in 1994 of another early turtle, *Australochelys.* Known only from a skull and a fragment of shell, *Australochelys* lived millions of years after *Proganochelys*, in the early Jurassic. It is the oldest turtle ever found in Africa, beating its nearest rival by 60 million years. While *Proganochelys* was apparently amphibious, like a modern snapping turtle, *Australochelys* seems to have lived in dry uplands. It may have been entirely terrestrial, like a tortoise.

Like *Proganochelys*, *Australochelys* retained many features found in other early reptiles. Its ear region was almost the same as in *Proganochelys*, but not quite. *Australochelys* had made a significant advance, a first step in the transition from an open, flexible skull to a closed, rigid one.

In *Proganochelys*, as in other early reptiles, the braincase attaches to the rest of the skull by a movable joint just in front of the ear. It could, at least in theory, move slightly within the animal's skull. With a movable braincase pushing or pulling against the middle ear, it may have been difficult to form a rigid bony box around it. Things would have been much easier if the braincase could have been locked in place first.

That, in fact, is the change we see in *Australochelys.* Instead of a movable hinge, its braincase is fused to the skull by a solid, sutured immovable joint. With that important first step taken, the evolution of the modern, rigid turtle skull may have been only a matter of time. The further evolution of a box around the middle ear may have given turtles such an advantage that it happened not once but twice, in pleurodires and in cryptodires.

Of course, *Australochelys* did not evolve its rigid braincase so that its descendants could ossify their ear capsules. Evolution does not plan ahead. There must have been some immediate advantage for *Australochelys*—or for turtles like it, because *Australochelys* is too young to be the direct ancestor of modern turtles—in having a rigidly attached braincase. What that advantage might have been, though, we do not, and perhaps cannot, know. Perhaps the more rigid skull the fused braincase-joint allowed may have affected the way *Australochelys* fed.

*Australochelys* shows that the very earliest

turtle lineages must have had a large geographic range, a lengthy geological history, and a broad range of lifestyles. Whatever their eventual fate, they were an evolutionary success.

### The *Proterochersis* (and *Palaeochersis*) Problem

For many years, paleontologists have assumed that *Proganochelys*, and, by extension, *Australochelys*, was a living fossil even in its own far-off day. This was because of the second-oldest-known turtle, *Proterochersis*. *Proterochersis* lived at almost the same time and in almost the same place as *Proganochelys*. The problem is that it seemed to be, not just a turtle, but a pleurodire. Like modern pleurodires, its pelvic girdle was firmly fused to its carapace.

If *Proterochersis* was really a pleurodire, then the great evolutionary divide between the cryptodires and the pleurodires must have happened even earlier, before the time of *Proganochelys*. *Proganochelys* could hardly have been one of the first turtles if, by its time, turtles had not only evolved but had split into pleurodires and cryptodires.

But was *Proterochersis* really a pleurodire? The conclusion that it was may actually reflect one of the most dangerous traps in paleontology: making deductions from specimens that are, more often than not, incomplete.

*Proterochersis*, which was described as long ago as 1913, is known only from its seemingly advanced shell and limb girdles. By contrast, *Australochelys* is known primarily from its fairly primitive skull. However, in 1995 Guillermo Rougier, of the American Museum of Natural History, and his colleagues announced the discovery of two skeletons of a new early turtle, *Palaeochersis*, from the late Triassic of northwestern Argentina, and at least one of the skeletons was almost complete.

*Palaeochersis*, like *Australochelys*, is certainly a primitive turtle. Like *Proganochelys*, it could not tuck its head into its shell. It did not have the neck armor that protected *Proganochelys*, but its neck may have been protected instead by a forward extension of its carapace, as in a number of living sidenecks and tortoises. It had an open, largely unfused skull like *Australochelys*. It was certainly not a pleurodire, at least not according to the way scientists have defined pleurodires, but, like *Proterochersis*, its pelvic girdle was sutured to its carapace. What is more, *Palaeochersis* displayed a few minor features that are otherwise known only from *Proterochersis*.

What does this discovery mean? Not, probably, that *Palaeochersis* and *Proterochersis* are close relatives. But its unexpected combination of a primitive skull and advanced shell may mean that a condition we had believed to be found only in pleurodires—a pelvic girdle sutured to the carapace—was actually widespread in primitive turtles before the true pleurodires evolved.

If *Palaeochersis*, with its pleurodire-type hip girdle, is not a pleurodire, then *Proterochersis*, supposedly the first of the "modern" turtles, may not be a pleurodire either. The two may be pleurodire ancestors, of course. But Rougier and his colleagues suggest that perhaps the ancestor of both pleurodires and cryptodires had a hip girdle fused to the carapace. If so, then at some point in the further evolution of the cryptodires it became unfused again, a change scientists call an *evolutionary reversal*.

If all this is true, it may change our whole picture of the timing of the evolution of early turtles. If *Proterochersis* is not a pleurodire, then we have no evidence that true pleurodires lived at the same time as *Proganochelys*. This means, in turn, that we have no reason to believe that pleurodires and cryptodires evolved as separate groups before *Proganochelys* appeared.

*Proganochelys* would, then, not have been the "living fossil" that we had thought. Nonetheless, it still cannot be either the oldest turtle, or the direct ancestor of any other known turtle. Even if *Proterochersis* is more like *Australochelys* and *Palaeochersis* than it is like living turtles, it is still more advanced than *Proganochelys*. Because it lived at practically the same time as *Proganochelys*, it cannot be its descendant.

## The Rise of Modern Turtles

If *Proterochersis* is not a pleurodire, then pleurodires do not appear in the fossil record until the late Jurassic, the time of *Platychelys* (if, in fact, *Platychelys* really is a pleurodire; we don't have its skull either). Another late Jurassic turtle, *Notoemys*, does seem to be a genuine pleurodire, though its relationships are still rather obscure.

The earliest known turtle that we can definitely call a cryptodire is *Kayentachelys*, a fossil from the Kayenta formation of northeastern Arizona. *Kayentachelys* lived in the early Jurassic, 185 million years ago—45 million years before the next-oldest cryptodire fossils. We know *Kayentachelys* was a cryptodire because it shows a key feature of the jaw hinge that only cryptodires have.

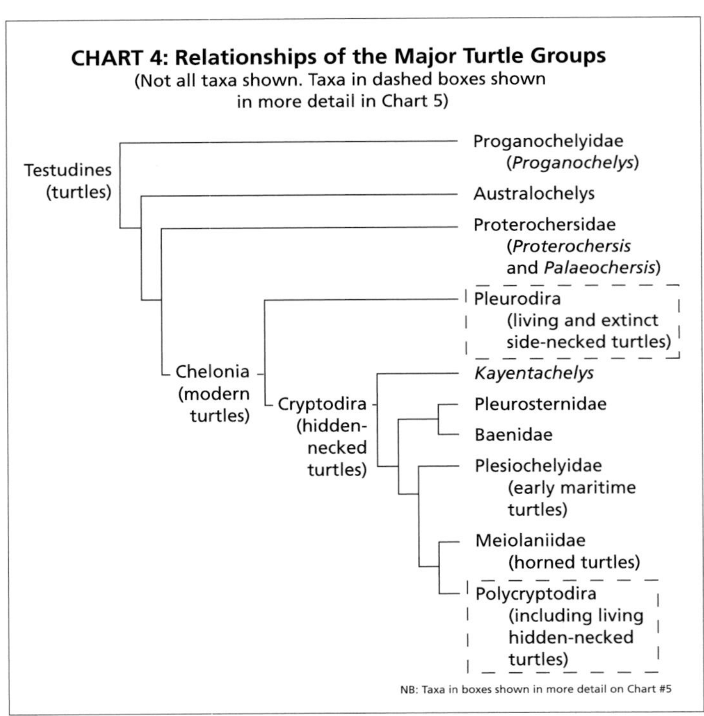


Reptiles (including birds) close their jaws by means of a powerful muscle, the *adductor mandibulae*; we mammals do not, because the bones that serve as the jaw hinge in reptiles are now part of our middle ear, and our muscles have changed accordingly. In early reptiles, and in *Proganochelys* and *Australochelys*, the *adductor mandibulae* attached by means of a tendon that met the lower jaw at a more or less shallow angle. In all modern turtles, this tendon passes instead around a bump of bone called the *trochlea*. The trochlea acts as a sort of pulley, forcing the tendon to meet the jaw at a right angle. This not only keeps the tendon out of the way of the enlarged, bony ear capsule; it also results in a much more powerful bite (an excellent reason to keep your fingers out of a turtle's mouth). Both pleurodires and cryptodires have a trochlea, but it must have evolved separately in each group, because the bones involved are not the same. In cryptodires the trochlea is part of two bones called the *proötic* and the *quadrate*, while in pleurodires it forms part of a different bone, the *pterygoid*. *Kayentachelys* has its trochlea on the proötic and quadrate, so *Kayentachelys* must have been a cryptodire.

This is not to say that *Kayentachelys* was, in every respect, a modern turtle. Bone had not yet completely enclosed its middle ear, which was still open from

Pictured here is the skull of *Kallokibotion*, a primitive cryptodire turtle from the Cretaceous of Transylvania.

below. In technical terms, *Kayentachelys* lacked the posteromedial process of the pterygoid, the piece of bone that forms the floor of the ear capsule in all other cryptodires. *Kayentachelys* still had teeth, confined, like those in *Proganochelys*, to its palate. This is interesting because it means that the ancestor of pleurodires and cryptodires must have had teeth too. Pleurodires and cryptodires apparently lost their remaining teeth independently.

For these and other reasons, *Kayentachelys* is regarded as forming a "sister-group" to all other known cryptodires, both living and extinct. That is, all the others are more closely related to each other than any of them is to *Kayentachelys*. *Kayentachelys* sits on one early limb of the cryptodire family tree; all the rest are on the other.

On that other limb, a number of side branches lead to families now extinct. One includes *Kallokibotion*, from the Cretaceous of Translvania. The Pleurosternidae and the Baenidae are sometimes grouped together, based on the path that their carotid arteries followed into their skulls. The Baenidae, which survived into the Eocene period, are known only from North America. The Pleurosternidae had a broader geographic range—they may have lived across the whole of the northern hemisphere—but they did not last as long, dying out at the end of the Cretaceous. Both families were quite diverse in their time. Two more extinct side branches lead to animals we will meet later in this chapter: the Jurassic marine turtles of the Plesiochelyidae and the weird, horned meiolaniids.

The living cryptodire families (and some extinct cousins), whose branches cluster together in the end of the limb, form the Polycryptodira. Though we may think of this as a "modern" group because many of its members are still with us, the Polycryptodira has been around for long time. Its oldest probable member, *Peltochelys*, from the early Cretaceous of Belgium, dates back 120 million years. Only 10 million years later, still in the early Cretaceous, it seems that the Polycryptodira was not only well established, but had already split into lineages that lead to the turtle families alive today. We know that because of two recent discoveries from that period: the oldest-known true sea turtle (that is, a member of the same group as modern sea turtles), *Santanachelys*, and a peculiar relative of the modern softshell turtles, *Sandownia harrisi*. Known from a skull and jaw from the Isle of Wight, England, and only described in 2000, *Sandownia* had an extremely strong skull and an extensive secondary palate forming a bony roof to the mouth, both strengthening adaptations that may have contributed to an unusually powerful crushing bite.

The living families of pleurodires boast an equally ancient pedigree. Like the cryptodires, they shared the Cretaceous with a number of vanished relatives. Though the Jurassic *Notoemys* probably sits on a side branch of the pleurodire tree, bearing the same relationship to modern pleurodire families as *Kayentachelys* does to modern cryptodire families, the next oldest unquestionable pleurodire is much closer to living forms. This is *Araripemys*, a 110-million-year-old fossil from Brazil, and a near contemporary of the early cryptodires *Santanachelys* and *Sandownia*. Though *Araripemys* belongs to a now-extinct family, the Araripemydidae, it appears to be more closely related to two of the living pleurodire families, the Pelomedusidae and Podocnemididae, than to the other living family, the Chelidae. If that is true, then the Chelidae must have already split from the other families by the early Cretaceous.

The discovery of another family of Cretaceous pleurodires, the Bothremydidae, has had a direct effect on the way living pleurodires are classified. The Pelomedusidae and Podocnemididae used to be considered part of the same family. The Bothremydidae, though, were actually closer to the Podocnemididae than to the Pelomedusidae. As a result, the Podocnemididae must be treated as a separate family, one that split from the Pelomedusidae as far back as the Cretaceous. The Bothremydidae include *Taphrosphys*, a possible seagoing pleurodire, and *Foxemys*, a recent discovery from southern France.

We do not have the space, here, to follow the living turtle families through the Age of Mammals to the present day. Their fossil record, though, is important. Three of the four cryptodire families that survive today only through a single representative—the leatherbacks (Dermochelyidae), the Central American river turtles (Dermatemydidae), and the pig-nose turtles (Carettochelyidae)—were once much more diverse. The Dermatemydidae and Carettochelyidae, each confined today to a single limited region, once ranged almost throughout the world. Pleurodires, too, once had a wider range. Though confined to the southern hemisphere today, in the Eocene members of the Podocnemididae lived in North America and Europe.

*Santanachelys gaffneyi* is the oldest-known cheloniid sea turtle. Its forelimbs are relatively unspecialized.

Instead of surveying their progress, though, we will end this chapter by looking at three highlights of turtle history: the rise of the sea turtles, the story of the meiolaniids, and the question of what, exactly, was the world's biggest turtle.

**Return to the Sea**

In the Mesozoic, the seas were invaded by reptiles from a number of different lineages: sauropterygians, ichthyosaurs (which may have been sauropterygians too), seagoing crocodiles, and mosasaurs, which were seagoing lizards or seagoing snakes. Marine turtles, too, arose in the Mesozoic. In fact, they arose at least three times.

Before the rise of modern sea turtles, during the Jurassic, a group of primitive cryptodires swam in the shallow seas off the coast of what is now Europe. These were the members of the family Plesiochelyidae, and they represented a radiation of marine turtles quite separate from the lineage that survives today. They were not particularly modified for an oceangoing life. Unlike modern sea turtles, their front limbs were no larger than their hindlimbs, so they could not have been the graceful, balletic swimmers the living species are (which does not mean they couldn't swim; living softshells, which swim very well, also have limbs of roughly equal size). They apparently vanished before the end of the Jurassic.

Modern sea turtles first appeared in the Cretaceous. A recent discovery from the famous Santana fossil beds of Brazil shows us what the earliest of them may have been like. *Santanachelys gaffneyi* was only described in 1998, by Ren Hirayama, a Japanese specialist in fossil sea turtles. *Santanachelys*, which lived about 110 million years ago, was very small by sea turtle standards: only about 20 cm (8 in) overall. Its front limbs had yet to become fully modified into long, flat, rigid swimming paddles. However, it has one crucial adaptation of modern sea turtles to life in the ocean: a space in the skull where the living species have large salt-removing glands. These glands are vital for an animal that must drink seawater, even inadvertently. Sea turtles apparently developed the ability to handle the salt load in seawater before they had fully evolved the elegant structures they use to swim through it.

Turtle paleontologists have not been able to agree on how many families of fossil sea turtles there are, mostly because of disagreements on how to handle some of the more primitive forms. They do agree that modern sea turtles and their fossil relatives, even including the highly distinctive leatherback, are each other's closest relatives, and belong in a single group, the Chelonioidea. The sea turtles of today are a pale remnant of the variety that existed in the past. In the Cretaceous, there was a wide range of specialized sea turtles, with a second, rather smaller flowering of forms in the Eocene. They included the very large European turtle *Allopleuron*, seemingly highly evolved for life in the open sea, with an elongate, streamlined shell and rigid hindlimbs that were more likely rudders than swimming paddles. Quite different were relatively primitive forms like *Toxochelys*, which had a rounded shell, retained the unspecialized, movable flippers of *Santanachelys*, and whose eye sockets faced upward, suggesting that, like modern snapping and softshell turtles, it spent much of its time lying on the bottom in shallow water.

*Calcarichelys gemima*, a protostegid sea turtle from the Cretaceous of Alabama, flees the 4 m (12 ft) mosasaur *Clidastes*.

Cretaceous sea turtles seemed to stay in their own corners of the ocean. Some, for example, are found in European deposits but not in North America, and vice versa. That may explain why there were so many of them; each region had its own sea turtle fauna. Apparently, they did not undertake the great migratory journeys of their living relatives, though perhaps we just haven't found the fossils of Cretaceous wanderers, or perhaps some of the fossil turtles we have found in different regions are not really separate species.

Both of the modern sea turtle families, the Cheloniidae (typical sea turtles) and the Dermochelyidae (leatherbacks), had representatives in the Cretaceous. The most famous turtle to swim Cretaceous seas, though, belonged to an extinct family, the Protostegidae. This was *Archelon ischyros*, largest of all sea turtles, an inhabitant of the shallow Niobrara Sea that once rolled over what are now the Great Plains of central North America. Just how big *Archelon* really was we will discuss at the end of this chapter, but it must have been an impressive creature. Its head alone was almost a meter long, culminating in a powerful, hooked beak. *Archelon* and its near relative *Protostega* were the most advanced of the protostegids, a line of turtles that started with some fairly small and unspecialized animals (*Santanachelys* may have been a protostegid), but whose evolution led to larger and larger size, more and more massive heads, huge limbs, and, like their close relatives the leatherbacks, a tendency to reduce the amount of bone in their carapaces and, eventually, to lose their bony carapace scutes altogether.

The protostegids may have been adapting to the pursuit of a specific group of food items. The seas of the Mesozoic were dominated by the ammonites, distant relatives of the living chambered nautilus. Like it, ammonites had spectacular shells, some of them truly immense. Unlike the deepwater nautilus, though, many of them drifted in the upper layers of the ocean. To pursue them, an ammonite hunter needed the ability to swim after them and, once it had captured them, to crush their shells. The protostegids seem to have been well adapted on both counts. *Archelon*'s enormous flippers, or, rather, the shoulder joints that supported them, appear to have been designed more for straight-line distance swimming than for deep diving, and its long, formidable jaws would certainly have been capable of dealing with an ammonite shell.

Short of finding a protostegid fossil with an ammonite in its stomach, we cannot prove that *Archelon* and its relatives were actually eating ammonites, but it seems a reasonable enough speculation. It may also explain their eventual extinction. Ammonites went into a decline towards the end of the Cretaceous, and the protostegids appear to have done the same thing. By the time of the final catastrophe at the end of the Age of Reptiles, there may have been only one protostegid left, an animal called *Atlantochelys*, pursuing the last of the ammonites.

A third line of turtles also entered the sea during the Cretaceous. Unlike the groups we have considered so far, these were not cryptodires but pleurodires, members of the extinct family Bothremydidae. They included *Taphrosphys*, a late Cretaceous turtle known from North America, South America, Africa, and Europe. This was not as big a range as it may first appear. In Cretaceous times, the Atlantic was a narrow ocean, just beginning to spread. *Taphrosphys* may have been able to move freely among the continents along its shores (though, of course, if that is true one wonders why the "true" Cretaceous sea turtles did not do the same thing). *Taphrosphys* does not seem to have been particularly modified for life in the sea, but despite that, its fossils are always found in

Plaster cast of a horn core of a meiolaniid turtle from the Miocene of Bullock Creek, Northern Territory, Australia.

marine deposits. It, and its relatives, may have been as close as the pleurodires ever came to producing a seagoing turtle.

### Ninja Turtles

In 1992, Dr. Eugene Gaffney had to find a new generic name for *Meiolania oweni*, a fossil turtle from the Pleistocene of Queensland, Australia. He named it *Ninjemys*, "in allusion to that totally rad, fearsome foursome epitomizing shelled success." Let it never be said that turtle paleontologists lack a sense of humor.

If ever there really were ninja turtles, surely they would have belonged to the Meiolaniidae. More bizarre creatures than *Ninjemys* and its kin it is difficult to imagine. Large, tortoise-like animals, most meiolaniids sported startling head ornaments, culminating in steerlike horns that either arched backward (in the members of the genus *Meiolania*) or jutted straight out to the side (as they did in *Ninjemys oweni*). The only hornless meiolaniid, *Warkalania*, bore horizontal corrugated ridges instead, while *Niolamia*, the oldest member of the family and the only one known from South America, supplemented its *Ninjemys*-like horns with a bony frill over the back of its neck. Meiolaniid tails were surrounded by rings of dermal armor and tipped with a bony club.

Aside from the oldest of them, *Niolamia*, from the Eocene of Argentina, meiolaniids are known only from eastern and northern Australia, Lord Howe Island in the Tasman Sea, and the islands of New Caledonia in the Southwest Pacific. They apparently survived there until only about 100,000 years ago. What is less clear is how they reached some of these areas in the first place. South America, Australia, and New Caledonia were all part of the great Southern continent of Gondwanaland during the Mesozoic, so the first meiolaniids could have walked from one to the other (there are fragmentary remains from the Cretaceous of South America that may be meiolaniid). Lord Howe Island, though, did not emerge from the Tasman Sea until the late Miocene, so the ancestors of its horned turtle, *Meiolania platyceps*, must have reached it over water, with possible stops along the way on islands that no longer exist. Perhaps they drifted there, as the giant tortoises of Aldabra and the Galápagos probably did in their turn (see Chapter 4).

The name *Meiolania* means "lesser ripper." This may seem like a strange name for even such peculiar animals as these, but it stems from one of the more famous mistakes in paleontology. When George Bennett found the first meiolaniid skull in Queensland, Australia, in 1879 (the skull that Gaffney would later rename *Ninjemys*), he identified it correctly as belonging to a turtle. However, when he sent the specimen to England, it came into the hands of Richard Owen, the famous paleontologist who coined the name "dinosaur" and bitterly opposed Charles Darwin's theory of evolution. Owen rejected

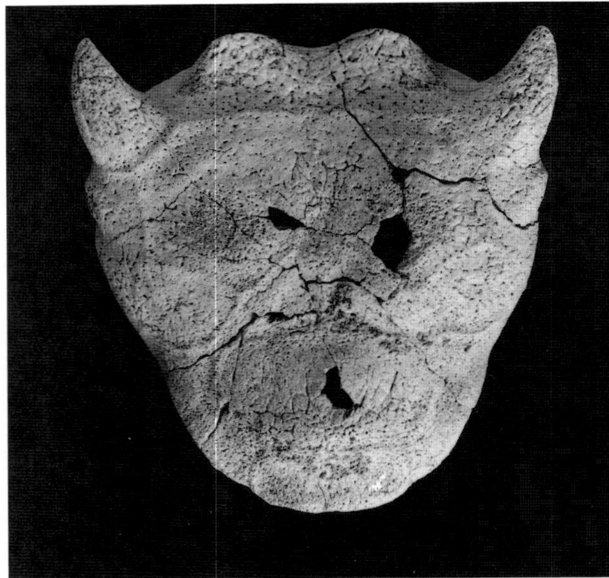

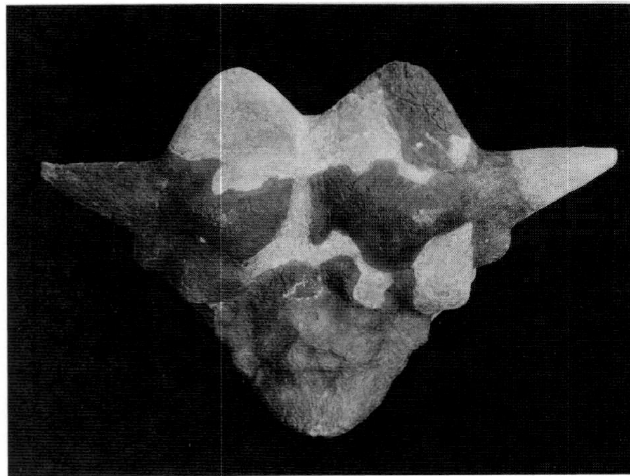

The horns of meiolaniid turtles either swept backward (*Meiolania platyceps*, above) or jutted straight out to the side (*Ninjemys oweni*, below; the lighter areas have been restored).

Bennett's conclusion, and decided instead that the skull belonged with the bones of a giant monitor lizard he had described in 1858 as *Megalania prisca*. *Megalania* means "great ripper," a name the lizard certainly deserved.

When Owen received specimens of the Lord Howe meiolaniid, a smaller animal, he decided that they were lizards, too, and named them *Meiolania*. A few years later, he added to his mistake by mixing in foot bones from *Diprotodon*, a giant wombat-like marsupial. By 1887, Darwin's great champion, Thomas Henry Huxley, was able to use newly discovered material from Lord Howe Island to show that Owen's "lizard" was, as Bennett had thought, a turtle. Owen grudgingly admitted that perhaps his remains belonged to a group of animals intermediate between lizards and turtles—a peculiar stance for someone who did not believe in evolution. Huxley tried to give the turtles a more appropriate name, *Ceratochelys*, which means "horned turtle," but was defeated by the rules of zoological nomenclature, which give priority to older names, however inept. The name *Megalania* is now only associated with the giant lizard named by Owen; Bennett's skull became *Meiolania oweni* (later to be changed to *Ninjemys oweni*); and the Lord Howe horned turtle remains *Meiolania platyceps*.

There has been considerable controversy over where the meiolaniids fit on the turtle family tree. They have been variously called pleurodires, cryptodires, or something neither pleurodire nor cryptodire but perhaps related to the earliest turtles. Like *Proganochelys*, but unlike modern turtles, meiolaniids had ribs attached to their neck vertebrae. Certainly meiolaniids were incapable of retracting their necks in either the pleurodire or cryptodire fashion. As Eugene Gaffney wrote, "Even a cursory examination of the horns on a *Meiolania* skull would show that either form of neck retraction would result in a kind of chelonian 'hara-kiri'." Dr. Gaffney has argued that meiolaniids are, nevertheless, cryptodires—though very peculiar ones—based on a number of characters including the pattern of arterial circulation in the head and the jaw muscle mechanism.

*Meiolania platyceps* from Lord Howe Island east of Australia, best known of the bizarre horned turtle family (Meiolaniidae).

We can only speculate on the uses to which meiolaniids would have put their formidable armament. In South America and Australia, they would have faced some large and dangerous predators, but not on Lord Howe Island, where they were probably the largest animals around. Perhaps they fought each other in spectacular battles over mates, but we will never know for sure.

The Lord Howe Island horned turtle, *Meiolania platyceps*, is the only one of its family for which we have reasonably complete skeletons. The others are known from skulls, tail armor, or even more fragmentary remains. At 2 m (6.5 ft) overall length, it was a substantial beast. If skulls and bits of horn are any guide, though, it was actually one of the smaller meiolaniids. Others may have been twice as large, making them among the largest turtles ever to have lived. The skull of *Ninjemys* measures almost 70 cm (2 ft) across from horn tip to horn tip. Without better specimens, though, they cannot, bizarre and wonderful as they may have been, be candidates for consideration in the final section of this chapter, which addresses that very question.

## Giants

We started this chapter with one of the most difficult questions one can ask about fossil turtles: how did they evolve? In following our story thus far we have dealt with other, equally complex and abstruse issues. Let us end the chapter, though, with a much more straightforward question: what was the biggest turtle that ever lived?

Most people, assuming they stopped to think about such things, would give the crown to *Archelon*, the giant protostegid sea turtle of the Cretaceous. *Archelon ischyros* was the largest sea turtle known to science (but it may have had a near rival in *Cratochelone berneyi*, known from fragmentary remains

*Meiolania platyceps* is the only meiolaniid for which we have nearly complete skeletons; this one has been reconstructed.

*Archelon ischyros* from the Cretaceous of North America, largest of the Protostegidae, and the biggest marine turtle known.

from the early Cretaceous of Australia). It was certainly an enormous animal, though its exceptionally long skull may give it a greater total length than its actual bulk implies. Turtle biologists tend to think of size in terms, not of total length, but of the length of the carapace (a more useful measurement for fossils anyway, as paleontologists don't always find the whole animal). In that department, *Archelon* may not have been much longer (though it was much wider) than a modern leatherback; its carapace, measured in a straight line, is about 1.9 m (6.3 ft) long.

Exactly how big *Archelon* was overall, though, seems a matter for some debate. Three meters or so (9–10 ft) may have been an average length. Two specimens found in South Dakota in 1996 and 1998 measured roughly 3.6 m (12 ft) from snout to tail. A specimen on exhibit in Vienna, collected in the mid-1970s, measures 4.5 m (15 ft) from snout to tail and 5.25 m (16.5 ft) from tip to tip of its outstretched flippers. It may have weighed 2200 kg (4500 pounds) when alive.

At one time, the most likely candidate for the largest land or freshwater turtle would have been a giant tortoise that roamed India and Indonesia about 2 million years ago, during the Pleistocene. From its fragmentary first remains, the naturalists who described it in 1844 calculated that its carapace was a staggering 3.6 m (12 ft) long. They named it *Colossochelys atlas* in consequence. Had this calculation proven to be correct, *Colossochelys* would unquestionably have remained the largest turtle known, even to this day. Unfortunately, later workers, with better material, revised the figure downward to 2.3 m (8 ft), and then to about 1.8 m (6 ft). That still makes it one of the largest tortoises ever found (though there is a similar-size fossil giant known from Spain). It was much bigger than any living form, but it is out of contention for the overall world record. To add insult to injury, it turns out, on further examination, not to have been different enough from living tortoises of the genus *Geochelone* to warrant its own special generic name. It is now known as *Geochelone sivalensis.*

With *Colossochelys atlas* (or *Geochelone sivalensis*) whittled down to size and shorn of its impressive name, the world record for the largest turtle that ever lived, based on carapace size—at least as far as we know—passes to another, even more remarkable creature. In 1972, a paleontological expedition from Harvard University unearthed the remains of several huge fossil turtles in northern Venezuela. Roger Conant Wood, who described them in 1976, gave these "gargantuan" turtles, as he called them, the name *Stupendemys geographicus.* The origin of the name *Stupendemys* should be obvious enough—"emys" simply means "turtle." *Geographicus* refers to the National Geographic Society, which funded Wood's research.

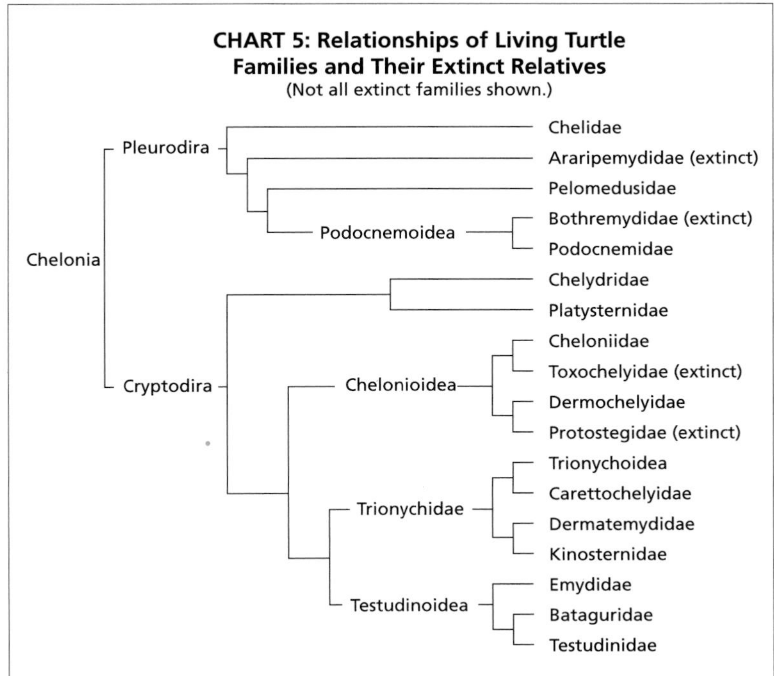


Gratitude, in science, makes good sense.

The largest *Stupendemys* carapace Wood examined was 2.3 m (8 ft) long. That would have been enough to give it the record right there, now that *Colossochelys* is out of the running, but in 1992 an even larger specimen turned up, 3.3 m (11 ft) long and 2.1 m (7 ft) wide. Unfortunately, we don't have a complete skeleton, so we cannot be certain exactly how this translates into the overall size of the animal. With a shell like that, *Stupendemys* would appear to be well in front of its rivals.

*Stupendemys* lived about 8 million years ago, in the late Miocene. It was a member of the same pleurodire family, Podocnemididae, as the largest living sideneck, the giant river turtle or arrau (*Podocnemis expansa*) of northern South America. The enormous, flattened shell of *Stupendemys* is very like that of other aquatic podocnemidids. Like the arrau, it probably spent almost all of its life in the water—probably in fresh water, though, like the Cretaceous sideneck *Taphrosphys*, it may have entered the ocean as well. Its peculiarly squat, massive humerus suggests that its limbs, unlike those of other pleurodires, may have been modified into flippers like those of a chelonioid sea turtle.

In recent years, remains of *Stupendemys* have been found in several parts of the Amazon basin, showing that it must have once had a very large range. With these remains, paleontologists have found others suggesting that *Stupendemys* may have only been one of a whole group of giant podocnemidid turtles in the region. One, known only from a single vertebra, may have been even larger than *Stupendemys.*

*Stupendemys*, in fact, lived in a land of giants. The Urumaco formation of northern Venezuela where it was first discovered, was a vast complex of fresh- and brackish-water marshes during the Miocene, like the Everglades or the Pantanal today. It has yielded fossils of giant crocodiles, and, surprisingly, giant rodents. One of them, *Phoberomys*, was a monster even more surprising than *Stupendemys*: the biggest rodent that ever lived. It was 3 m (10 ft) long, 1.3 m (4 ft) tall, and weighed perhaps 500 kg (1100 lb)—fit company, perhaps, for the largest of the turtles.

CHAPTER THREE

# *Turtles Round the World I: Side-Necks and Hidden-Necks*

THERE ARE ONLY about 280 to 300 species of turtle alive today. That is a small number indeed compared to the numbers of extant lizards (over 4,500) or snakes (nearly 3,000). Yet, that small number covers almost as great a span of the globe as either of those larger groups, and, if the sea turtles are included, a much greater one. Turtles live on every continent except Antarctica, roam all but the coldest seas, and have reached a range of isolated oceanic islands from the West Indies to the Galápagos and the islands of the Indian Ocean.

Of the three families of pleurodires and 11 families of cryptodires, only the tortoises (Testudinidae), come close to having a global reach on land, and even that family is missing from Australia. Cryptodires are mostly northern (except for the sea turtles), though tortoises have done very well in Africa, and one family, the Carettochelyidae, survives today only south of the equator in New Guinea and tropical northern Australia. Pleurodires are entirely southern, confined to South America, Africa, Madagascar, and Australia, with a single, tiny toehold in Asia—though, once again, fossils show that they once enjoyed a wider range.

The northern long-necked turtle (*Chelodina rugosa*), like all but one of Australia's freshwater turtles, belongs to the Chelidae.

**How the Pleurodires Got to Where They Are**

The living pleurodires fall naturally into two groups, one containing the more "advanced" Chelidae, and the other the remaining two families, the Pelomedusidae and the Podocnemididae. The Chelidae live in Australia and South America. The Podocnemididae has all but one of its members in South America, but that one exception lives, of all places, only on the island of Madagascar. The third family, the Pelomedusidae, is the only pleurodire group in Africa.

This distribution looks extremely peculiar, until you remember that the southern land masses where these families live were once united into a single supercontinent. This great vanished block of earth, which received the name Gondwanaland long after it had sundered, contained the pieces of the earth's crust that were to become Antarctica, South America, Africa, Madagascar, Australia, New Zealand, Arabia and India. In the early Jurassic it even included part of the future North America. Gondwanaland did not break up until the Jurassic, probably after the pleurodires and cryptodires had gone their separate evolutionary ways. Pleurodire turtles were free to spread throughout Gondwanaland, perhaps even crossing the narrow stretches of sea that opened as the supercontinent split.

What is odd is that none of the Gondwanaland remnants carries all three living pleurodire families. There are no chelids in Africa, no podocnemidids in Australia, and no pelomedusids in South America. There are no living pleurodires in India, and no land turtles of any kind in New Zealand.

The explanation for this may be a phenomenon biogeographers call *vicariance*. A vicariant event, like the breakup of Gondwanaland, splits up what was once a continuous ecosystem. It isolates animals and plants into separate areas where they follow their own independent histories. In some of the separate bits, one group may go extinct and another may survive; in other areas, for one reason or another, the opposite may happen. We know that happened to the podocnemidids because fossils related to the living genus *Podocnemis* have been found in Africa, India, and even in Europe and North America. No fossil chelids, though, have ever been found in Africa. Could it be that they never occurred there? What is their history?

Gondwanaland did not break up all at once. By the mid-Cretaceous, 125 million years ago, Africa had drifted away to the north, but South America remained close to, and, perhaps, joined with, a land mass that still included Antarctica and Australia. Antarctica was not a polar wasteland but a rich expanse of territory boasting dense, cool forests, its own special dinosaurs, and turtles. Their presence there is testified to by a single fossil bone, a neural dating from the Miocene. The chelids, which can tolerate cold better than the other pleurodires, may have evolved in Antarctica. They could have walked into Australia and crossed the narrow ocean gap (if indeed there was one) that may have separated Antarctica from Patagonia. By the Cretaceous, though, it was too late for them to walk, or swim, into Africa.

Did the chelids make the crossing to South America once or twice? In both Australia and South America, chelids come in two distinct body types: long-necked and short-necked. For a long time, herpetologists assumed that the Australian longnecks of the genus *Chelodina* were close relatives of the South American longnecks of the genus *Hydromedusa*. If that is true, it would mean that the present distribution of the Chelidae is the result of two separate vicariant events. The first would have split the family into long-necked and short-necked forms. Each of these lines would have then spread on its own to Australia and South America, before another vicariant event split them into Australian and South American groups, and Antarctica lost all its turtles beneath miles of ice.

It now appears, however, that *Chelodina* and *Hydromedusa* are not each other's closest relatives after all. They differ in a number of anatomical features, including details of the skull and forward extension of plastron that shields the head and neck when they are tucked out of the way. Recent molecular studies, though unclear on other details, support the idea that the two longnecks are not closely related. Instead, *Chelodina* appears to be most closely related to the Australian short-necked chelids, while *Hydromedusa* seems closer to the other South American species. If this is true, we now need only one vicariant event to explain the modern distribution of the Chelidae, the one that separated the Australian and South American branches of the family. It also means that chelids evolved long necks at least twice, once on each continent. Since both genera use their necks in the same way—to make a rapid, snakelike strike at their prey—the fact that they have hit upon the same evolutionary solution should not be too surprising.

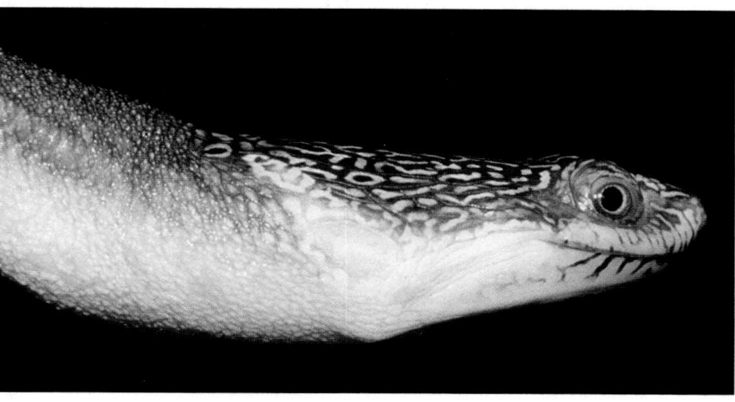

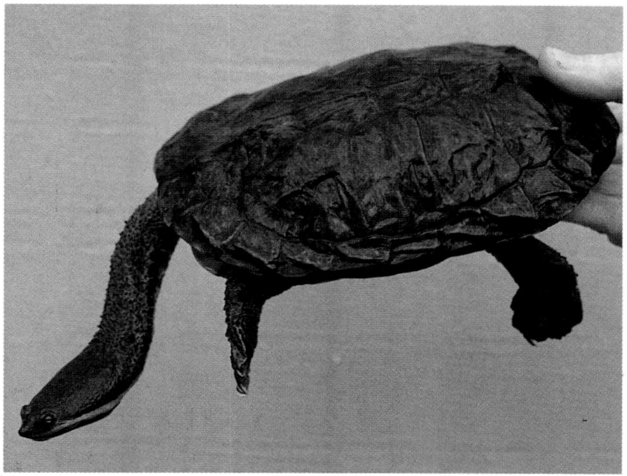

Australasian (e.g. Parker's long-necked turtle *Chelodina parkeri* from New Guinea, above) and South American (e.g. the South American snake-necked turtle [*Hydromedusa tectifera*], below) long-necks probably evolved long necks independently.

## Chelidae: Australo-American Sideneck Turtles

There are roughly 50 species in the family Chelidae, the majority of them—some 30 or so—located in Australia and New Guinea, with the remainder in South America. Chelids are flat-headed and often rather flat-shelled animals, with soft skin or small scales covering their heads instead of the large, platelike scales that cap the other sidenecks. Chelids

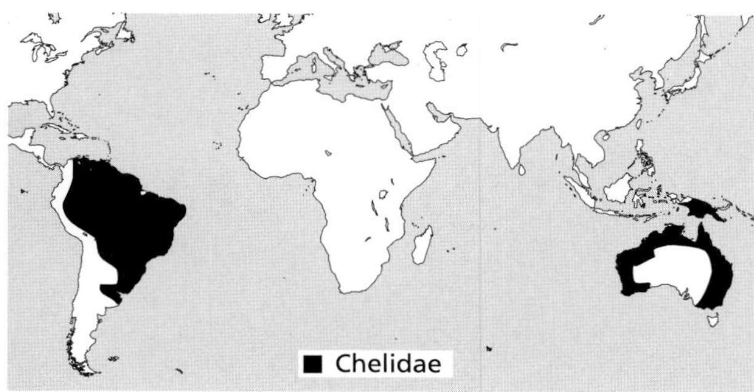


spend most of their lives in fresh water. Some, like the Mary River turtle (*Elusor macrurus*) and the Fitzroy River turtle (*Rheodytes leukops*) of Australia, almost never leave it. The northern long-necked turtle (*Chelodina rugosa*) even lays its eggs underwater—a truly remarkable feat (see Chapter 7). On the other hand, the South American twist-necked turtle (*Platemys platycephala*) and Zulia toad-headed sideneck (*Phrynops zuliae*) often emerge to hunt for food on the forest floor.

The Australasian chelids first came to the knowledge of western scientists during Captain James Cook's 1770 expedition to Australia, when Joseph Banks collected a snake-necked turtle (*Chelodina longicollis*) somewhere near Botany Bay, south of the present-day city of Sydney. To a northerner, the *Chelodina* turtles are startling animals, with elongate, seemingly overstretched necks and eel-like heads. When a snake-necked turtle extends itself, its head and neck add some 60 to 65 percent to its overall length. The oblong turtle (*Chelodina oblonga*) of Australia's southwestern corner is even more extreme, with a head and neck stretching to 90 percent of the length of its carapace—an impressive figure for a turtle whose shell may reach 31 cm (12 in).

The snake-neck is one of a dozen or so long-necked turtles from Australia and New Guinea. In the southeast, it shares much of its range with the broad-shelled turtle (*Chelodina expansa*); both live throughout Australia's

The Arnhem land long-necked turtle (*Chelodina burrungandjii*) was only recognized as a separate species in 2000.

largest river system, the Murray–Darling, in lagoons, lakes, and swamps and in the rivers themselves. New *Chelodina*s are still being described: the Arnhem Land long-necked turtle (*Chelodina burrungandjii*) was only named in 2000, by Scott Thomson, Rod Kennett, and Arthur Georges, and another new species, from farther west in the Kimberlies, awaits a scientific name.

*Chelodina* turtles are active and very efficient predators. Fish, freshwater shrimp, tadpoles, and frogs are the chief items on their menu, but the oblong turtle will occasionally seize a waterbird. The Arnhem Land longneck, unusually for its genus, also eats some plant material. When necessary, most will undertake lengthy overland migrations to reach better feeding sites. When a snake-necked turtle crawls from one pond to another, it shelters its head and neck from the hot sun beneath the overhanging lip of its carapace. Broad-shelled and oblong turtles, though, seem to disdain such measures, and stretch out their heads and necks to their full extent as they march along.

The short-necked Australasian turtles are grouped into five genera, three of them—*Elusor*, *Pseudemydura*, and *Rheodytes*—with a single species each. The Mary River turtle (*Elusor macrurus*) is the largest chelid in Australia, with a carapace 40 cm (16 in) long. Males have a long, heavy tail as thick as a human wrist—the largest tail on any freshwater turtle, much bigger than its head and neck. What it uses this immense tail for we have no clear idea, but it probably has something to do with mating. The male's cloacal opening is a long slit more than halfway down the underside of the tail, so possibly the male uses its tail to bring the opening into position for inseminating the female. At least, that is the rule for other male turtles boasting less impressive equipment.

Remarkably, the Mary River turtle remained unknown to science until 1990. Starting in the early 1960s, pet shops in Adelaide, Bristol, Melbourne, and Sydney began receiving large shipments of hatchling turtles that no one could identify. Although the ship-

The recently discovered Mary River turtle (*Elusor macrurus*) is confined to a single river system in Queensland, Australia.

A hatchling Mary River turtle (*Elusor macrurus*) sports the "egg-tooth" that it used to break out of its shell.

ments continued until around 1984, when it became illegal to sell hatchling turtles in the state of Victoria, no one knew where they were coming from. For 25 years, Australian turtle enthusiast John Cann explored river systems from one end of Australia to the other (and in New Guinea, just in case), searching for the turtle he called "shortneck alpha."

Finally, a contact in Maryborough, Queensland, John Greenalgh, sent Cann a note saying, "I've got one." Cann dashed to Maryborough, only to find that the turtle swimming in a water-filled metal drum on his host's property was perfectly ordinary. Cann was furious, until Greenalgh said, "Well, if that's not him, let's have a look in this drum." The next drum held a full-grown female of Cann's mystery turtle.

In October 1990 Cann found four specimens himself, in the murky waters of the Mary River, south of Maryborough, an area already well known to zoologists as one of only two rivers where the Australian lungfish (*Neoceratodus forsteri*) is found. Cann described it in 1994, with American biologist John Legler, as *Elusor macrurus*, meaning roughly "the hard-to-find animal with the big tail." As far as we know, it lives only in deep pools along 100 km (60 mi) of the river's middle reaches.

The Fitzroy River turtle or white-eyed river diver (*Rheodytes leukops*), one of the most thoroughly aquatic of all turtles, is another fairly recent discovery by John Legler and John Cann. This time only seven years elapsed from 1973, the first time a turtle hobbyist showed Cann a specimen he could not identify, until its official description in 1980. Like the Mary River turtle, *Rheodytes* is confined to a single river system in Queensland, this time the drainage of the Fitzroy River, named after the captain of Charles Darwin's ship the *Beagle*. The Fitzroy River itself has become polluted from the tailings of upstream mining operations, but, fortunately, the turtle lives in the other major rivers of its drainage system. Its hatchlings are distinctive, with shells whose serrated edges look as though they had been trimmed with a pair of pinking shears.

The western swamp turtle (*Pseudemydura umbrina*) is not only the rarest turtle in

Another recently discovered and highly distinctive Australian chelid, the Fitzroy River turtle (*Rheodytes leukops*).

The Critically Endangered western swamp turtle (*Pseudemydura umbrina*) lives only in swampland near Perth, Australia.

Collecting Fitzroy River turtles (*Rheodytes leukops*) in their only home, the waters of the Fitzroy River drainage.

Australia, it is one of the rarest in the world. Though described in 1901, it was forgotten until 1953 when a Perth schoolboy named Robert Boyd exhibited one at a pet show. One of the judges, a famous Western Australian naturalist named Vincent Serventy, spotted the animal, had no idea what it was, and passed it on to Ludwig Glauert, curator of the Western Australian Museum. Glauert described it as a new species, but other scientists soon realized that it was the same animal that had been named at the beginning of the century.

No one has ever found the western swamp turtle anywhere but in two tiny patches of swampland about 30 km (18.6 mi) north of Perth. Both patches were purchased by the Western Australian government and set aside as nature reserves, Ellen Brook (65 hectares/163 acres) and Twin Swamps (155 hectares/387 acres), but unfortunately the population at Twin Swamps disappeared during the 1980s. That left only 20 or 30 individuals at Ellen Brook and a number of others in captivity. The situation was dire, but a combination of successful captive breeding, publicity, and habitat management has brought the turtle's numbers up to over 150. The western swamp turtle was reintroduced to Twin Swamps, now protected by an electrified fence, in 1994.

The other Australasian chelids are the river turtles of the genus *Emydura* and the Australian snappers of the genus *Elseya*. The Macquarie turtle (*Emydura macquarii*) is fairly typical of the *Emyduras*. It shares the Murray–Darling river system with its long-

A foxproof fence now protects Twin Swamps, one of only two sites for the western swamp turtle (*Pseudemydura umbrina*).

The Jardine River turtle (*Emydura subglobosa*) of New Guinea and tropical Australia is one of the more colorful chelids.

necked cousins, the snake-necked turtle and the broad-shelled turtle. Unlike its predatory relations, the Macquarie turtle is an omnivore, eating pretty much whatever it can get, including plant matter, insects, carrion, and fish. Like most of its congeners (members of the same genus), the Macquarie is a plain-looking turtle, with a pale stripe running from the corner of its mouth down the side of its neck. Some of the tropical *Emyduras*, though, like the Jardine River turtle (*Emydura subglobosa*) of New Guinea and the northern Cape York Peninsula of Australia, are quite colorful animals, with stripes and patches of yellow and red.

The Victoria River or northern Australian snapper (*Elseya dentata*) lives (in a variety of forms, some of which may actually be different species) in river systems across the tropical north and northeast of Australia. It may live up to its name if handled carelessly; a large one can deliver a severe bite. Each of the marginal scutes on its shell is tapered into a point, giving the edge of the carapace a saw-toothed look; hence its specific name, *dentata*, which means "toothed." The Victoria River snapper is almost entirely a plant eater. Other *Elseya* turtles may be more carnivorous, like the saw-shelled turtle (*Elseya latisternum*) of northern and eastern Australia, or, like Bell's turtle (*Elseya bellii*), omnivorous. They are equally effective biters, and the saw-shelled turtle can emit a powerful musky stink.

The rest of the chelids live across the Pacific, in South America. There are three genera of South American short-necked chelids: *Platemys* (with three species); *Acanthochelys* (with two species, which used to be included in *Platemys*); and *Phrynops* (with 12 species). The species in *Platemys* and *Acanthochelys* are quite small turtles. Most are less than 20 cm (8 in) in carapace length. The twist-necked turtle (*Platemys platycephala*) of northern South America lives in shallow water in marshes, ponds, and rainforest creeks. It is apparently a poor swimmer, but will leave the water and travel overland at times. Russell Mittermeier of Conservation International has found several in puddles on forest paths in Suriname, and they have been captured while swimming by night among the trees in the flooded rainforest.

The saw-toothed edge of its carapace gives the Victoria River snapper (*Elseya dentata*) its specific name.

The twist-necked turtle (*Platemys platycephala*) lives in shallow water in rainforests of northern South America.

*Phrynops* turtles are a variable lot. Recent molecular evidence suggests that they should be split into three separate genera. The red Amazon sideneck (*Phrynops rufipes*) of northwestern Brazil and southeastern Colombia seems to be a food specialist, something unusual among turtles. It lives only in closed-canopy rainforest streams, where it feeds on palm fruits that fall into the water. Although once thought to be one of the rarest turtles in the world, a recent study of the species by William E. Magnusson and his colleagues has suggested that it may be one of the most abundant turtles of the Amazon.

Toad-headed sidenecks have large, broad heads with powerful, bulging jaw muscles. The common toad-headed sideneck (*Phrynops nasutus*) of the Guianas and the Amazon basin has a head width equal to one-quarter of the length of its carapace. What it does with this enormous head we are not quite sure. Its jaw surfaces are poorly developed and, perhaps fortunately, it does not bite when handled.

We do not know much about the two South American longnecks. The South American snake-necked turtle (*Hydromedusa tectifera*) lives from southeastern Brazil to Paraguay, northeastern Argentina, and Uruguay. Maximilian's snake-necked turtle (*Hydromedusa maximiliani*) is confined to the coastal Atlantic rainforest of Brazil, one of the most endangered habitats on earth. While most South American freshwater turtles dwell in large, deep, often muddy rivers, Maximilian's snake-neck lives in clear, shallow, rather cool rocky streams where it hunts insects, crustaceans, and other prey. It seems to require unpolluted water flowing through undisturbed montane rainforest, a preference that may affect its chances of survival in the future. There is little undisturbed montane rainforest left in southeastern Brazil.

There is a third South American longneck, farther north in the Amazon and Orinoco drainages: the matamata (*Chelus fimbriatus*). According to some recent molecular studies, it may be only distantly related to *Hydromedusa*. Whatever its affinities, it is the best-known, and certainly the most bizarre, of South American turtles. Once you have seen a matamata, you will surely never forget it. It looks more like a pile of dead leaves than a turtle—useful camouflage for an ambush predator that spends most of its time lying in wait for prey at the bottom of murky streams. Its carapace is flattened and rough, each scute raised into a conical boss. Its head is so broad and flat that it looks squashed, and is decorated with fringed, fleshy filaments on the underside and along the sides of its neck. Its snout is drawn out into a long, thin snorkel whose tip only needs to touch the water surface to keep the animal supplied with air. The matamata is the largest chelid in South America, though perhaps not quite as big as the two biggest Australian species, the broad-shelled turtle and the Mary River turtle.

The matamata is such an aquatic animal that it rarely even comes to the surface. It is a weak swimmer that may only emerge to nest, close to the water's edge. Despite its seeming sluggishness, though, it is a very effective hunter, with a remarkable feeding technique (see Chapter 6). Its strange appearance has had a paradoxical effect on its relationship with people. It is highly prized in the pet trade, but, as Peter Pritchard and Pedro Trebbau reported in *The Turtles of Venezuela:*

South Americans may avoid eating the matamata turtle (*Chelus fimbriatus*) because of its bizarre appearance.

> . . . although there are no reports of *C. fimbriatus* being unpalatable or poisonous, they are so peculiar in appearance that they are not eaten even in many areas were turtles of all other types—even mud turtles—are highly favored. In Venezuela, I found no evidence that this species was eaten along the middle Rio Orinoco, even though the more numerous and "normal"-looking *Podocnemis* were eaten in large numbers . . . Fiasson (1945) wrote that the flesh of the matamata was as tasty as that of *P. unifilis*, but that the former was not sought after due to its repulsive appearance.

## Podocnemididae: South American River Turtles and Their Malagasy Cousin

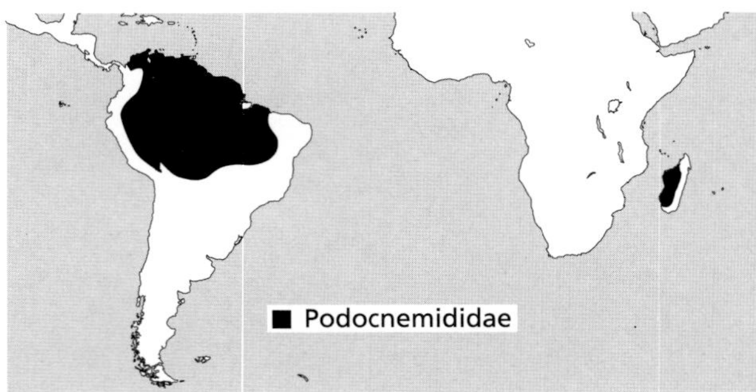


The other seven pleurodire turtles of South America—the big-headed Amazon River turtle (*Peltocephalus dumerilianus*) and the six species of *Podocnemis*—belong to the Podocnemididae. Most are river turtles, with broad, rounded shells streamlined for swimming in a current. Several enter backwaters and ponds, and the savanna side-necked turtle (*Podocnemis vogli*) is confined to the llanos of Venezuela and Colombia, an Everglades-like maze of swamps, streams, and pools.

The genus *Podocnemis* includes a species of great scientific interest, ecological significance, and economic importance, the arrau or giant South American river turtle (*Podocnemis expansa*). It is the largest living pleurodire. The carapace of a male arrau may be 40–50 cm (16–20 in) long. Females are significantly larger, averaging between 60 and 70 cm (24–28 in) in carapace length. An average female arrau may weigh between 20 and 25 kg (45–55 lb). In earlier times, before the largest specimens had been hunted out of existence, some may have weighed up to 90 kg (200 lb).

The arrau lives, or lived, throughout much of the Amazon, Orinoco and Essequibo basins. It spends most of its life in the larger rivers, swimming into the flooded forest during the wet season to feed on fallen fruits. Males almost never leave the water, and

The giant arrau (*Podocnemis expansa*), photographed in the Tamshiyacu-Tahuayo Community Reserve, Loreto, Peru.

females only do so to nest, or to bask on the edges of sandbanks in the weeks before nesting begins. The nesting season of the arrau ushers in a great natural spectacle. Legends along the Orinoco tell of a magical "turtle lady" who guards the arrau and shows them the best places to lay their eggs. Perhaps responding to her call, huge numbers of turtles crawl ashore at night on traditional sandy nesting beaches. In the past, they may have numbered in the hundreds of thousands.

Such a high concentration of protein has proved irresistible to humans. For centuries, the meat and eggs of the arrau were a staple for river tribes. Some villages kept turtles in pens for year-round use. With the arrival of European settlement, exploitation of the arrau increased enormously. Oil from its eggs was used as, among other things, fuel for lamps, and even as insect repellent. By the mid-20th century, arrau populations had collapsed throughout much of their range (see Chapter 10). Only in the last few years has the

Brazilian government begun a special effort to protect arrau nesting beaches—an effort that has apparently paid off, resulting in a 1200 percent increase in egg production in the 13 years prior to 1997. A similar program is also in place in Venezuela, with some modest success. The World Conservation Union (IUCN) currently classifies the arrau as Lower Risk/Conservation Dependent: a species that requires continuing conservation efforts if its situation is not to deteriorate.

The other *Podocnemis* turtles are considerably smaller than the arrau, and have not been the target of such an intense harvest. They are nonetheless heavily hunted. IUCN considers one, the Magdalena River turtle (*Podocnemis lewyana*) of the Magdalena River basin in Colombia, to be Endangered. The extremely attractive yellow-spotted river turtle (*Podocnemis unifilis*), reported to have the tastiest meat of its genus, has become rare in many areas, especially as the arrau has become difficult to find and hunters have turned their attention to its smaller cousins. As the yellow-spotted river turtle has declined in its turn, hunters have switched to a still smaller species, the six-tubercled Amazon River turtle (*Podocnemis sextuberculata*). Russell Mittermeier has compared this downward progression to a similar shift to smaller and smaller species by the whaling industry as whalers decimated the larger animals.

Far away from the other podocnemidids, across the Atlantic and past the continent of Africa, lives the Malagasy big-headed turtle (*Erymnochelys madagascariensis*), a large, gray turtle that may weigh up to 15 kg (33 lb). A Gondwanaland relic, it may be the only living turtle whose ancestors walked into Madagascar instead of floating there from Africa. Today, *Erymnochelys* lives in open water in slow-moving rivers, backwaters, and lakes in the lowlands of western Madagascar. We do not know very much about it, but fishermen in the area report that they usually only find *Erymnochelys* during the rainy season. In May, at the beginning of the cool, dry winter,

In Amazonian Peru, yellow-spotted river turtles (*Podocnemis unifilis*), and their eggs, are intensely hunted.

The Malagasy big-headed turtle (*Erymnochelys madagascariensis*) is the only Old World member of the Podocnemididae.

it disappears, but we do not know exactly where. A popular food item among the local people, *Erymnochelys* is heavily fished and may be endangered, though it still occurs over a wide, and little-studied, area.

**Pelomedusidae: African Mud Terrapins**

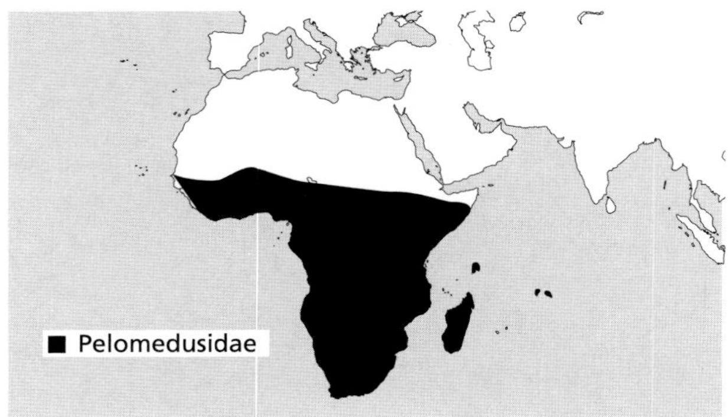


The Pelomedusidae, the only living sidenecks on the African continent, make up a fairly small family with only two genera and 16 or so species. At least 14 species of *Pelusios* terrapins live in sub-Saharan Africa. They prefer permanent bodies of water—a risky choice because crocodiles prefer the same places. Perhaps in response, most *Pelusios* species can protect their head and front limbs behind a hinged plastron—the only pleurodires to do so. They operate the hinge via a lever, connected to a powerful muscle derived, it seems, from one of their breathing muscles.

Their habitats range from the tropical rainforest streams, where the African forest terrapin (*Pelusios gabonensis*) lives, to the deep waters of the Okavango Swamp and the upper Zambezi, home to the Okavango hinged terrapin (*Pelusios bechuanicus*). Broadley's mud terrapin (*Pelusios broadleyi*) is known only from the southeastern shores of Lake Rudolph in northern Kenya. The pan hinged terrapin or East African black mud turtle (*Pelusios subniger*) lives in permanent waters over most of its East African range, but in southeastern Africa it will also live (as its common name implies) in pans and other temporary water bodies. So will the yellow-bellied hinged terrapin (*Pelusios castanoides*) and the West African mud terrapin (*Pelusios castaneus*). When their pools dry out, they bury themselves in the mud to await the next rainy season.

The first sideneck I ever saw in the wild was a helmeted terrapin (*Pelomedusa subrufa*) grimly making its way across the dry plains of the Serengeti. This common and widespread species ranges throughout Africa in dry, open country, from south of the Sahara to the Cape. It has even crossed the Red Sea to establish a population in Yemen, making it the only pleurodire in Asia. It is a thin-shelled turtle, usually missing from permanent bodies of water where crocodiles live. This vulnerability, though, may have served it in good stead. Its preference for temporary, crocodile-free rain pools has probably allowed it to spread to many areas that might seem unsuitable for a semiaquatic turtle. In South Africa, it even reaches the dry Central Karroo. Like the pan hinged terrapin, it buries itself in mud when its ponds dry.

The helmeted terrapin is a carnivore and scavenger, and may even pick parasites from the skins of rhinoceroses that enter its water holes. In his *Field Guide to Snakes and Other Reptiles of Southern Africa*, Bill Branch notes that helmeted terrapins "are belligerent, long-necked and ever-willing to bite or to eject their cloacal contents." He concludes, not unnaturally, that "they do not make the best pets."

A few pelomedusids have spread beyond Africa to the surrounding islands. The helmeted terrapin, the pan hinged terrapin, and the yellow-bellied hinged terrapin have reached Madagascar. The pan hinged terrapin

The helmeted terrapin (*Pelomedusa subrufa*) ranges widely through Africa, and also lives in Yemen and Madagascar.

also occurs on Mauritius. With the yellow-bellied hinged terrapin, it has reached the Seychelles, home of the Seychelles mud terrapin (*Pelusios seychellensis*), the only pleurodire to be confined to small oceanic islands. It is possible that the turtles reached these oceanic islands with human help. Even the rare , and doubtfully distinct, Seychelles mud terrapin may have a population in Africa. *Pelusios* was reported to have occurred on Diego Garcia in the Chagos Islands, a place it could not have reached by natural means. Humans certainly brought the pan hinged terrapin to the Americas; there is a small introduced population around Miami, Florida, and another on the island of Guadeloupe in the West Indies.

The pelomedusids on Madagascar must be fairly recent arrivals. Most Malagasy reptiles are found nowhere else. While the only other sideneck on the island, the Malagasy big-headed turtle, is an ancient relic whose nearest relatives live in South America, the three Madagascar pelomedusids do not even belong to endemic species (that is, species found nowhere else). If they have not displaced their much more ancient cousin, as they may have done on the African continent, it is probably because they occupy different habitats. *Erymnochelys* prefers open water in permanent rivers and lakes. The yellow-bellied hinged terrapin occupies densely vegetated areas, while the helmeted terrapin lives mostly in temporary ponds. The pan hinged terrapin does not even meet *Erymnochelys*; in Madagascar it is only found along part of the east coast, while the Malagasy big-headed turtle is confined to the west.

## The Cryptodires

The 11 families of cryptodires include four—

the Platysternidae (big-headed turtles), Dermochelyidae (leatherback turtles), Carettochelyidae (pig-nose turtles) and Dermatemydidae (Central American river turtles)—with but a single living species each. By contrast, there are some 45 species of tortoises (Testudinidae), and more than 65 species of Asian river turtles (Bataguridae), in 23 genera.

There has been some disagreement on exactly how the various cryptodire families are related to one another, but the most widely accepted view today is that the most primitive of them—the ones on the lowest branch of the cryptodire family tree—are the snapping turtles (Chelydridae), with their possible Asian ally, the big-headed turtle. The two families of sea turtles, Cheloniidae and Dermochelyidae, are the living representatives of a second branch, the Chelonioidea. The remaining limb of the cryptodire tree splits into two main branches. One, the Trionychoidea, includes, on one sub-branch, the softshell turtles (Trionychidae) and their relative the pig-nose turtle (Carettochelyidae), and, on the other sub-branch, the mud and musk turtles (Kinosternidae) and their relative the Central American river turtle (Dermatemydidae). The other main branch, the Testudinoidea, includes the Asian river turtles (Bataguridae), the tortoises (Testudinidae), and the American pond turtles (Emydidae). These three groups are closely related to one another, but there has been some difficulty in determining just how (see Chapter 4).

The center of cryptodire distribution is the northern hemisphere. Except for the wide-ranging sea turtles, only a single cryptodire species, the pig-nose turtle (*Carettochelys insculpta*), reaches Australia, and then only in its tropical north. Only two of the 11 families, the Trionychidae and Testudinidae, reached sub-Saharan Africa, though the tortoises have their highest diversity there. By contrast, six of the nine families of land and freshwater cryptodires are represented in North America.

Southeastern North America and eastern Asia are particularly rich in cryptodires, with diverse arrays of, respectively, emydids and batagurids. Both regions have a long history of climatic and ecological stability, free from waves of mountain-building and the advancing and retreating sheets of ice that scraped away many of the plants and animals of the other northern lands. Today, they are uniquely rich, not just in turtles, but in groups as diverse as magnolias and salamanders.

## Chelydridae: Snapping Turtles

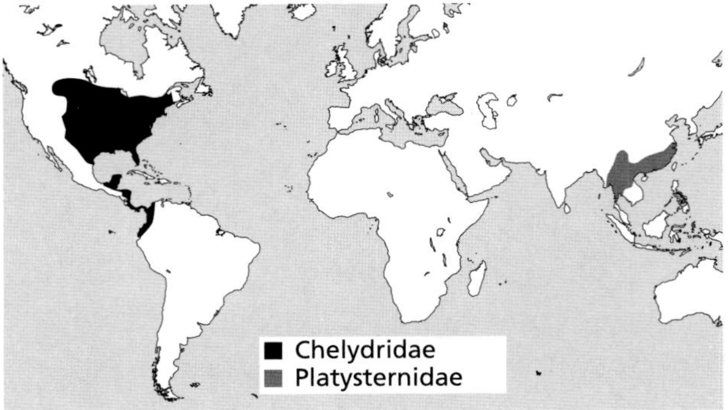


There are only two species in the Chelydridae, the common snapping turtle (*Chelydra serpentina*) and the alligator snapping turtle (*Macrochelys* [formerly called *Macroclemys*] *temminckii*). They are distinctive, fearsome-looking animals, with oversize heads and hooked jaws. Their carapaces are ornamented with three keeled ridges, though in an old common snapper the ridges may be almost worn away. The chelydrid plastron is very much reduced, shrunk to a blunt, bloated cross in the center of the belly. The arms of the cross still reach out to connect with the carapace by a narrow bridge.

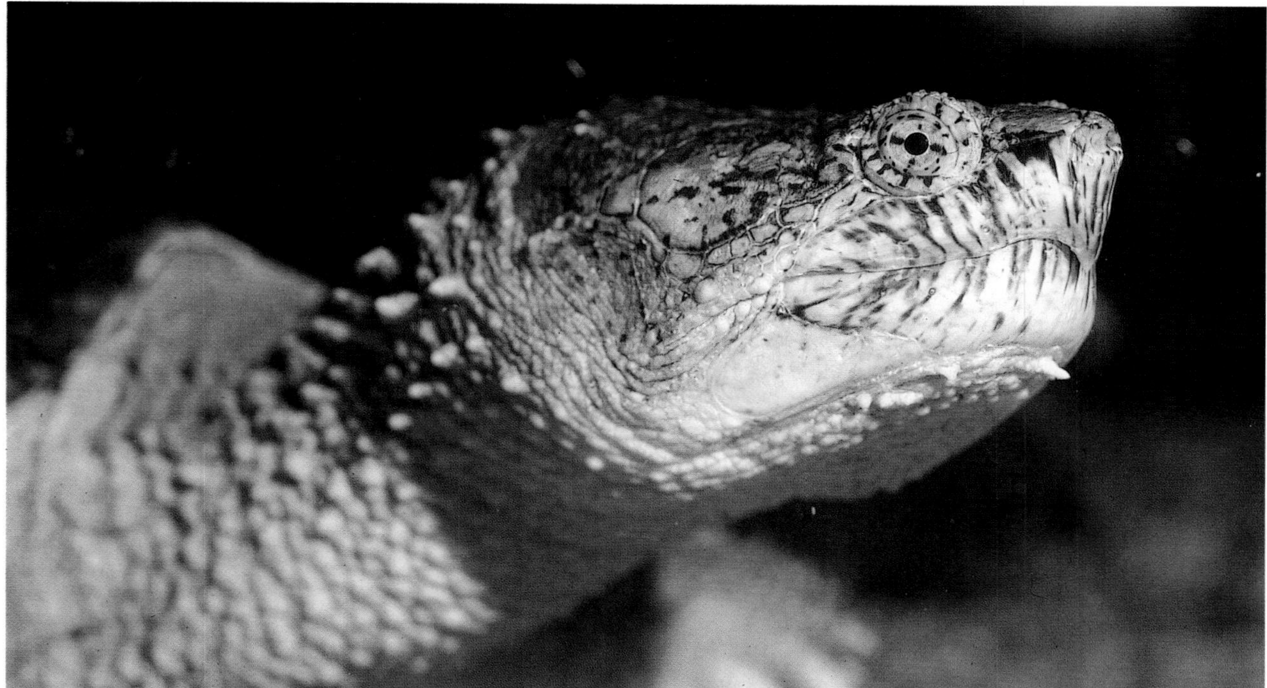

This common snapping turtle (*Chelydra serpentina*) was photographed in Shenandoah National Park, Virginia.

Chelydrids cannot fully withdraw into their shells, though this may have less to do with their supposed primitive condition than with the size of their heads.

Chelydrid tails are the longest on any turtle. A young snapper's tail is as long or longer than its carapace. Some people will tell you that the tail is the safest thing to grab should you wish to pick up a snapping turtle—advice you should not take, because picking up a large snapper in this way may injure or even kill it. Picking up a snapper by the sides of its shell is risky too; it has a long neck, and is quite capable of reaching around and biting you if you try. Given its famously pugnacious disposition, you probably should not pick one up at all. Its strike is amazingly fast, the edges of its jaws are sharp cutting tools, and it can also emit a foul-smelling musk from glands along the lower edge of its carapace and bridge. This is not an animal to treat lightly.

Like many another allegedly primitive creature, the common snapping turtle is tough, widespread, and successful. It ranges from southern Canada westward to the Rocky Mountains and southward through Central America, even penetrating into South America as far as western Ecuador—one of the widest ranges of any turtle. Snappers can live in almost any freshwater habitat they come across, even polluted ones, and they will enter brackish coastal waterways. They have a preference for soft, muddy bottoms in shallow water. Here, they bury themselves, waiting for prey, only their eyes and nostrils showing, occasionally stretching upward for a breath without emerging from their hiding place. A common snapper will eat just about any animal it can subdue, from a muskrat to a snail, in addition to carrion and various sorts of water plants. On the other hand, humans have developed a taste for snapping turtles, a taste that has taken a severe toll on some populations.

Though they are among the most aquatic of turtles, common snappers will occasionally climb out of the water to bask, or even make surprisingly long overland journeys; one was found on the top of Big Black Mountain, Kentucky, more than 2 km (1.24 mi) away from the nearest decent-sized stream.

The other representative of the Chelydridae, the alligator snapping turtle, is a truly remarkable creature. The alligator snapper is confined to the southeastern United States, from Kansas and Illinois south to northwestern Florida, Louisiana, and Texas, where it lives in the river systems flowing into the Gulf of Mexico. It prefers the deeper water of large rivers, though it will venture into lakes, ponds, oxbows, and bayous. Even more aquatic than its smaller cousin, it rarely leaves the water except to nest. The alligator snapper is one of the heaviest freshwater turtles in the world. It may reach 113 kg (248 lb).

The most obvious way to identify an alligator snapper, other than by sheer size and bulk, is the position of its eyes. In a common snapper, the eyes are positioned high and forward, but in the alligator snapper, they lie on the sides of its head. The keels on an alligator snapper carapace are even more pronounced than on the common snapper, and are more likely to persist even in old individuals. But to see the alligator snapper's most unusual feature, you would need to see it open its mouth—it is usually glad to oblige.

On the surface of its tongue is a little, whitish forked structure that sits on a rounded, muscular base. At times, it fills with blood, stretches out and comes to look like nothing more than a succulent pink worm. The resemblance is not an accident. What the turtle does with this peculiar object was first described in this account, quoted by Archie Carr in his classic *Handbook of Turtles*:

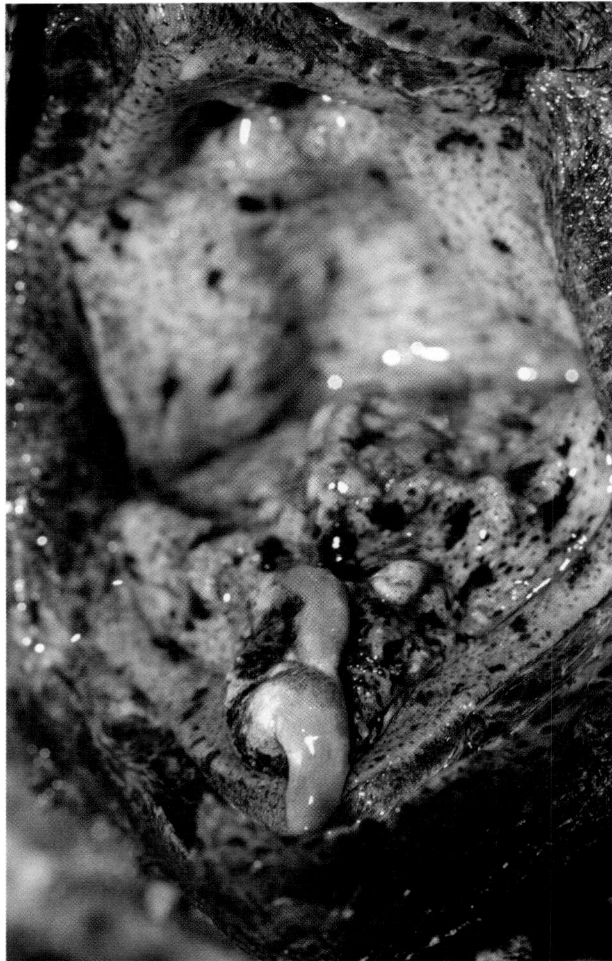

A close-up of the pink, wormlike lure on the tongue of an alligator snapping turtle (*Macrochelys temmincki*).

> Several baby *Macrochelys*, three to four inches in carapace length, were kept in an aquarium and supplied with live fish. The young turtles would hide between rocks in a corner of the aquarium and open their mouths widely. The muscular base of the lure would then pull down, first on one side and then on the other, imparting a wiggling motion to the two portions of the appendage. Sometimes the turtle would "fish" for hours without success, but often a *Molliensia* or a *Gambusia*

would swim into the open jaws and bite at the "bait." The turtle's jaws would immediately snap shut on the fish, which was next manipulated into position and then swallowed whole.

Peter Pritchard appears to have settled a long-standing controversy about the fearsomeness of the alligator snapper's bite. Although some writers have claimed that an alligator snapper is barely capable of biting a pencil in two, Pritchard taunted a 75 kg (165 lb) specimen with the handle of a brand-new household broom. The turtle bit clean through it.

The head of the big-headed turtle (*Platysternon megacephalum*) of Asia is heavily armored with tough, horny scutes.

**Platysternidae: The Big-Headed Turtle**

The big-headed turtle (*Platysternon megacephalum*), sole member of its family, lives in rocky mountain streams from southern China southward into Thailand. It deserves its name. Its head is so large that it looks out of place on its body—it may be half the width of the carapace.

Since the big-headed turtle cannot, for obvious reasons, withdraw its head into its shell, it carries armor instead. The top of its head is entirely covered by a single tough, horny scute, and its skull is completely roofed over with bone, without the emarginations for the jaw muscles found in most other turtles. The horny sheath lining the edges of its jaws extends up over its face, almost reaching the scute on the top of its head, leaving only a narrow band of exposed skin between eye and nostril.

Almost nothing is known about the big-headed turtle in the wild. It appears to be nocturnal and carnivorous, burrowing into gravel or under rocks in the bottom of a stream by day, and emerging to hunt for small animals in the streambed at night. As befits an animal that has to clamber over slippery boulders, the big-headed turtle is a remarkably good climber. It can scale sheer, nearly vertical slopes and even climb trees. Its bridge is flexible, perhaps as an adaptation to allow it to use its limbs more freely as it climbs. It can even use its long, whiplike tail as a climbing prop; its tail is strong enough to support the turtle's entire weight.

We are not exactly certain where this peculiar animal belongs on the turtle family tree. Eugene Gaffney of the American Museum of Natural History placed it in the Chelydridae. Indeed, it has some resemblance to snapping turtles, though it is far smaller, with a usual carapace length of only 18.4 cm (7.2 in). It has a similar combination of large head, long tail, and ridged carapace (though it has only a single, slight median ridge instead of three pronounced ones). Its shell structure, though, is very different; its plastron, for example, is much broader and more extensive. There are also a number of differences in its skeleton. Though its DNA supports chelydrid affinities, its chromosome patterns do not. Its similarities to the snappers are probably the result of convergence.

Where, then, does the big-headed turtle belong? We don't know; though instead of lying near the base of the cryptodire family tree, it may be better placed near the top, close to the Asian river turtles (Bataguridae). In the meantime, it is best kept in its own family—if only to display how little we know about it.

### Cheloniidae: Sea Turtles (Except the Leatherback)

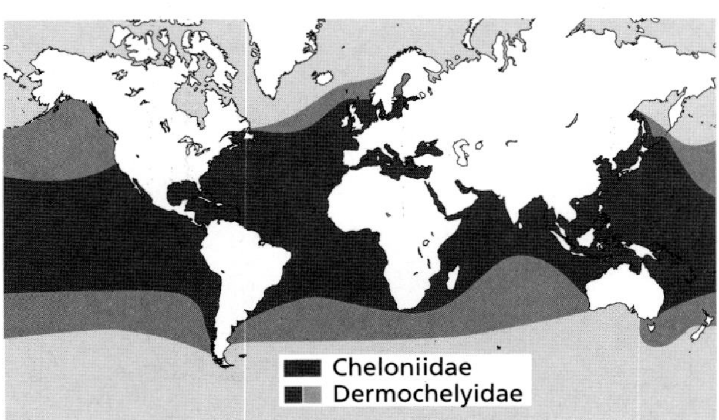


All of the living sea turtles, except the leatherback, are placed in the family Cheloniidae. Sea turtles are so distinctive, and so well known, that they hardly need describing. Beautifully adapted for life in the sea, they range throughout the warm waters of the world. Four of the six usually recognized species are almost global in range; only the leatherback, which swims into colder waters, has a broader distribution. The two exceptions are the flatback (*Natator depressus*) of the coastal waters of northern and northeastern Australia, and Kemp's ridley (*Lepidochelys kempii*), which is confined to the Atlantic and nests almost exclusively on the southern coast of the state of Tamaulipas, Mexico. The other ridley, the olive (*Lepidochelys olivacea*), ranges widely in the Indian and Pacific Oceans, but in the Atlantic rarely swims north of the Southeastern Caribbean Sea.

The six species of cheloniid sea turtles are the green (*Chelonia mydas*), the flatback, the hawksbill (*Eretmochelys imbricata*), the loggerhead (*Caretta caretta*), Kemp's ridley, and the olive ridley. Some scientists recognize a seventh, the black sea turtle (usually called *Chelonia agassizii*). Arguments have waxed hot and heavy in recent years about whether this animal, which lives in the Indian and Pacific Oceans, is really a separate species, a well-marked form of the green turtle, or merely a localized color phase. There have been accusations that some who recognize the black sea turtle do so only to make it easier to attract attention to it for conservation purposes.

Sea turtles fly through the water, beating their elongate flippers in a forward-tilting figure eight. Unlike a flying bird, a sea turtle can use the denser medium of water to drive it forward on the upstroke as well as on the downstroke. Underwater flight, with the flippers moving simultaneously up and down, is the normal way that sea turtles swim. It is not the only one; in shallow water, sea turtles will sometimes use their limbs in alternate strokes like other turtles.

The transformation of their limbs is far from the only adaptation sea turtles developed for life in the ocean. Their bodies have become streamlined for swimming speed and efficiency. In doing so, they have lost the sheltering overhang at the front of the carapace that protects most other turtles when they withdraw their heads into their shells. Without the ridge, water can flow smoothly over the shell as a turtle swims forward,

An Australian flatback sea turtle (*Natator depressus*) heads back to the sea while volunteers remove and count its eggs.

instead of being swept into turbulent eddies that could create friction and drag. Consequently, sea turtles cannot pull themselves into their shells, though the hawksbill does retain some slight ability to do so. Perhaps to compensate, they, like the big-headed turtle, have solid, roofed-over skulls.

Their reshaping for life in the sea has left sea turtles ungainly on land. A male sea turtle may never touch dry land again after he joins his siblings in digging himself out of his nesting burrow and scampers down to the water's edge. Females usually only return to land to lay their eggs. In a very few places, including the white sand beaches of French Frigate Shoals and the black sand beaches of the Big Island of Hawaii, green turtles haul themselves out on the beach to bask in the sun. You can watch them at Punaluu State Park, if you observe the signs and keep a respectful distance.

Sea turtles are mostly omnivores, but the different species have their own dietary preferences. Green turtles are primarily vegetarians. They will take animal food when they can get it, including sponges and jellyfish, but in most parts of their range they graze on sea

Female olive ridleys (*Lepidochelys olivacea*) come ashore during a mass nesting, or *arribada*, at Ostional, Costa Rica.

grass and other marine plants. They even keep sea grass gardens (see Chapter 8). Flatbacks are more omnivorous than greens. The hawksbill is a coral reef specialist that eats both animal and plant matter, but seems particularly partial to sponges and other soft invertebrates. Loggerheads are more carnivorous, and the two ridleys almost exclusively so.

Ridley nesting beaches are, or were, one of the great spectacles of the natural world. While other sea turtles come ashore singly or in small groups to lay their eggs, ridleys emerge in vast numbers. The greatest of these concentrations, or *arribadas*, occurs today on the beaches of the state of Orissa in India. Tens of thousands of olive ridley turtles may come ashore in a single night. How long this phenomenon will continue no one knows. Trawlers dragging off the Orissa coast for shrimp and fish drown thousands of turtles every year—one of the heaviest annual kills of any endangered species.

All sea turtles have been hunted for centuries for their meat and eggs. The green sea turtle's vegetarian preference, apparently, and unfortunately, has made it particularly tasty to humans. Hawksbills have been singled out for even greater economic attention, not for food—its flesh can be toxic—but for the beautiful, translucent scutes of its carapace, the only source of commercial tortoiseshell. Hawksbills have been hunted so intensively that today the Marine Turtle Specialist Group of the World Conservation Union (IUCN) classifies it as Critically Endangered throughout its global range. International commercial trade in tortoiseshell is now banned, though, as we shall see in Chapter 10, that ban is under attack.

It is one of the paradoxes of sea turtle biology that while we probably know more about their nesting habits than we do about those of any other turtle, many aspects of their lives—particularly during the years between hatching and sexual maturity, which sea turtles pass in the open ocean—remains a mystery. Today, though sea turtles are flagship species for conservation around the world—certainly more so, in the public eye, than any other reptile—we are a long way from understanding them, or from giving them the protection they need.

A leatherback (*Dermochelys coriacea*), largest and strangest of turtles, photographed off Mexico's Pacific coast.

### Dermochelyidae: The Leatherback

The leatherback or luth (*Dermochelys coriacea*), largest and strangest of turtles, is the only living member of the Dermochelyidae. It is so unusual that scientists once placed it in a group separate from all other turtles. Today we include it in the same major group of cryptodires, the Chelonioidea, as the other sea turtles, but as the last survivor of an evolutionary line that went its own way far back in the Cretaceous.

An animal whose shell is made of thousands of tiny separate pieces of bone (see Chapter 1) is not likely to produce very complete fossils, and *Dermochelys* has a poor fossil record. The Family Dermochelyidae, though, was once more diverse than it is today. There appear to have been at least six dermochelyid species in the Eocene (56–35 million years ago), representing three different lines within the family. Why these other lines became extinct we do not know, but by the end of the Pliocene, two million years ago, only the modern line seems to have survived.

Its size alone singles out the leatherback. A leatherback's carapace may be up to 180 cm (71 in) in length. Its forelimbs, too, are exceptionally long; a leatherback with its flippers stretched out may measure over 270 cm (9 ft) across, from tip to tip.

Even if it were smaller, the leatherback would still be an amazing creature. Roger Conant Wood and his colleagues have called it the most remarkably specialized turtle in the world. No other turtle has carried shell reduction to such an extreme degree. The leatherback has no surface β-keratin at all: no claws, no scales, no scutes, not even cutting plates lining its jaws—structures hardly necessary for an animal that lives almost entirely on jellyfish. It is the fastest-growing of all turtles, increasing its body weight by roughly 8000 times between hatching and sexual maturity, within possibly as few as nine years.

The leatherback is the most wide-ranging reptile in the world. It has traveled north to Alaska and Iceland, and south of the Cape of Good Hope. It can live, and thrive, in seas far colder, and dive far deeper—more than 1,500 m (4920 ft)—than any other sea turtle. It accomplishes all this with physical and physiological adaptations more like those of a marine mammal than a turtle. An adult leatherback can even maintain its body temperature well above that of the surrounding waters.

In the Atlantic, leatherbacks seems to be doing reasonably well; they still return in numbers to their nesting beaches in the Guianas. The Pacific population, however, collapsed during the last five years of the 20th century. The once-famous nesting beaches of Terangganu in Malaysia are all but deserted, and the much-studied nesting populations at Playa Grande in Costa Rica and along the western coast of Mexico have declined by an order of magnitude. The causes range from over-harvest of eggs on the nesting beaches to accidental capture in fishing gear—especially by long-liners, drift nets, and squid nets on the high seas.

The smooth softshell (*Apalone mutica*) is one of only three species in the family Trionychidae found in the New World.

### Trionychidae: Soft-Shelled Turtles

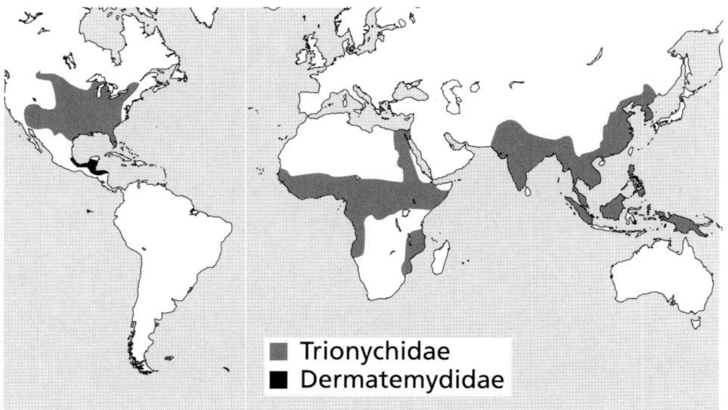


Is hard to mistake a softshell for anything else. Their narrow, elongate heads, tipped with snorkel-snouted nostrils, are unlike those of any other turtle. Their long, broadly webbed three-clawed feet—the name "Trionychidae" means the three-clawed ones—emerge from a leather-covered carapace so round and, usually, flat that its owners have been compared to animated pancakes. They have no scutes and no peripheral bones except in a single genus *Lissemys*. Even in *Lissemys*, the bones are small and, rather than forming a continuous ring around the shell, are embedded separately in the leathery skin at the margin of the carapace. They may, in fact, have a different embryonic origin from the peripheral bones found in other turtles.

Most softshells are medium-sized to fairly large animals. Some are very large indeed. The Indian narrow-headed softshell (*Chitra indica*) reaches 120 kg (265 lb), and its endangered relative, the Kanburi narrow-headed softshell (*Chitra chitra*) of Thailand, may weigh 150 kg (330 lb). The Asian giant softshell (*Pelochelys cantorii*), a huge, bloated animal with a relatively broad head, ranges from India south and east over a huge area to the lowlands of northern New Guinea. New Guinea's southern lowlands are home to a second species, the New Guinea giant softshell (*Pelochelys bibroni*). Other than their distant relative, the pig-nose turtle (*Caret-*

*tochelys insculpta*), giant softshells are New Guinea's only land or freshwater cryptodires.

Softshells spend much of the time submerged in the water or buried in the mud of the bottom, only the tip of their snorkel poking occasionally above the surface. They are excellent, powerful swimmers and active hunters, agile and often wary. They are also, frequently, scavengers, and some species are omnivores. Softshells have long necks that they can shoot out and twist around with great speed and agility. That, combined with an often aggressive disposition, sharp raking claws, and, in many species, sharp cutting edges to the jaws, makes a softshell an animal that should be definitely handled with care (if at all).

Though softshells are primarily creatures of fresh water, at least two species enter the sea and even nest on ocean beaches. The Nile softshell (*Trionyx triunguis*) has turned up off the coast of West Africa, in Senegal and Gabon, and was once common enough along the coasts of the eastern Mediterranean to be considered a pest by Turkish fishermen, though the Mediterranean population of this turtle is now considered to be Critically Endangered. The Asian giant softshell has been found nesting on the same sandy beaches in Orissa, India, that the olive ridley uses for its massive *arribadas*.

Today, softshells are primarily an Old World family, concentrated in Africa and, particularly, in Asia. Fossils shows that they once reached Europe as well, and they survived in Australia until only about 500,000 years ago. In the New World, only a single genus with three species lives in North America. There are none in South America, though a single fossil bone from the Pliocene of Venezuela shows that they once lived there as well.

The six flapshells of the subfamily Cyclanorbinae—four in sub-Saharan Africa and two, the Indian flapshell (*Lissemys punctata*) and the Burmese flapshell (*Lissemys scutata*), in Asia—are fairly small to medium-size turtles. When a flapshell withdraws into its shell, it covers its hindlimbs with fleshy flaps of skin, called *femoral valves*, attached to the rear of the plastron. The fossil softshells of Australia may have been part of the flapshell group.

An Indian flapshell (*Lissemys punctata*), a member of the softshell subfamily Cyclanorbinae, burrows in the mud.

None of the flapshells is particularly well known. They are shy animals, and seem to spend much of their time hidden in mud or sand at the bottom of rivers or ponds, waiting for prey. The African flapshells are grouped in two genera, *Cycloderma* and *Cyclanorbis*, with two species in each. The Indian flapshell, the best-known of the subfamily, is an omnivore. It will come on shore at night to feed on carrion.

The other 20 species of the family, the typ-

The Malayan softshell (*Dogania subplana*) is unusually small-bodied and large-headed for a member of the Trionychidae.

ical softshells, make up the subfamily Trionychinae. Except for the Asian giants *Chitra* and *Pelochelys*, they were all once included in the genus *Trionyx*. However, in 1987 Peter Meylan split *Trionyx* up into nine separate genera. The split is, perhaps, a better reflection of how varied these turtles are.

In a family of generally large, small-headed animals, the Malayan softshell (*Dogania subplana*) stands out. It is both fairly small (its carapace reaches just over 30 cm [12 in]) and markedly large-headed. Its skull may be over half as long as its carapace, and its unusually flexible shell bulges visibly when the turtle pulls it out of sight (see Chapter 1). Its diet is fairly unusual for a softshell, including algae and fruits that fall into the forest streams where it lives. The Malayan softshell ranges from Myanmar and Thailand southward and eastward to Borneo, Java, and the Philippines.

The best-known and most thoroughly studied of the softshells are the three American species in the genus *Apalone*. The smooth softshell (*Apalone mutica*) lives primarily in the watersheds of the Ohio, Mississippi, and Missouri Rivers. The spiny softshell (*Apalone spinifera*), named for the spiny projections along the front edge of its carapace, has a much broader range, from southern Ontario to northern Florida and Mexico, with isolated populations in Lake Champlain, Montana, Wyoming, and the Colorado River basin. The third American species, the Florida softshell (*Apalone ferox*), is found, as its name implies, throughout Florida, north to southern South Carolina, Georgia, and Alabama.

Smooth and spiny softshells are primarily river animals; the smooth, in particular, prefers large rivers with at least moderate currents. It is primarily an insect-eater; insects amounted to 75 percent of the diet of smooth softshells in Iowa. It eats a wide range of other animals as well, including frogs, young birds, and small mammals, but will also eat plant material. The spiny softshell, by contrast, is a carnivore in both sexes, while the Florida softshell is an omnivore, though one with definite carnivorous preferences. The Florida softshell prefers canals, ditches, swamps, and similar areas, and is more likely to be seen out of the water than its relatives. It is usually a much easier animal to watch.

The only species left in *Trionyx* is the Nile softshell. It is the largest member of the family after the Asian giants; a big one may weigh 45 kg (100 lb). It ranges throughout most of the river systems of the continent except for the south and northwest, and has apparently followed the Nile downstream to reach its saltwater range in the Mediterranean. Like the much smaller Malayan softshell, it is an omnivore, varying its diet of mollusks, frogs, fish, and insects with palm nuts and dates. In return, it is eaten by human beings in many parts of its range.

In Asia, the Euphrates softshell (*Rafetus euphraticus*) lives in warm, shallow, slow-moving waters in the basins of the Tigris and Euphrates. A declining species, it is suffering from pollution and the destruction of some of its river habitat by the rising waters behind the Ataturk Dam in Anatolia. The only other species of *Rafetus*, the little-known Swinhoe's softshell (*Rafetus swinhoei*), lives far away to the east, in southern China and Vietnam.

The rest of the Asian softshells range from India to China and Borneo. The Asiatic softshell (*Amyda cartilaginea*), a handsome turtle decorated with an attractive pattern of yellow freckles, has a broad range stretching from Myanmar south to Borneo and Java. It even reaches Lombok, an island that lies just on the other side of Wallace's Line, the famous boundary between the Oriental and Australasian zoogeographic regions. The Asiatic softshell is not only valued for food, but for the supposed medicinal qualities of its flesh and cartilage; as a result, its meat fetches very high prices, and the species may be in decline. In 1979, a second species, possibly invalid, was described from Thailand, with the jawbreaking name *Amyda nakornsrithammarajensis.*

The Indian softshell (*Aspideretes gangenticus*) has been used to scavenge corpses and garbage in the Ganges River.

The wattle-necked softshell (*Palea steindachneri*) and the Chinese softshell (*Pelodiscus sinensis*) are both native to China and Vietnam, but have been spread elsewhere by human hands. Both species are now established in Hawaii, and the Chinese softshell has also been introduced into the Bonin Islands, Timor, and Japan. They were probably brought to Hawaii by Chinese or Japanese immigrants in the late 1800s, though both species were imported into Hawaii for food until the outbreak of the Second World War.

The four species of *Aspideretes* live in India and surrounding countries, with a near relative, the Burmese peacock softshell (*Nillsonia formosa*), in Myanmar. This species and the Indian peacock softshell (*Aspideretes hurum*) get their names because the carapace of young animals is decorated with four striking eyespots like the markings on a peacock's train, dark circular splotches bordered with yellow and ringed with black. The markings, which fade with age, may startle a predator long enough to help the baby turtle escape—or may not; no one is quite sure.

The black or bostami softshell (*Aspideretes nigricans*) survives for religious reasons. The entire population of some 400 animals is held in semi-captivity in an enclosed pond, or tank, about five miles from Chittagong, Bangladesh, where visitors and pilgrims feed them bread, bananas, and offal. The tank is attached to an Islamic shrine devoted to a famous saint, Sultan al-Arefin Hazrat Bayazid Bistami (or Bayazid Bostami, 777–874). The turtles have been there at least since 1875, and possibly for far longer, nesting on the banks of the pond (though the area available for nesting is shrinking, and only the east bank and a raised area near the west are still suitable for the turtles). According to an early report:

> The Mahommedans will neither kill them nor permit them to be killed; they believe they are in some way connected with the saint. The tank is surrounded by steps leading down to a platform a few inches under water, and the turtles are so tame that they come to feed when called, placing their forefeet on the edge of the platform or even climbing upon it and stretching their necks out of the water... Some even allowed us to touch them, and ate pieces of chicken from wooden skewers held in our hands.

**Carettochelyidae: The Pig-Nose Turtle**

The pig-nose turtle (*Carettochelys insculpta*) belongs in its own family, and looks it. It is one of the most distinctive of all turtles. At first glance, it looks like an undersize sea turtle (its carapace rarely reaches more than 50 cm [20 in], and is usually only 30–34 cm [12–13 in] at 14 to 16 years of age, when the turtle reaches maturity). Its front limbs have been transformed into soft, flexible flippers and its hindlimbs into paddles. Like a sea turtle, it flies through the water, moving both front limbs at the same time instead of alternately as other freshwater turtles do. Each flipper and paddle retains two free claws on its leading edge. The turtle uses the claws on its flippers for ripping food apart; the only time it appears to use its hind claws is while digging its nest. Males seem to use their front claws while mating: one researcher in Australia found "mating scars" on both sides of the neck in 60 percent of adult females, but never in males or juveniles.

The pig-nose's snout—the source of its rather unattractive Australian common name—is extended into something that looks almost like the beginnings of a trunk. Its shell is covered by a thin, fragile skin dotted with tiny pits (another one of its common names is "pitted-shelled turtle"). Unlike the softshells, though, *Carettochelys* still has a heavy, complete bony shell. It still retains, for example, a set of 10 peripherals around each side of the edge of the carapace, bones that the softshells have lost.

Fossil carettochelyids have been found in North America, Europe, and Asia, in deposits

The pig-nose turtle (*Carettochelys insculpta*) has sea turtle–like flippers, but is more closely related to the softshells.

dating from the Cretaceous to the Oligocene. The living pig-nose turtle is confined, instead, to the savannas of southern New Guinea and northern Australia, where it dwells in clear, shallow rivers, swamps, water holes, and billabongs. Along the shores where the pig-nose swims grows a dense gallery forest of broad-leafed trees, including figs, whose fruits and leaves fall into the water to become food for the turtles. The turtles also feed on aquatic plants, and on animal food, including prawns, insects, snails, and carrion. Like sea turtles, the pig-nose only comes to land when the females haul themselves ashore to dig their nests in a sandbank.

For a long time, scientists thought that the pig-nose turtle lived only in the Fly River region of southern New Guinea, where they discovered it in the 1880s. It was not until 1970 that Harold Cogger, then of the Australian Museum, published the first report proving that the species lived in the Northern Territory of Australia as well. The Aboriginals, of course, had known it was there all along. Rock paintings near Kakadu National Park clearly showing the pig-nose turtle—paintings, ironically, that had been photographed and studied by Europeans for over a century—date back an estimated 7000 years.

The pig-nose was once thought to be one of the rarest turtles in the world. Now that we have come to know it better, it seems that there may be more than we thought, particularly in remote areas. There actually seem to be fairly high densities of pig-nose turtles in some rivers in New Guinea, and in the Alligator River of Australia. Nonetheless, some populations of pig-nose turtle in western Papua New Guinea appear to have been severely overhunted, and it does not pay to become complacent about any species with such a restricted range, particularly one as unusual as this.

### Dermatemydidae: The Central American River Turtle

On the other side of the world from the home of the pig-nose turtle lives another rare, localized species. It, too, is the last survivor of what was once a wide-ranging family, with fossil representatives from North America, Europe, and eastern Asia. The Central American river turtle (*Dermatemys mawii*) is found only in the Atlantic lowlands of southern Mexico, Belize, and northern Guatemala. One of its fossil relatives, *Zangerlia testudinimorpha* from Mongolia, may have been a tortoise-like land animal, but *Dermatemys* is almost helpless on land. Out of the water, it is barely able to lift its head. It probably never emerges except to nest. Even then, it uses the rainy season floods to lift it to its nest sites. The annual rise of the waters carries the turtle to places it could not reach over land.

*Dermatemys* is a fairly large turtle, with a carapace to 65 cm (26 in) long. Its feet are webbed; like some softshells, its head appears small for such a big turtle. Its shell is unusual; though its bones are relatively thick, the scutes covering them are thin, delicate, and easily injured. In old turtles the boundaries between the scutes are often obliterated.

The Central American river turtle is almost entirely a herbivore. It lives on water plants, and on fallen leaves and fruits. Captive juveniles will take meat, and perhaps in the wild they are more omnivorous than their parents.

Because this turtle is so helpless out of the water, it is surprising to find that in Guatemala it sometimes occurs in isolated seasonal pools called *aguadas*. How does it get there? The answer may have less to do with its natural history than with its flavor. The Central American river turtle is highly prized for its meat, and as much in demand in local markets. It is possible that the turtles in the *aguadas* are placed there, as if in a holding pen, until they are ready to be eaten or sold.

Years of overhunting are taking their toll on the Central American river turtle. Hunting has almost eliminated it from Mexico, and populations elsewhere are declining. Belize has begun a program to protect it, but even so, the Central American river turtle may be, today, one of the most endangered turtles in the Americas.

The Endangered Central American River turtle (*Dermatemys mawii*) is the only living member of its family.

## Kinosternidae: Mud Turtles and Musk Turtles

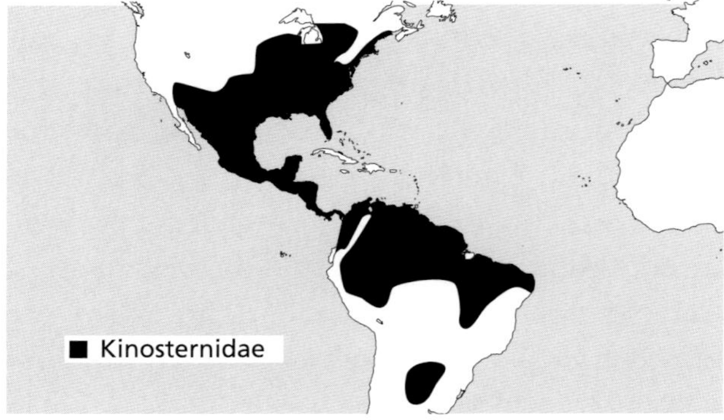


The 27 or so species of mud turtles and musk turtles are confined to the Americas, where they range from southern Canada to northern Argentina. The center of their distribution, though, is Mexico and northern Central America, home to 17 species. One was described only in 1997, the Jalisco mud turtle (*Kinosternon chimalhuaca*). Even their fossils have only been found in the New World; the oldest known, *Xenochelys formosa*, is from the Oligocene of South Dakota.

Musk turtles take their name from a series of four glands along the edges of the shell. One pair opens at the posterior end of the bridge; the other, about halfway between the front of the bridge and the front of the carapace. If you pick up a musk turtle (and if it doesn't bite you first), these glands can give off a yellowish fluid with a very unpleasant odor, pungent enough to have given the common musk turtle both the scientific name *Sternotherus odoratus* and the unflattering nickname "stinkpot." The fluid consists of a number of acids, two of which (5–phenylpentanoic acid and 7–phenylalkanoic acid) are responsible for the odor. The smell probably deters predators, but no one has ever been able to establish how well it works.

All members of the family except for the chopontil or narrow-bridged musk turtle (*Claudius angustatus*) have at least one hinge across the plastron. Apparently this originally evolved as a way to open the front of the shell as the turtle reacts to a threat by pulling its head back into the shell and opening its jaws wide at the same time (see Chapter 1). In the mud turtles (*Kinosternon*), the mechanism for opening the shell has switched to one that closes it. Some of the more terrestrial mud turtles can raise both the front and back lobes of the plastron, using hinges in front of and behind the bridge, and effectively seal themselves inside their shells.

The typical mud and musk turtles (Subfamily Kinosterninae) are small to medium-size, rather dull-colored animals. The young of some species have two yellow stripes on the side of the face that tend to break up and fade

An eastern mud turtle (*Kinosternon subrubrum*) in the Back Bay National Wildlife Refuge, Virginia.

with age. Although they spend much of their lives in the water, mud turtles and musk turtles are poor swimmers. Instead, they walk along the bottom, searching for worms and insects. They are most active at night, particularly during the warmer months of the year, something that may explain why they are so seldom seen, even in areas where they may be among the most abundant turtles.

The 19 species of mud turtles (*Kinosternon*) are much more likely to spend extended periods of time on land. The Sonoran mud turtle (*Kinosternon sonoriense*) of the American Southwest and the scorpion mud turtle (*Kinosternon scorpioides*) of northern South America travel overland from one pond to another, sometimes setting off in groups. Creaser's mud turtle (*Kinosternon creaseri*), a localized species of the northern central Yucatán Peninsula in Mexico, lives in isolated temporary pools and caves, and must be able to wander over land to reach new homes. Other mud turtles, including the eastern mud turtle (*Kinosternon subrubrum*) of eastern North America, the striped mud turtle (*Kinos-*

*ternon baurii*) of the Atlantic seaboard from Virginia to Florida, and the Central American white-lipped mud turtle (*Kinosternon leucostomum*), may wander extensively on land.

The eastern mud turtle is particularly terrestrial. Sometimes, especially in the spring and fall, it hunts worms and caterpillars on land. It frequently overwinters there, in burrows in the soil, rotten logs, or piles of debris. This species, and to a lesser extent the striped mud turtle, can also tolerate brackish or salt water. Eastern mud turtles may be abundant in tidal marshes.

Musk turtles (*Sternotherus* spp.) are, in general, more tied to the water than are mud turtles; their shells are often covered with algae and infested with leeches. Common musk turtles live in a wide range of watery habitats in eastern North America, from southern Ontario to Florida and Texas, though they prefer waters with a slow current and a soft bottom. At times, they can be found far from water, sheltering under leaves or debris. Much more localized is the flattened musk turtle (*Sternotherus depressus*), a species restricted to a single river system, the Black Warrior, in west-central Alabama. The same area is also the only known home of at least one species of salamander, three fishes, one mollusk, and six caddisflies.

Musk turtles are surprisingly good climbers. Common musk turtles and loggerhead musk turtles (*Sternotherus minor*), a large-headed species of the American Southeast, will occasionally scale cypress knees or small tree trunks to bask. They may climb 2 m (6.6 ft) or more out of the water. The razorback musk turtle (*Sternotherus carinatus*), which ranges from Oklahoma and eastern Texas to the lower Mississippi, basks more than the other species. The flattest-shelled of the lot, the flattened musk turtle, rarely leaves the water.

Mud and musk turtles are mostly omnivores, with a preference for items like insects and snails. Flattened musk turtles and loggerhead musk turtles of the nominate race (*Sternotherus minor minor*) shift their diet, as they age, from insects to clams. As they do so, their jaw muscles grow heavy, and their jaws develop expanded crushing surfaces. The Mexican rough-footed mud turtle (*Kinosternon hirtipes*) is almost entirely carnivorous; one researcher baited them in successfully with canned sardines preserved in soybean oil.

The other subfamily, the Staurotypinae, includes only three species, all restricted to Mexico and northern Central America: the two giant musk turtles of the genus *Staurotypus*, and the narrow-bridged musk turtle. Anyone familiar only with the small musk turtles of the United States might be startled to encounter a Mexican or northern giant musk turtle (*Staurotypus triporcatus*), and might be forgiven for mistaking it for a snapping turtle. It even shares a snapping turtle's reduced, cross-shaped plastron and triple ridges running the length of the carapace. As in an alligator snapper, these ridges become more prominent with age.

The northern giant musk turtle lives in the Gulf and Caribbean lowlands of southern Mexico south to western Honduras. With a carapace that may reach almost 40 cm (16 in), it is the largest in the family and an impressive turtle by any reckoning. It has a particularly large head; its jaws have powerful crushing surfaces that it uses to devour, among other things, smaller turtles, especially mud turtles that have the misfortune to share its range. As Jonathan Campbell writes in *Amphibians and Reptiles of Northern Guatemala, the Yucatán, and Belize*, "This turtle is capable of delivering a severe bite and readily defends itself. It

The attractive stripe-neck musk turtle (*Sternotherus minor peltifer*) ranges from Tennessee to the Florida Panhandle.

should never be handled just for the hell of it."

The Chiapas giant musk turtle (*Staurotypus salvinii*) of the Pacific drainage of southern Mexico, Guatemala, and El Salvador, is almost as large, with a carapace reaching 25 cm (10 in). It seems to share its congener's disposition. Carl Ernst and Roger Barbour report, in *Turtles of the World*, that "This species is well known for its vile temper and sharp jaws!"

By contrast, the narrow-bridged musk turtle is one of the smallest members of the whole family. Its carapace average is only about 12 cm (4.7 in) for males and slightly less for females. Its range is almost the same as that of its larger cousin, the northern giant musk turtle, though it does not reach as far as Honduras. Its plastron is even more reduced than that of the giant musk turtles, and it is the only member of the family to lack a plastral hinge. It has a large head and formidably hooked jaws, with a pair of sharp toothlike cusps on the upper jaw below the eye—a feature found in no other turtle. These cusps may help it to hold on to slippery prey, including frogs which gather in great numbers in the temporary ponds it prefers.

During the rainy season, the narrow-bridged musk turtle swims into seasonally flooded fields to forage among the grasslands. What it does in the dry season we are not altogether sure; until very recently, this was one of the least known turtles in its family. It was thought to be very rare, but biologists simply did not know where to find it. Fishermen do, and in coastal Veracruz, Mexico, they have been catching narrow-bridged musk turtles for centuries.

CHAPTER FOUR

# *Turtles Round the World II: Terrapins and Tortoises*

CLUSTERED TOGETHER ON A branch of the cryptodire tree are the three largest families of turtles: the Emydidae (American pond turtles), Bataguridae (Asian river turtles), and Testudinidae (tortoises). Exactly how they cluster there, though, has been a matter for some debate. Until fairly recently, scientists did not realize that the Emydidae and Bataguridae—the animals that Europeans call terrapins—represent separate branches of the tree. To make matters more confusing, there are Asian river turtles in Central and South America and one American pond turtle in Europe. Asian river and American pond turtles look very much like each other. There are box turtles in both families: *Terrapene* spp. in North America and *Cuora* spp. in Asia. Both have domed shells and hinged plastra, and it was once thought that they were each other's closest relatives. It took deeper analysis—particularly, a comparison of the structure of the bones and cartilages of the lower jaw and palate, plus a few other skeletal details—to show that this was a case of parallel evolution.

The Asian river turtles are actually more closely related to the tortoises than they are

The best-known emydid: a painted turtle, spattered with duckweed, at Bison Range National Wildlife Refuge, Montana.

Asian box turtles, including the flowerback box turtle (*Cuora galbinifrons*, above) are members of the Bataguridae, only distantly related to American box turtles of the Emydidae, like the eastern box turtle (*Terrapene carolina*, below).

to their American near look-alikes. The tortoises may have arisen from within the Bataguridae. Some Asian pond turtles may be more closely related to tortoises than to others within their own family. If so, the tortoise "family," to be consistent, would have to be reduced to a subgroup of the Bataguridae.

For our purposes, though, it is easier to think of these turtles as members of three separate, but closely related, families, and to survey them one at a time. We will start with the turtles most familiar to North Americans, the members of the Emydidae.

The ringed map turtle (*Graptemys oculifera*) is restricted to the Pearl River system of Mississippi and Louisiana.

## Emydidae: American Pond Turtles

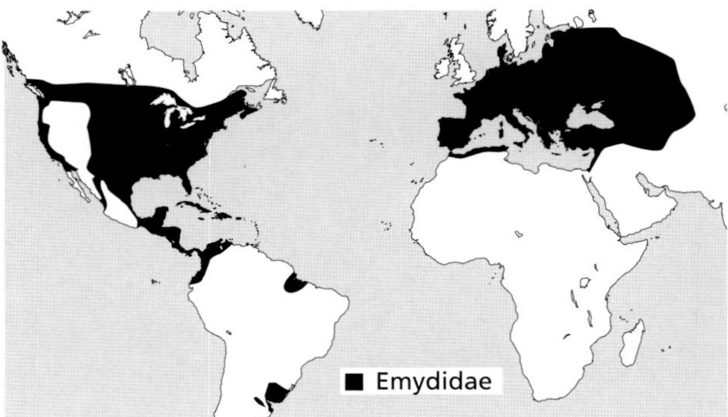


For most North Americans, the Emydidae are the "typical" turtles. They are familiar not only because they are varied and abundant, but because they are often easy to watch as they bask in the open on floating logs or emergent stones. Many of us—at least those over a certain age—can remember having a baby emydid, most probably a red-eared slider (*Trachemys scripta elegans*), as a (usually) short-lived childhood pet.

Emydids range through the Americas from southern Canada to northern Argentina, including the West Indies. Along much of the Pacific coast there is only one species, the western pond turtle (*Clemmys marmorata*), a misnamed animal that prefers streams and rivers to ponds. Only three species reach

The declining western pond turtle (*Clemmys marmorata*) is the only freshwater turtle along much of the U.S. Pacific coast.

The bog turtle (*Clemmys muhlenbergii*), the smallest turtle in North America, has suffered from habitat loss and poaching.

The European pond turtle (*Emys orbicularis*), of Europe, Western Asia, and North Africa, is the only emydid in the Old World.

South America: the extremely broad-ranging and variable slider (*Trachemys scripta*), which ranges from Virginia and Lake Michigan south to Venezuela, the Brazilian slider (*Trachemys dorbigni*) of Brazil, Uruguay, and Argentina, and the recently described slider *Trachemys adiutrix* from the state of Maranhao in eastern Brazil (there may be a fourth South American slider, as yet undescribed). The chief emydid stronghold, though, is the eastern United States. Thirty-five species live east of the Mississippi River. Some have tiny distributions: the Alabama red-bellied turtle (*Pseudemys alabamensis*) only lives in two counties around Mobile Bay, Alabama, and the ringed map turtle (*Graptemys oculifera*) occurs only in the Pearl River system of Mississippi and Louisiana. Many of the other map turtles, too, are confined to single river systems. By contrast, the painted turtle (*Chrysemys picta*) ranges clear across the continent, the only North American turtle to do so.

The western pond turtle is a comparatively dull creature that seems to have escaped the pet collector's furore. The spotted turtle (*Clemmys guttata*), bog turtle (*Clemmys muhlenbergii*) and wood turtle (*Clemmys insculpta*) have not been so fortunate; their beauty has made them the most sought-after turtles in North America. All *Clemmys* spp. suffer from habitat loss, and the three eastern species from (mostly illegal) collecting for the pet trade. They have disappeared from a number of parts of their former range.

The spotted turtle ranges from southern Ontario south to northern Florida, mostly along the eastern seaboard of the United States. The spots on its carapace that make it so attractive and easy to recognize are actually places where the scutes are transparent, revealing deposits of yellow pigment lying beneath. Some spotted turtles lack them, or may lose them with age. The spotted turtle prefers bogs, woodland streams and other areas with clear, shallow water, muddy bottoms, and lots of vegetation.

The attractive little bog turtle, a bright orange blotch daubing the side of its head, is a habitat specialist. It favors sphagnum bogs,

Blanding's turtle (*Emydoidea blandingii*) is a northern North American species, able to tolerate low temperatures.

tamarack and spruce swamps, and marshy meadows. It is the smallest turtle in North America, with a carapace reaching only 11 cm (4.4 in) or so. Increasingly rare, it survives in scattered and fragmented colonies in the northeastern United States south to nortern-most Georgia.

The wood turtle of the northeast and the Great Lakes region, a larger animal than the other two, is mostly brown with bright orange, yellow, or red on its throat, neck, tail, and the underside of its forelimbs. In contrast to the smooth, almost polished shell of the spotted turtle, the scutes of the wood turtle's carapace are decorated with radiating, raised ridges crossed by well-marked growth rings. They give the shell the appearance of a cluster of limpets, and justify the scientific name *Clemmys insculpta. Insculpta* means carved or engraved. Though it can be found in other places, the wood turtle prefers forested streams with open areas along their banks. During summer, it may spend quite a bit of its time on land, especially in the eastern part of its range.

*Clemmys* lacks a hinge across its plastron. Its more advanced relatives, the European pond turtle (*Emys orbicularis*), Blanding's turtle (*Emydoidea blandingii*), and the American box turtles, have developed hinges, but to varying degrees of effectiveness. Box turtles can seal themselves tightly within their shells, both fore and aft. Some Blanding's turtles can do the same, but others seem hardly able to raise the plastron at all. Most European pond turtles cannot close their shells completely.

The European pond turtle, the only member of the Emydidae in the Old World, is a shy carnivore that prefers still or slow-flowing waters with muddy or sandy bottoms and abundant vegetation overhanging their banks. It has a low, rounded carapace that may be olive brown, brown, or black, heavily dotted and streaked with yellow. It ranges northeast to Latvia and Lithuania, south and east to Iran and the Caspian Sea, and across the Straits of Gibraltar to Morocco, Algeria, and Tunisia. North of Spain, Italy, and the Balkans, it is the only freshwater turtle of any kind. Its subfossil remains have been found as far north as Sweden, and in the British Isles.

Blanding's turtle is one of the most northerly of all turtle species. Its range is concentrated around the Great Lakes and the Midwest, reaching no farther south than south-central Indiana and Illinois. There are isolated colonies in New England, Nova Scotia, and southern New York State. It tolerates fairly low temperatures; Blanding's turtles have even been seen swimming under the ice of a frozen Indiana pond in November. It is a larger animal than the *Clemmys* turtles, with a higher, more domed carapace and a long neck highlighted by a bright yellow chin and throat. Like its close relative, the European pond turtle, Blanding's turtle is a carnivore, usually catching its food underwater.

American box turtles recall miniature

tortoises, with high, domed shells, heavyset limbs, and a preference for a land-based life. The eastern box turtle (*Terrapene carolina*) of the eastern United States usually lives in open deciduous or mixed woodlands, though it sometimes can be found in marshy meadows or, in Florida, in palmetto thickets. Though primarily a land animal, it needs access to water, where it may spend time sitting and soaking around the edge of a pond. It may be particularly active after rain. The Gulf Coast (*Terrapene carolina major*) and the Florida (*Terrapene carolina bauri*) races, in particular, spend considerable time in the water. Eastern box turtles eat a wide variety of plant and animal foods; a turtle may stay within a ripening raspberry or blackberry patch until it has eaten every berry it can reach. The ornate box turtle (*Terrapene ornata*) of the American Midwest south to Texas and northern Mexico, as befits an animal of the prairies, is more likely to be found in sandy plains or open grassland country, while the much less well known spotted box turtle (*Terrapene nelsoni*) is restricted to hilly savanna or dry woodland in scattered localities in western Mexico.

The fourth species, the Coahuilan box turtle (*Terrapene coahuila*), is, however, not terrestrial but semiaquatic. It is confined to a single isolated wetland basin, no more than 800 sq km (320 sq mi) in extent, the Cuatro Ciénegas (Four Pools), in the otherwise arid state of Coahuila, Mexico. Herpetologists think that it is the descendant of an isolated population of eastern box turtles that retreated to small, spring-fed pools as the surrounding country became too dry even for their terrestrial lifestyles. Today, the Coahuilan box turtle is effectively trapped within its wetland refuge. The pools continue to shrink as canals drain the water away for irrigation, and the species is endangered. It is not the only Cuatro Ciénegas turtle in this situation; it shares the pools with two endangered turtle subspecies, the Cuatro Ciénegas slider (*Trachemys scripta taylori*) and the black spiny softshell (*Apalone spinifera ater*, sometimes considered a full species).

*Clemmys*, *Emys*, *Emydoidea*, and *Terrapene* form the smaller of two major evolutionary lines within the Emydidae. The other, referred to as the *Chrysemys* complex, includes the "typical" log-basking pond and river turtles: the painted turtle, sliders (*Trachemys*), cooters (*Pseudemys*), the chicken turtle (*Deirochelys reticularia*), and the map turtles (*Graptemys*). It also includes the only turtle to be entirely confined to brackish water, the diamondback terrapin (*Malaclemys terrapin*).

Groups of painted turtles clustering on a log, lying on top of each other, pushing each other out of the way in an attempt to reach the sun, are a common sight on many a northern pond. The painted turtle may be the most thoroughly studied species of freshwater turtle in the world. It occupies an extremely broad range, from Nova Scotia to Vancouver Island, southward through the eastern and Midwestern United States to the Gulf of Mexico, and in scattered localities in the Southwest and northern Mexico. Its four races differ in details of color pattern, but all have attractively marked carapaces with varying amounts of red on the edge and, in the southern form, a bright red or yellow stripe running down the midline. The painted turtle is an omnivore, foraging along the bottom or skimming fine particles of food from the water's surface. Like Blanding's turtle, it tolerates cold temperatures; it will swim in ice-covered ponds, or even emerge to bask before the ice on the water surface has quite melted.

There are seven (or eight) species of slider, but only two live north of the Mexican bor-

Introductions have made the red-eared slider (*Trachemys scripta elegans*) the world's most widespread freshwater turtle.

der: *Trachemys scripta*, which is simply called the slider, and the Big Bend slider (*Trachemys gaigeae*) of central New Mexico and the U.S.–Mexican border country along the upper Rio Grande. The Brazilian slider is South American, and four further species are confined to the West Indies.

One race of slider has been taken up by our own species and turned, willy-nilly, into perhaps the best-known, most widespread, and least endangered turtle in the world. The tiny, brilliant patch of red on the side of the head of *Trachemys scripta elegans*, the red-eared slider, singled it out to be, once, the quintessential dime-store turtle. Bred for the pet trade and for food, and exported around the world in huge numbers, accidental populations of red-eared sliders soon began to appear in all sorts of unlikely places. Today, there are wild red-eared sliders in South Africa, Seychelles, Britain, Spain, France, Italy, Japan, Korea, Taiwan, West Malaysia, Singapore, and Australia.

A peninsula cooter (*Pseudemys peninsularis*) surfaces to breathe in Everglades National Park, Florida.

In December 1997, the European Parliament banned the importation of red-eared sliders on the grounds that they pose a serious and lasting threat to native species, including the European pond turtle. However, European turtle societies reported that dealers were planning to get around the ban by either switching to a different race of slider,

As in many emydids, the juvenile Florida cooter (*Pseudemys floridana*) is more colorful than the adult.

Map turtles like this juvenile Ouachita map turtle (*Graptemys ouachitensis*) may depend on floating logs for basking.

or by smuggling red-ears in through European countries that are not members of the European Union (EU). Though red-eared sliders no longer appear in numbers in North American pet shops and department stores, they are still being exported to China in ever-increasing numbers: exports jumped from 10,000 in 1996 to 1,832,400 in 1997.

The eight or nine species of cooter, including the red-bellied turtle (*Pseudemys rubriventris*) and the Florida red-bellied turtle (*Pseudemys nelsoni*), resemble the painted turtle and the sliders. Adult males, though, have a distinctive feature: extremely long, straight foreclaws that they vibrate along the sides of a female's face during courtship. Though cooters start life as omnivores, the adults are almost entirely herbivorous.

The chicken turtle of the southeastern United States, though superficially similar to the cooters and often sharing the same bodies of water, is a quite different animal, a long-necked, narrow-headed hunter specializing in aquatic insects and crustaceans. It is named for its particularly succulent flesh.

Map turtles are river dwellers, shy animals that rarely leave their homes for the land. They get their name from the intricate series of yellowish lines on the carapace of several species, lines that may recall the contours on a map. The so-called sawbacks, including the ringed map turtle, have a series of black, spinelike projections running down the middle of the carapace. Some map turtles show a remarkable difference in size and head shape between males and females, paralleling a difference in their diets (see Chapter 6).

The eastern map turtle (*Graptemys geographica*), is broad-ranging, from southern Ontario and Quebec south to Arkansas and Alabama. Their preference for rivers, though, has meant that a number of map turtle populations have become isolated in different drainage basins along the Gulf Coast, where they have evolved into localized species with highly restricted distributions. Several, like the

The diamondback terrapin (*Malaclemys terrapin*) is a highly variable species. This is the Mississippi race *pileata*.

ringed map turtle, are confined to single river systems, or even portions of those systems.

Map turtles make extensive use of, and even seem to require, logs and other deadwood as basking sites. At night, they roost clinging to their perches below the waterline. Removal of deadwood from their home rivers has been held partly to blame, in a study by Peter V. Lindeman, for the decline of two threatened species, the ringed map turtle and the yellow-blotched map turtle (*Graptemys flavimaculata*).

The diamondback terrapin is a habitat specialist, restricted to salt marshes, estuaries, and tidal creeks along the whole of the eastern and Gulf coasts of the United States from Cape Cod to Texas. It is the only emydid turtle with a salt-excreting gland (see Chapter 5). It is a dietary specialist too, concentrating in some areas on hard-shelled snails. In Virginia, it prefers mud snails (*Nassarius obsoletus*), though it also takes periwinkles, worms, crabs, and other shellfish. In turn, a diamondback can become so encrusted with barnacles that its shell may erode.

The diamondback was once a food staple so cheap that 18th-century tidewater slaves actually went on strike to protest the amount of terrapin in their diet. In the late 19th century, though, the diamondback made an unfortunate transition from despised staple to gourmet delicacy. It remained one until the 1930s, subjected to an intense commercial hunt: between 1880 and 1936, an estimated 200,000 diamondbacks were processed into meat in Maryland alone. Even though the commercial hunt has largely collapsed, turtles continue to drown in large numbers in pot traps designed for crabs. The remaining populations face coastal development, disturbance on their nesting beaches, boat injuries, and pollution. Terrapins are still being hunted, mostly for sale to Chinese restaurants in New York City. The diamondback terrapin continues to decline.

### Bataguridae: Asian River Turtles

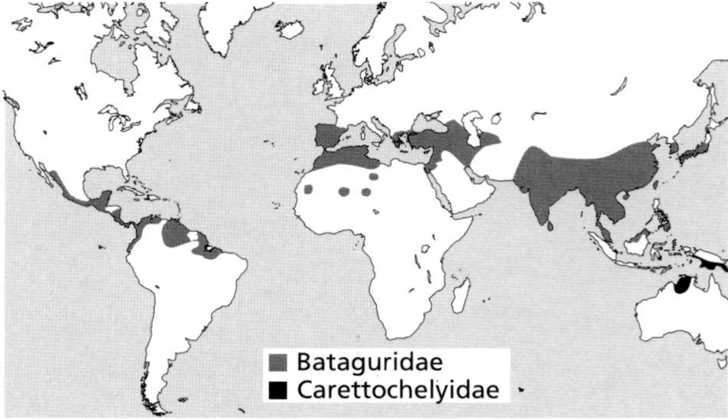


The Bataguridae is the largest and most diverse of turtle families. It has more than 65 species, including some of the handsomest turtles. They are divided up among roughly 23 genera, though their exact classification is a matter for some debate. New ones are still being discovered. The Sulawesi forest turtle (*Leucocephalon yuwonoi*) was described in 1995 from seven specimens purchased on the island of Sulawesi in Indonesia by a Jakarta

The Sulawesi forest turtle (*Leucocephalon yuwonoi*) was described in 1995; it is already extremely rare in the wild.

Neotropical wood turtles like the Sonoran wood turtle (*Rhinoclemmys pulcherrima*) are the only batagurids in the New World.

animal dealer, Frank Bambang Yuwono. Sulawesi is east of Wallace's Line, the eastern border of the Oriental zoogeographic region, making its forest turtle the only batagurid confined to Australasia. Only one other, the Malayan box turtle (*Cuora amboinensis*), crosses the line, reaching Sulawesi and beyond, to the Moluccas.

There has been a recent rediscovery, too: in May 1994, colleagues of herpetologist Oscar Shiu bought a pair of live Arakan forest turtles (*Geoemyda* [or *Heosemys*] *depressa*) from a villager in western Yunnan province, China. They were first seen by western scientists since the species was described, based on a specimen from nearby Myanmar, in 1875. We still do not know if the newfound animals, which continue to appear, represent a Chinese population, or if they are smuggled over the border.

While new turtles are being discovered, old ones may be disappearing. Most of the batagurids live in southern Asia. No other region of the world is being so rapidly stripped of its turtles, both to supply food markets in China and the pet trade in Europe, North America, Japan, and elsewhere. Of the 25 full species of turtle that IUCN listed in 2000 as Critically Endangered, its highest category of risk, 13—more than half—were batagurids. A few have not been seen for years.

There are so many batagurids in Asia that it will probably be less confusing to start our survey on the other side of the world, with *Rhinoclemmys*, the neotropical wood turtles. *Rhinoclemmys* is an anomaly, the only batagurid genus in the New World. There are eight or nine species in the genus, ranging from Mexico south to Ecuador and Brazil. They are attractive turtles, with low, rounded shells mostly colored in warm brownish tones. They have no hinge on the plastron, but when a *Rhinoclemmys* withdraws its head, two folds of skin close over it like curtains, hiding it completely from view. What protection this provides it is hard to say.

Some *Rhinoclemmys* species, like the Black wood turtle (*Rhinoclemmys funerea*) of Central America, are more or less aquatic, but others, including the Mexican spotted wood turtle (*Rhinoclemmys rubida*) of the coastal lowlands of western Mexico, spend much of their time on dry land. This difference in habitat preference is mirrored in the texture

The most easterly representative of the genus *Mauremys* is the Japanese turtle (*Mauremys japonica*).

The spotted pond turtle (*Geoclemys hamiltonii*) is a rare, carnivorous inhabitant of the Indus and Ganges river drainages.

of their shells. The aquatic species, like many other freshwater turtles, have low, smooth shells. The shells of the terrestrial species, free from the need to be streamlined for swimming, are often rough and sculptured.

The only other batagurids found outside Asia, including Indonesia and other Asian island states, are the ordinary-looking turtles that belong to the genus *Mauremys.* Two of its seven species live in Europe. The Mediterranean turtle (*Mauremys leprosa*) ranges from Spain and Portugal across North Africa to Libya, Senegal, and Niger; and the Caspian turtle (*Mauremys caspica*) extends from the Middle East and the countries around the Caspian Sea westward to Cyprus, Crete, Greece, and the Balkans.

A huge gap separates these turtles from their nearest relatives, the Japanese turtle (*Mauremys japonica*), the rare Vietnamese leaf turtle (*Mauremys annamensis*) of central Vietnam, and the Asian yellow pond turtle (*Mauremys mutica*) of southern China, Vietnam, and Taiwan. The gap was partly filled with a description in 1997 of a new species from northern Myanmar and nearby China, *Mauremys pritchardi* (named, I am pleased to say, in honor of Peter Pritchard, one of the consultant editors of this book). That's still leaves thousands of kilometers of *Mauremys*-free territory in between, much of it barren desert today, where the genus must once have existed but has now disappeared.

As vast area of perfectly good turtle habitat does lie between Iran and China, on the Indian subcontinent. Four genera—*Geoclemys* and *Hardella*, with one species each, *Melanochelys* with two, and *Kachuga* with seven—are largely confined to the territory between Pakistan and Myanmar.

The spotted pond turtle (*Geoclemys hamiltonii*) is a scarce inhabitant of the Indus and Ganges river drainages, where it lives in quiet, shallow oxbows and sloughs, occasionally entering the rivers themselves. A relatively large and striking animal, it is heavily marked on its head, neck, and carapace with prominent orange, yellow, or white spots that may fade with age. It is a thoroughgoing carnivore. In the wild it seems to dine principally on

The high, arched carapace of the Indian roofed turtle (*Kachuga tecta*) recalls the roof of a house.

snails. By contrast, the crowned river turtle (*Hardella thurjii*), a highly aquatic species that lives in the same river basins as well as in the Brahmaputra drainage, is a vegetarian, though a small one was once seen polishing off a frog.

The little-known tricarinate or keeled hill turtle (*Melanochelys tricarinata*) of Assam and Bangladesh is apparently an entirely terrestrial species of hill forests. Much better known is the Indian black turtle (*Melanochelys trijuga*), a widespread and variable species that even reaches the Maldives and the Chagos Islands in the Indian Ocean. It may spend considerable time on dry land, even living in burrows on Sri Lanka. In the Royal Chitwan National Park, Nepal, Indian black turtles not only live in the same terai grasslands as the great one-horned rhinoceros; they even incubate their eggs by burying them in rhinoceros latrines.

The high carapace of the Indian roofed turtle (*Kachuga tecta*) with its strong median ridge and flattened sides, does look vaguely like the roof of a house. This is another species that lives in quiet streams and ponds in the Ganges, Indus, and Brahmaputra drainages, though in Bangladesh it also lives in brackish coastal waters. The other six species in the genus *Kachuga* vary considerably in size, from the rare Assam roofed turtle (*Kachuga sylhetensis*) of eastern Bangladesh and Assam, whose carapace only reaches 20 cm (8 in) or so, to the almost extinct Burmese roofed turtle (*Kachuga trivittata*), a turtle of large, deep rivers in the Irrawaddy and Salween drainages of Myanmar, whose females may be three times that size. Its males are somewhat smaller.

Only four batagurid genera—including *Cuora*, the Asian box turtles—are found in both the Indian subcontinent and Southeast Asia. The Burmese eyed turtle (*Morenia ocellata*) of southern Myanmar just barely makes it into the northwestern part of the Malay Peninsula, while the Indian eyed turtle (*Morenia petersi*) lives in northeastern India. They get their name from eyespot markings on their carapaces. Eyed turtles are strictly aquatic. A naturalist who studied them early in the 20th century reported that during the rainy season, Burmese eyed turtles ventured out into flooded plains where they became trapped when the waters receded. Local people took advantage of the situation, gathering up the stranded turtles for the pot.

The Asian leaf turtle (*Cyclemys dentata*), a semiaquatic denizen of shallow streams from northern India to Borneo in the Philippines, looks rather like one of the more sculptured members of the New World genus *Rhinoclemmys*—or, if you prefer, they both look like a pile of dead leaves on the forest floor, and probably for the same reason, camouflage. Older individuals develop a hinge across the plastron, but though the hinge allows the anterior lobe of the plastron to move a bit, the animal cannot completely close its shell. Anywhere from one to four other species have also been placed, perhaps

One of the largest of the batagurids, the river terrapin (*Batagur baska*) is now considered to be Critically Endangered.

questionably, in *Cyclemys*, including a new one, *Cyclemys atripons* from southeast Thailand, Cambodia, and Vietnam, first described in 1997.

Among the largest and handsomest of the batagurids are the river terrapin or tuntong (*Batagur baska*), which ranges from eastern India to Malaysia and Vietnam, and its close relative, the painted terrapin (*Callagur borneoensis*), found from southern Thailand to Sumatra and Borneo. A painted terrapin may have a carapace length of 60 cm (24 in) and a weight of 25 kg (55 lb). Both terrapins live in tidal river estuaries and mangrove forests, and can tolerate brackish water (up to 50 percent salinity in the case of the painted terrapin). They eat the fruits and leaves of mangrove as well as some shellfish. The river terrapin is the more carnivorous of the two, consuming mollusks, crustaceans, and fish; it follows the high tide upstream into the rivers to forage, returning to the estuaries as the tide falls. Both migrate upstream to nest on sandbanks, but the painted terrapin also lays its eggs on sea beaches. Adults of both species have a pronounced, upturned snout that allows them to breathe while keeping their bodies below the surface of the water.

Most unusually for any turtle, the males of both terrapins go through a striking color change in the breeding season. River terrapins are normally a dull olive-brown above and cream-colored below, but as males come into the breeding season, their heads, necks, legs, and sometimes their whole bodies turn a deep, glistening jet black. Their irises change from yellowish to pure white. The breeding color of the painted terrapin is more garish. The heads of breeding males turn almost entirely pure white, with a lozenge of bright red, outlined with black, running from the tip of the snout to the top of the forehead. Their carapaces turn pale olive-gray, with three broad black streaks running the length of the shell and black blotches decorating the marginals.

Both the river terrapin and the painted terrapin are suffering from the destruction of mangrove forests. Historically, and even today, over-harvest of eggs has been the most important factor causing population crashes of both these species in West Malaysia. Both are hunted; living painted terrapins are particularly in demand as good-luck charms. Today, IUCN lists both the river terrapin and the painted terrapin as Critically Endangered.

The remaining 11 genera in the family are confined, or nearly confined, to the Far East. They include three species of Chinese pond turtles (*Chinemys*), the black-breasted leaf turtle (*Geoemyda spengleri*), the Malayan flat-shelled turtle (*Notochelys platynota*), the Chinese stripe-necked turtle (*Ocadia sinensis*), the keeled box turtle (*Pyxidea mouhotii*), and the black marsh turtle (*Siebenrockiella crassicollis*). They also include the species of

The spines on the carapace and the beautiful patterns on the plastron that make juvenile spiny turtles (*Heosemys spinosa*, left) so attractive to the pet trade have worn away in old adults (right).

*Geoemyda* and *Heosemys*, many of which have been traded back and forth among these genera and others like *Cyclemys* as herpetologists try to work out just how they are interrelated.

The unusual and handsome spiny turtle (*Heosemys spinosa*) lives among the leaf litter on forest floors almost throughout Southeast Asia. In juveniles, the plastron is covered with beautiful patterns of radiating lines. Almost every scute on the carapace ends in a small spine. The marginal scutes extend outward to form a ring of spikes circling the shell. The spines may increase the turtle's resemblance to a pile of dead leaves, and put off predators such as snakes, though they attract predators such as collectors for the pet trade. As the turtle ages, the spines wear down, until adults are left with only a few serrations at the back of the carapace. The pattern on the plastron, too, blurs, smears, and largely disappears with age.

At the other end of the scale of batagurid appearance is the huge Malayan giant turtle (*Orlitia borneensis*) of the Malay Peninsula, Borneo, and Sumatra. With a carapace that may be 80 cm (32 in) long, this is the largest turtle in the family. It has a large, blunt head and a dark carapace that is as smoothly rounded as the spiny turtle's is prickly. Like the arrau of South America, it nests on riverbanks and river islands; like many large river turtles, it is taken by humans for food. The Malayan giant turtle is an omnivore, but it prefers fish and will even catch the occasional snake.

It is always difficult to know why some ani-

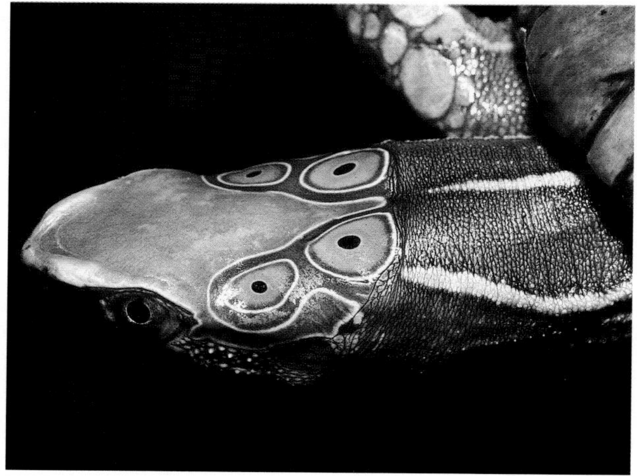

The four-eyed turtle (*Sacalia quadriocellata*) has four startling eyespots on its head, behind its real eyes.

Devout Buddhists in Asia release orange-headed temple turtles (*Heosemys grandis*) into temple ponds.

mals have the color patterns they do. The attractive little Chinese eyed turtles of the genus *Sacalia* have startling circular eye-spots—two in Beale's turtle (*Sacalia bealei*) and four in the four-eyed turtle (*Sacalia quadriocellata*)—on the top of their heads, behind their real eyes. At least, they look startling to us, and we usually assume that eye-spots may have the same effect on potential predators. We don't know if they serve that purpose for *Sacalia*, or play some other role, or no role at all. The three species live in southern China, with one, the four-eyed turtle, extending into northern Vietnam.

The Malayan snail-eating turtle(*Malayemys subtrijuga*) is not quite the gastropod specialist that its name implies, but freshwater snails do appear to be the juveniles' primary food. Adults vary their diet with other types of shellfish and, perhaps, other small animals as well. A small turtle of the lowlands of Thailand, Indochina, and West Malaysia, it has a distinct preference for rice fields; in southern Thailand, its local name means "ricefield turtle." Unlike most other turtles, it may have benefited, at least in Thailand, from the conversion of natural wetlands to agriculture (as may the otherwise-declining Asiatic softshell).

Although large numbers of Malayan snail-eating turtles are sold for food in Thailand, the species may also be benefiting from religious beliefs. It is one of a number of turtles, including the orange-headed temple turtle (*Heosemys grandis*) and the yellow-headed temple turtle (*Hieremys annandalei*), that devout Buddhists release into temple ponds. As Lim Boo Liat and Indraneil Das explain in *Turtles of Borneo and Peninsular Malaysia*, "According to the tenets of Buddhism, a live gift offered to the temple gains merit for the donor in his next life. Unfortunately, the obligation ceases after the turtle's life has been saved, and as a result, hundreds of turtles are left to survive on tidbits offered by worshipers." However, in the words of the Ven. Lama Tenzin Kalsang, Spiritual Director of the Tengye Ling Tibetan Buddhist Temple in Toronto, Canada: "Saving the life of turtles (or any being for that matter) creates virtuous karma and, further, the person would create still more virtuous karma by feeding and caring for them."

Asian box turtles (*Cuora* spp.) make up the largest genus in the Bataguridae. There are anywhere from eight to 11 or 12 species, depending on whom you read. Many of them are extremely beautiful animals, highly valued in the pet trade. The flowerback or Indochinese box turtle (*Cuora galbinifrons*) has a carapace textured and colored like inlaid wood, with contrasting patterns of mahogany, tan, and cream. The Malayan box turtle has a head and neck patterned with bright yellow stripes, while the crown of the golden-headed box turtle (*Cuora aurocapitata*), a Chinese species first described in 1988, is a rich golden-yellow.

They share the ability of American box turtles to seal off their shells by raising their plas-

Zhou's box turtle (*Cuora zhoui*) is known only from Chinese markets, where it has not been seen for years.

tral lobes, but the mechanism that permits them to do so was independently evolved, using different muscles and ligaments. Asian and American box turtles are not even that similar in lifestyle. While American box turtles are almost entirely diurnal, Asian box turtles are often active at night. The flowerback and yellow-margined box turtles (*Cuora flavomarginata*) are highly terrestrial, with the high, domed shells you would expect in land turtles (they are sometimes placed in their own genus, *Cistoclemmys*). Most of the others, however, are at least semiaquatic, with lower, flatter carapaces. The most widespread and common species, the Malayan box turtle, lives in wetlands ranging from mangrove swamps to rice fields. It has webbed toes and is a good swimmer. It feeds, sleeps, migrates, and possibly mates in the water, though it will hunt for plants, fungi, and worms on land.

*Cuora* has become a flagship for concern about the trade in Asian turtles. They are valuable items: a golden-headed box turtle may fetch up to US$900 in Chinese markets. Heavy hunting pressure has eliminated the three-striped or golden box turtle (*Cuora trifasciata*), of southeastern coastal China, Hainan Island, northern Vietnam, and Laos, from rivers in the foothills. Today it appears to be confined to mountain streams, where it is, presumably, more difficult to catch. The Yunnan box turtle (*Cuora yunnanensis*) is known from only a old few museum specimens. McCord's box turtle (*Cuora mccordi*) and Zhou's box turtle (*Cuora zhoui*) are only known from markets in Guanxi and, for Zhou's box turtle, Yunnan, where they have not been seen for some years. There are fears that both may be extinct in the wild.

Altogether, six species of *Cuora*, including the flowerback and golden-headed box turtles, are listed as Critically Endangered. In April 2000, all species of *Cuora* were listed under the Convention on International Trade in Endangered Species of Wild Fauna and Flora (CITES)—a valuable first step, but only the beginning of what must be done if Asian box turtles are to survive.

### Testudinidae: Tortoises

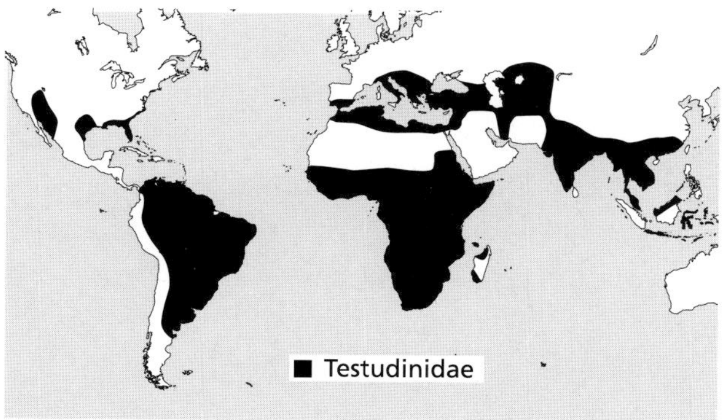


The appearance of tortoises, like that of sea turtles, is so well known as to hardly need description. It, too, represents a series of adaptations to an extreme habitat—in this case, not to the sea but to dry land. There are rainforest tortoises in Southeast Asia and South America, but most tortoises live in places that, if they are not actual deserts, at least go through seasons where water is scarce and the vegetation sere. They are beautifully adapted to survive there.

The high, domed shells of many tortoises decrease their surface to volume ratio, helping them to resist dry conditions. Their heavy, elephantine limbs, flattened in the front but rotund and columnar in the rear, are designed for walking (and, in some species, digging), not swimming. Their limbs are also designed for protection. When a tortoise retracts its head, its front limbs completely close over it, knee to knee. Large, shieldlike scales on the outer surface of its front legs protect both the tortoise's limbs and the head hidden behind it. At the rear, only the hardened soles of the feet are exposed for a predator to scrabble at.

Tortoises are, of course, not designed for speed (though they may be more agile than their popular image implies). With all that armor, they usually don't have to be. Perhaps as a consequence, the family contains a higher proportion of vegetarians than any other turtle group. Another consequence, as Peter Pritchard points out in his *Encyclopedia of Turtles*, is that a tortoise "is about the only vertebrate that one can walk up to in the wild and pick up without difficulty." That, of course, has not served them well with people, who have found them, over the centuries, not only easy to pick up but easy to store (they can live, and thus stay fresh, for a long time without being fed) and highly satisfying to eat.

Tortoises, to their detriment, make such attractive pets that tens of thousands of them, of many species, have been taken from the wild for sale into the international pet trade. Overcollecting has decimated or eliminated or nearly eliminated many a tortoise population. For some species, like the Egyptian tortoise (*Testudo kleinmanni*), overcollection represents the chief threat to their survival. Today, international bans are in place barring legal commercial trade in a number of tortoises. The once-huge imports of *Testudo* tortoises into the EU have been brought under control by import bans passed during the 1980s (though importers responded, not by cutting back on the tortoise trade, but by switching to other species). However, collecting, both legal and illegal, goes on. So does the increasing use of tortoises for food and medicine in many parts of the world, and the destruction of tortoise habitat through overgrazing, fire, and other causes. The world's tortoises remain under severe threat.

The classification of tortoises is something of a mess. Herpetologists do not seem to be able to agree on how many genera there are, which species belong in them, or what the species should be called. The genus *Geochelone*, for instance, has been broken up into at least six pieces. A Tunisian population of the widespread spur-thighed tortoise (*Testudo graeca*) of the Mediterranean has been named as a separate genus and species, *Furculachelys nabeulensis*. Though some pet owners and dealers use the name, most herpetologists think it simply refers to variant individuals of the ordinary spur-thighed tortoise. If you read more about tortoises, you may come across some of the animals we are going to meet in this book under different, and perhaps unrecognizable, scientific names.

Southern Africa is the tortoise capital of the world. Five genera and 14 species of tortoise live there. Of these, three genera, *Psammobates*, *Chersina*, and *Homopus*, are found nowhere else. Representatives of all three live within a day's drive of the Cape of Good Hope.

The Cape is southern Africa's Mediterranean zone. Its native vegetation is *fynbos*, a bewildering and magnificent variety of proteas, heaths, bulbous plants and others. Fynbos is a threatened habitat. In the far southwestern Cape, a dry, inland variety of fynbos called renosterbosveld (rhinoceros bushveld), restricted to acidic, sandy soils, has been particularly hard hit by frequent wildfires and by clearing for agriculture and development. This is the only home of the rarest and most beautiful tortoise in Africa, the geometric tortoise (*Psammobates geometricus*). Ninety-six percent of its habitat has been destroyed. Today, wild tortoises survive in only 13 tiny patches ranging from less than 10 to no more than 30 hectares (25–74 acres).

*Psammobates* means "sand-loving." The geometric tortoise's two more widespread relatives, the highly variable tent tortoise (*Psammobates tentorius*) and the somewhat more northerly serrated or Kalahari tent tortoise (*Psammobates oculiferus*), are arid-country animals, absent from South Africa's more humid east. The serrated tortoise lives among the dunes and dry, rocky bushveld of the Kalahari Desert.

All three species are fairly small, with a maximum carapace length of 20 cm (8 in) or less. Males are even smaller than females. In the geometric and Kalahari tent tortoises, each scute on their domed shells, except the marginals, is mounded into a peak. They are beautiful animals, their scutes marked with radiating sunbursts of yellow or even pinkish lines against a blackish background. This pattern is also found on some other tortoises, including the radiated tortoise of Madagascar and the star tortoise of India. Its function is not beauty but camouflage. The contrasting pattern of black and yellow mimics the dappled sunlight under a desert bush, and may hide the tortoise sheltering there.

The angulate, or bowsprit, tortoise (*Chersina angulata*) is entirely restricted to the lowlands of the southwestern Cape. It has fared much better there than the geometric tortoise, probably because it has adapted to a much wider range of habitats. Angulate tortoises range from semidesert to areas of high rainfall, though they prefer sandy coastal regions. Jutting out beneath the angulate tortoise's neck is a blunt, upcurved process that resembles, particularly in males, the bowsprit of a ship. The bowsprit is formed by the fused gular scutes and underlying bone of the plastron extended forward into a ramming, pushing, and heaving weapon. Rival males use it to beat back and, with luck, overturn their opponents. As might be expected in a species featuring contests of strength between males, females are the smaller sex—the opposite of the situation in *Psammobates.*

The tent tortoise (*Psammobates tentorius*) is a member of a genus of three species entirely confined to southern Africa.

I once found one of these attractive tortoises making its way across a road in the karoo, a few hours' drive north of Cape Town. Fortunately, I was not treated to the defensive response Bill Branch describes in his *Field Guide to Snakes and Other Reptiles of Southern Africa*: "It readily ejects the liquid contents of its bowels when handled, often spraying them up to one metre, and this with surprising accuracy." This defense does not always work; Liz McMahon and Michael Fraser, illustrator and author of *A Fynbos Year*, watched a Verreaux's eagle (*Aquila verreauxii*), a large and spectacular predator, drop an angulate tortoise from a height of 30 m (100 ft) onto the rocks below "before gliding down to eat the shattered remains."

The padlopers or pygmy tortoises (*Homopus*) are the smallest tortoises in the world. Their Cape representative is the parrot-beaked tortoise or common padloper (*Homopus areolatus*), a medium-size (but still very small) member of the genus with a carapace rarely more than 10–11 cm (4–4.3 in). It lives beneath dense cover in a variety of habitats from deciduous woodland to coastal fynbos thickets, where it hides under rocks or in abandoned animal burrows. Its carapace is flattened and colorful, each scute sculptured with concentric ridges and marked with warm reddish-brown splotches in the center, ringed with bright olive or yellow and bordered in turn with black. In the breeding season, the heads of males turn a deep bright orange-red, remaining that way for several weeks before the color gradually fades.

The other four padlopers range from the greater padloper (*Homopus femoralis*) of the eastern Cape and the Free State, which tips the scales at 200–300 g (7–11 oz), to the tiny speckled padloper (*Homopus signatus*) of Little Namaqualand in the northwestern Cape, which may weigh as little as 85 g (3 oz) even as an adult. Less broad in their selection of habitats than the common padloper, they prefer rocky areas. The speckled padloper is entirely confined to boulder-strewn outcrops, or *kopjes*. In this restricted habitat it may be quite common. Several individuals may shelter under a single slab of rock. They emerge in the early morning to feed on small succulent plants growing among the rocks. The karoo or Boulenger's padloper (*Homopus boulengeri*) is so active on cool, cloud-threatened summer days that it is known in Afrikaans as "donderweerskilpad"—the thunderstorm tortoise.

The most extreme rockpile specialist of them all, and the most unusual tortoise in the world, is the pancake tortoise (*Malacochersus tornieri*) of semidesert and grassland areas of Kenya and Tanzania. The pancake is a small animal, about 17 cm (6.5 in) in carapace length, with a shell so varied in color and pattern, from pale yellow to black and from plain to star patterned, that it is hard to find two alike. Pancake tortoises are agile climbers, easily able to right themselves if they fall on their backs. We have already discussed the peculiarities of the pancake tortoise's shell (see Chapter 1), that flattened, flexible construct that allows the animal to wedge itself into a crevice among the rocks. The tortoise spends most of its time hiding there, rarely straying far away. It may spend the whole of the dry season wedged in its crevice.

Just any crevice will not do. A suitable one must provide protection from overheating, be a safe haven from predators, and protect the animal against drying out. The crevice may be horizontal or vertical, but whatever its orientation it must be deep, easily accessible from the ground below, and it must narrow, as it penetrates the rocks, to a height of 5 cm (2 in) or less. The surrounding rocks must be weathered and eroded enough for plants to take root, sheltering the crevice and probably providing the tortoise with food

The rare pancake tortoise (*Malacochersus tornieri*) lives in Kenya and Tanzania, where it shelters in rocky crevices.

close at hand. Such ideal locations may be few and far between. Even where they occur, normally only one or two tortoises live in each, though sometimes larger groups—usually up to six, but rarely as many as 10—will share a crevice, particularly in the dry season. Even in the right sort of country, pancake tortoises are likely to be thin on the ground, clustered in scattered colonies separated by the flat, unsuitable plains that stretch from kopje to kopje. Increasingly, habitat alteration by humans has meant that the right sort of country is hard to find. Combine that with poaching for the pet trade and a natural breeding cycle that produces only one egg two or three times a year, and it should not be surprising that the pancake tortoise is a species in decline.

Hinged or hinge-back tortoises (*Kinixys*) are rather elongate, medium-size, omnivorous tortoises with almost rectangular shells. Some individuals never develop the hinge across the carapace that gives the group its name. Depending on which herpetologist you talk to, there are anywhere from three or four to six species of hinged tortoise, found throughout most of sub-Saharan Africa. They cover a broad ecological range: the Natal hinged tortoise (*Kinixys natalensis*) is a dry-country animal; the wide-ranging Bell's hinged tortoise (*Kinixys belliana*) lives in savanna and grassland but favors humid conditions; while the serrated hinged tortoise (*Kinixys erosa*) of West Africa prefers moist forests. The serrated hinged tortoise—somewhat unusually for the family—is often found in water, where it is a fair swimmer.

Bell's hinged tortoise is one of five surviving tortoises on Madagascar, the only one not confined to that island. It must be a comparatively recent arrival, so recent that herpetologists think human immigrants brought it there from Africa between 1000 and 1500 years ago.

Of the other four Malagasy tortoises, two are included in the endemic genus *Pyxis*: the spider tortoise (*Pyxis arachnoides*) and the much rarer flat-tailed tortoise or kapidolo (*Pyxis planicauda*). These are very small tortoises, almost as small as the padlopers, restricted to Madagascar's dry west and south. The spider tortoise, named for a pattern of yellow lines on its carapace that recalls a spider's web, lives only in a narrow belt along the southwest coast no more than 50 km (30 mi) wide.

The flat-tailed tortoise, which does indeed have a flattened tail ending in a large nail-like scale, is confined to an area of some 500 sq km (190 sq mi) in the Morondava region of western Madagascar. Here, it lives in a type of deciduous forest that is being rapidly destroyed. Even in those few areas where the forest is intact, it is an uncommon animal. Very little is known about its natural history. It spends the dry season buried in leaf litter, emerging to feed (probably on fallen fruits) and to mate between January and March, when the rains come.

The carapace of the radiated tortoise (*Geochelone radiata*) of Madagascar mimics dappled sunlight as it hides under a bush.

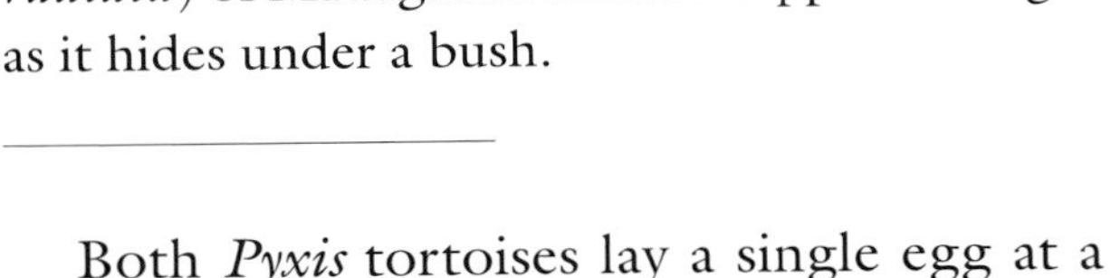

The spider tortoise (*Pyxis arachnoides*) is confined to a narrow belt of dry forest along the southwest coast of Madagascar.

A flat-tailed tortoise (*Pyxis planicauda*), photographed in the Kirindy Forest of western Madagascar, one of the few places where it lives.

Both *Pyxis* tortoises lay a single egg at a time, and cannot recover quickly, if at all, from a fall in numbers.

Madagascar's other two endemic tortoises are usually included in the widespread genus *Geochelone*, but might be better placed in a genus of their own, *Astrochelys.* They may be closer to the *Pyxis* tortoises than to the African species of *Geochelone*. A DNA study published in 1999 suggests that all four of Madagascar's endemic tortoises are the descendants of one species, and possibly of a single pregnant female that rafted to the island from Africa between 14 and 22 million years ago.

Like South Africa's geometric tortoise, the carapace of the radiated tortoise (*Geochelone radiata*) is patterned with starbursts of yellow lines. The radiated tortoise, though, grows much larger, reaching a carapace length of 40 cm (16 in). It lives in dry woodlands, including the unique spiny forests where pot-bellied baobabs crouch and thorn-studded *Didieria* trees tower over the sand like enormous flexible candelabra. The radiated tortoise is still common in many parts of southern Madagascar despite hunting for the pot, and heavy (and illegal) collecting for the pet trade and for souvenirs. In some areas, taboos, or *fadys*, have protected the tortoise, at least from local people. The *fady* means little, though, to outsiders.

A male radiated tortoise has an extension at the front of its plastron, like a smaller version of the angulate tortoise's bowsprit. It seems to use this to attempt to subdue or even overturn the female during mating. The radiated tortoise's ram pales into insignificance, though, beside the enormous upcurved hook, or *ampondo*, jutting out from the plastron of the ploughshare tortoise or angonoka (*Geochelone yniphora*), a fairly hefty creature with a carapace length of almost 45 cm (18 in). The male ploughshare's ploughshare is so long that the animal has to tilt its head to one side to see around it, or to feed.

The angonoka possesses the dubious distinction of being the rarest tortoise species in the world. Its total wild population is fewer than 1000 (and may indeed be less than half that number), broken up into at least two isolated subpopulations around Baly Bay in western Madagascar. It is entirely confined to a few small patches of deciduous forest and bamboo scrub, totaling no more than 40–80 sq km (15.5–31 sq mi).

The remaining angonoka habitat is threatened by brushfires set by local people to stimulate growth of grasses for their cattle, to drive cattle from the forest, and to keep bush pigs away from cultivated lands. Natural fires must occur there occasionally, but the fires set by people are much more frequent. Even if the tortoises survive them, their habitat, which cannot recover from too frequent burning, may not. The ploughshare tortoise is now the subject of an intensive conservation program, Project Angonoka, coordinated by the Durrell Wildlife Conservation Trust in cooperation with the Malagasy Water and Forests Authority at the Ampijoroa Forestry Station, within the Ankarafantsika Strict Nature Reserve in northwestern Madagascar.

Back on the African continent, two very large tortoises in the genus *Geochelone* live south of the Sahara. The spurred tortoise (*Geochelone sulcata*) ranges along the southern edge of the Sahara, from Senegal to Ethiopia, among plant-covered desert dunes, bush country, and acacia woodland. It appears well adapted to near-desert conditions; a study in Mali found that neither high temperatures nor intense sunshine restricted the spurred tortoise's range. Other than the giants of Aldabra and the Galápagos, this is the largest tortoise in the world. Captive specimens have reached a carapace length of 83 cm (33.5 in) and a weight of 105.5 kg (233 lb), though in the wild a spurred tortoise that size would be quite unusual.

The world's rarest tortoise is the angonoka or ploughshare tortoise (*Geochelone yniphora*) of Madagascar.

The leopard tortoise (*Geochelone pardalis*) is the second-largest of the world's mainland tortoises. It varies in size in different parts of its range. The largest recorded specimen in South Africa was a 40 kg (88 lb) individual that lived in the Addo Elephant National Park. The local rangers named him "Domkrag" (auto-jack) because of his habit of crawling underneath vehicles and attempt-

The spurred tortoise (*Geochelone sulcata*) of the southern edge of the Sahara is the world's largest continental tortoise.

ing to hoist them up (not to mention his tendency to ram visitors from behind).

The leopard tortoise occurs from Djibouti and the southern Sudan through eastern and southern Africa to Angola, Namibia, and South Africa. Over this broad area it is a common animal, occurring in a wide range of country from fynbos in the south to dry scrub in the northeast, and from humid forests to the deserts of Namibia. It can even tolerate South Africa's mountains, where it may face freezing winters and occasional snow. It gets its name not from its habits—it is a vegetarian, though it will nibble on animal droppings—but from its pattern of black spots on a pale ground. Like the patterns of many turtles, the spots may fade and even disappear with age.

If we cross the Sahara to the north, we enter the range of the Mediterranean tortoises of the genus *Testudo*. Leaving aside the controversial "*Furculachelys*" *nabeulensis*, there are five to seven species in the genus, ranging from Spain, the Balkans, and North Africa eastward to Pakistan, Afghanistan, and northwestern China. These are the animals the Romans first called *testudo*, tortoise. Aesop's hare almost certainly raced against a species of *Testudo*; there are three to choose from in Greece. So does Achilles in Zeno's paradox, the famous mathematical "proof" that, if a tortoise has enough of a head start, the hero will never overtake it. The shell of a *Testudo* became, so told the Greeks, the first lyre, fashioned by the young god Hermes; the word "testudo" also means, in Latin, "lyre."

All *Testudo* tortoises but one, the Central Asian tortoise (*Testudo horsfieldii*, sometimes placed in its own genus, *Agrionemys*), have a weak hinge across the plastron. As its name implies, the Central Asian tortoise is the most easterly of the lot, a creature of steppes and rocky deserts from the Caspian Sea to the eastern edge of the genus's range in farwestern China. The most westerly, reaching into Spain, is the spur-thighed tortoise, named for a large conical scale on its hind leg, and Hermann's tortoise (*Testudo hermanni*). The spur-thighed tortoise ranges eastward to Iran, though the form in the Middle East and the Caucasus may be a separate species. The Egyptian tortoise (*Testudo kleinmanni*) of North Africa is an endangered species as small as the *Pyxis* tortoises of Madagascar or the larger padlopers. *Testudo* tortoises are almost entirely vegetarian, although Hermann's tortoise will sometimes consume a variety of invertebrates.

Traveling eastward and southward into Asia, we reach the range of two more species of *Geochelone*: the Indian star tortoise (*Geochelone elegans*) of India and Sri Lanka, and the Burmese star tortoise (*Geochelone platynota*), a very rare and still-exploited species from Myanmar. As their names imply, these tortoises are decorated with the same

The spur-thighed or Greek tortoise (*Testudo graeca*) was once traded in huge numbers for the pet market.

sort of sunburst patterns that ornament the geometric tortoise and the radiated tortoise.

There are three species of *Indotestudo*, the elongated tortoise (*Indotestudo elongata*), which ranges from Nepal to Vietnam and the Malay Peninsula, the Travancore tortoise (*Indotestudo travancorica*) of southwestern India (and the of Sulawesi tortoise (*Indotestudo forsteni*) of Central Indonesia. These tortoises avoid the rocky wastelands favored by the species of *Testudo*. In Thailand and Malaysia, the elongated tortoise lives in hill forests where it feeds on everything from slugs and carrion to leaves and fruit. It seems particularly fond of mushrooms. During the breeding season, both sexes, but particularly the males, turn pink around their nostrils and eyes. The elongated tortoise has suffered from human-caused fires that destroy its forest habitat (and the habitat of other south Asian tortoises), and populations of this once-common species have crashed in western Thailand since the mid-1980s.

The largest tortoise in Asia is the Asian brown tortoise (*Manouria emys*), which can reach a carapace length of up to 60 cm (24 in). *Manouria* may be the most primitive of tortoise genera. In its humid forest habitat, broad, flattened shell, and other features, it resembles the batagurids from which tortoises probably arose. The Asian brown tortoise lives in lowland rainforests (up to 1000 m [3280 ft]), from Assam and Bangladesh to Sumatra and Borneo. The other species of *Manouria*, the impressed tortoise (*Manouria impressa*), is a highly localized and little-known animal of middle-altitude primary evergreen and bamboo forests, found in scattered regions of Myanmar, West Malaysia, Cambodia, Laos, Vietnam, and portions of China.

Like the flat-tailed tortoise of Madagascar, the Asian brown tortoise spends much of its time burrowing in leaf litter or damp soil near streams. It is an omnivore, but prefers plant food, including bamboo shoots, mushrooms, and fallen fruits. This species has religious significance for Buddhists. It is often kept in temple gardens, and sometimes individuals are found with Buddhist inscriptions carved in their shells.

Nonetheless, both *Manouria* tortoises are widely sought after for food. The Asian brown tortoise has been a trade item for centuries; its remains have been found in 3000-year-old archeological deposits in China, far north of their normal range. Even its empty shell may be put to use: in Bangladesh its carapace is occasionally used as a feeding trough for chickens, pigs, and dogs. The flesh of both species is considered to have medicinal properties, and the impressed tortoise is a prized rarity in the pet trade. Their habitats, too, have been seriously degraded, and both tortoises have become so rare that even professional tortoise hunters have difficulty finding them.

Though North America is rich in turtles,

The Travancore tortoise (*Indotestudo travancorica*) lives in forests of southwestern India. This animal is in breeding color.

The largest tortoise in Asia, and possibly one of the most primitive of tortoises, is the Asian brown tortoise (*Manouria emys*).

except in eastern Panama it has only a single genus of tortoise, *Gopherus*. The desert tortoise (*Gopherus agassizii*) lives in deserts from southern Nevada and southwestern California south into northwestern Mexico. The Texas tortoise (*Gopherus berlandieri*) lives from southern Texas southward to northeastern Mexico, and the southeastern coastal plain, from Louisiana through Florida to South Carolina, is the home of the gopher tortoise (*Gopherus polyphemus*). The fourth, and largest, species is the Bolson tortoise (*Gopherus flavomarginatus*). The Bolson tortoise may have been much more widespread long ago, during the Pliocene, but today it lives only in the Bolson de Mapimí, an isolated basin in north-central Mexico.

Most *Gopherus* tortoises, like the rodents for which they are named, are consummate burrowers. Their front limbs are highly effective spades, even flatter and more rigid at the wrist than in other tortoises. The gopher tortoise is almost entirely restricted to deep sandy soils, suitable for digging its extensive burrow systems. Its burrow—a life's work that the tortoise usually starts soon after it hatches—may be 9 m (30 ft) long and 3.6 m (12 ft) deep. The burrow provides refuge from heat, cold, predators, and the brushfires that regularly sweep its fire-adapted pineland habitat. Only the Texas tortoise does not dig a burrow, but uses its front limbs and the edge of its shell to scoop out a shallow resting place, called a *pallet*, under a bush or a cactus.

Desert and gopher tortoises have been listed as Threatened under the U.S. *Endangered Species Act*. The desert tortoise has declined substantially in recent decades in the face of pressures ranging from overcollecting and habitat destruction to collisions with automobiles and off-road vehicles, vandalism, and disease. The Texas tortoise is rather more secure, though its semidesert scrub habitat has been reduced by 90 percent in the Rio Grande valley since the 1930s. The specialized habitat of the gopher tortoise has been largely lost to mining, agriculture, and other human uses, and its populations have dropped by 80 percent in the last century.

The Bolson tortoise is listed as Endangered. It is still being eaten by local people, and during the 1980s its habitat was cut into fragments by roads driven through its range for oil and gas exploration. The remaining tortoises may now be restricted to an area of not much more than 1000 sq km (400 sq mi), cut into six separate areas.

The three species of tortoise on the continent of South America have traditionally been included in the Old World genus *Geochelone*. However, it is rather hard to believe—unless we suppose that they floated across the Atlantic from Africa—that South American tortoises could belong to the same genus as the leopard tortoise. Many herpetologists today place them in a separate genus, *Chelonoidis*, along with their giant relatives from the Galápagos.

The red-footed tortoise (*Geochelone carbonaria*) ranges through much of the continent, mostly east of the Andes, from Colombia to northern Argentina. It has a North American toehold in eastern Panama. The yellow-footed tortoise (*Geochelone denticulata*), a considerably larger animal with a carapace occasionally up to 82 cm (32 in) long, is a species of northern and central South America centering on the Amazon basin. Both are also found on Trinidad, and humans have introduced the red-footed onto a number of other islands in the Caribbean.

Unlike the other *Gopherus* tortoises, the Texas tortoise (*Gopherus berlandieri*) digs a shallow pallet instead of a burrow.

The red-footed may be the most brilliantly colored of tortoises. Its carapace is deep, shiny black, each of its vertebral and costal scutes tipped with a splotch of yellow. Its head is marked with yellow and orange, and, depending on the subspecies, there are bright scarlet spots on its limbs and tail. Not surprisingly, it is a popular pet. In the wild, the red-footed tortoise lives in forests and woodlands with nearby grassland, or in moist savannas. It spends most of its time in "food patches" such as clearings left by fallen trees in the forest, where it eats plants, carrion, and living insects such as termites.

The yellow-footed tortoise is a strict rainforest dweller, never found in any other habitat. It is even more of an omnivore than its

This yellow-footed tortoise (*Geochelone denticulata*) was photographed in the rainforest of Manu National Park, Peru.

cousin, eating pretty much anything it can get. Both species are widely eaten by humans, and in Venezuela the red-footed tortoise is esteemed as a traditional Holy Week delicacy.

In contrast to the other two South American tortoises, the chaco tortoise (*Geochelone chilensis*) is a species of dry lowlands found from southwestern Bolivia to northern Patagonia ("chaco" is a type of dry forest country that dominates much of Paraguay and its neighbors). It is the smallest of the three, with a carapace length of 20 cm (8 in), rarely reaching 43 cm (17 in). Its ecology and lifestyle recall the desert tortoise. Like the desert tortoise, the chaco tortoise is a burrower, excavating winter dens that may penetrate several meters into the soil. It may face some of the same threats. In Argentina, much of the southern end of this tortoise's range is heavily grazed by goats, sheep, cattle, and horses. Overgrazing has had a severe impact on the desert tortoise, and the same problem may be harming the tortoises that live at the other end of the Americas.

**Island Giants**

Tortoises, landlubbers though they may be, have a remarkable ability to reach oceanic islands. There were once tortoises, related to the living South American species, on at least nine West Indian islands. In 2000, a fossil tortoise from Bermuda, thought to be some 300,000 years old, was described as *Hesperotestudo bermudae*. Today, the largest tortoises in the world are island animals: the giants of Aldabra in the Indian Ocean and, famously, the Galápagos Islands in the eastern Pacific.

The giant tortoises most people see in zoos come not from the Galápagos, but from Aldabra, a remote and virtually uninhabited atoll within the Republic of Seychelles in the Indian Ocean north of Madagascar. The proper scientific name of the Aldabra tortoise is a source of considerable argument among specialists: you can find it labeled as *Geochelone gigantea*, *Aldabrachelys gigantea*, *Aldabrachelys dussumieri*, *Aldabrachelys elephantina*, *Dipsochelys elephantina*, or *Dipsochelys dussumieri*. Here, we will call it *Geochelone gigantea*.

The Aldabra tortoise is the last certain survivor of a whole complex of giants. Close cousins of the Aldabra tortoise once lived on Madagascar and the Comoros. Another genus of giant tortoises, *Cylindraspis*, was restricted to the Mascarenes, home of many species that vanished beneath the first onslaught of Western exploration. The last Mascarene tortoise died between 1804 and 1850. Most, if not all, of the captive tortoises now found on other islands in the region are the descendants of (or are themselves) animals directly imported from Aldabra. The other species were thought to be as dead as that most famous of Mascarene animals, the dodo.

Recently, Justin Gerlach and Laura Canning have argued that a number of venerable and odd-looking captives in the granitic Seychelles are not Aldabra tortoises, but the aged survivors of otherwise-extinct populations native to the Seychelles themselves. They have assigned them to two separate species, *Dipsochelys hololissa* and *Dipsochelys arnoldi*. Their conclusion remains, though, highly controversial. Although most islands in the granitic Seychelles (located some 600 miles northeast of Aldabra) once hosted substantial populations of giant tortoises, there is no unequivocal evidence that there was ever a species of granitic islands tortoise distinct from that now on Aldabra. Prior to 1900, only two specimens of "*D. hololissa*" were collected from the granitic islands, and these were destroyed by fire during the Second World War. Only four specimens of "*D. arnoldi*" were collected from Seychelles (in c.1840 from an unknown locality in the central Seychelles) and one female which lived on St. Helena from 1776 to 1874.

The chaco tortoise (*Geochelone chilensis*) is probably the closest living relative of the giant Galápagos tortoises.

Many authorities believe that the unusual features of the captive tortoises in the Seychelles could have been the result of improper feeding when the animals were young; we know that a diet too high in protein can cause a captive Aldabra tortoise to grow up with a distorted shell. Moreover, even under natural conditions, diet can influence the shape of a tortoise's shell. In the drier realms of eastern Aldabra, adult tortoises are significantly smaller, more rounded, and smoother than are their conspecifics living in the more heavily vegetated northern and western regions of the same atoll.

The tortoise population of Aldabra was nearly exterminated by the year 1900. Fortunately, the government implemented protective measures, and by the late 1960 to early 1970s, a census showed the population to number some 150,000 animals. Since then, however, the vegetation on which the tortoises depend for both food and for shelter from the hot mid-day sun has suffered from, among other things, over-grazing by tortoises, a significant increase in the feral goat population on Aldabra, and persistent drought during the 1980s and 1990s. In 1997, a new census by David Bourn (who also did the census in the early 1970s) showed that the tortoise population had dropped to some 100,000 animals. Most of the mortality occurred among the dense tortoise populations of the eastern part of the atoll, and had, in fact, been predicted based on data collected in the earlier census. Meanwhile, the population has been increasing elsewhere on the atoll, where tortoise density is lower.

How did tortoises get to Aldabra, or to the other islands in the Indian Ocean that were home to giant tortoises until humans exterminated them? Did the Aldabra giants and their relatives start out as smaller animals that rafted

The Aldabra tortoise (*Geochelone gigantea*) is last survivor of the giant tortoises of the Indian Ocean Islands.

on floating debris to their eventual destinations? It is not at all unusual for a small island immigrant to become a giant. There have been giant island insects, flightless birds, and lizards. Giant tortoises, then, should be no surprise.

The Indian Ocean tortoises, though, most likely started out as giants that floated to their ultimate homes. The fossil record shows that within the last 100,000 years—far too short a time for small tortoises to evolve into giants—Aldabra was completely inundated during periods of sea level rise, and subsequently recolonized by tortoises that must have been giants already.

Giant tortoises once existed on the mainland in many parts of the world. If they are not there now, we may have ourselves to blame: one of the best specimens of *Geochelone crassiscutata*, a monster tortoise from Florida with a carapace 1.2 m (4 ft) long, is the shell of an animal that was killed with a wooden spear and roasted over a fire. There are still some pretty big tortoises wandering around the continents today. A spurred tortoise may be every bit as large, or larger, than the two extinct tortoises from the island of Rodrigues; their carapaces measured only 85 cm (33 in) and 42 cm (17 in), respectively. It is not that much smaller than even the Aldabra tortoise, whose record carapace length in the wild is 106 cm (40 in).

The Aldabra tortoise is probably the direct descendant of an even bigger species from Madagascar, one that may have invaded the atoll via the Comoros at least three times. The occasional Aldabra tortoise will even put to sea today, where it bobs in the waves like a cork. If the Aldabra tortoise and its extinct cousins could float from island to island, their ultimate African ancestor may, too, have floated away from the continent.

What about the other surviving giants, the tortoises of the Galápagos? A 1999 study by Adalgisa Caccone of Yale University and her colleagues, using mitochondrial DNA sequences, has shown that these tortoises' closest living relative is actually the smallest species on South America, the chaco tortoise. This is not really a surprising result. The chaco tortoise, like the Galápagos tortoises, is an animal of dry scrub, while the other two South American tortoises live in humid forests and savannas. Did the Galápagos tortoises, then, become giants only after they arrived on the islands?

Probably not. The chaco tortoise once had giant fossil relatives on the South American mainland. The molecular evidence suggests that the Galápagos tortoises arose as a separate evolutionary line between 6 and 12 million years ago. That probably happened on the continent; until 5 million years ago, the present Galápagos Islands did not exist (though other islands, now submerged, probably did). It appears that for a tortoise to survive the crossing to an oceanic island, it may be better to be a giant to start with.

The most famous encounter between man and tortoise took place on San Cristóbal (Chatham) Island in the Galápagos, on September 17, 1835. What the tortoises thought of it we have no idea, but the man involved wrote the following:

> As I was walking along I met two large tortoises, each of which must have weighed at least two hundred pounds: one was eating a piece of cactus, and as I approached, it stared at me and slowly walked away; the other gave a deep hiss, and drew in its head. These huge reptiles, surrounded by the black lava, the leafless shrubs, and large cacti, seemed to my fancy like some antediluvian animals.

The writer, of course, was Charles Darwin, in the 2nd edition of *The Voyage of the Beagle*. Contrary to popular myth, he did not see the tortoises, shout "Eureka!," and immediately develop the theory of evolution. Though Darwin found the tortoises fascinating and recorded quite a bit about their natural history, he was not even aware of the most significant fact about them until he was almost ready to leave the islands. He records that the vice-governor, Mr. Lawson, told him:

> that the tortoises differed from the different islands, and that he could with certainty tell from which island any one was brought. I did not for some time pay sufficient attention to this statement, and I had already partially mingled together the collections from two of the islands. I never dreamed that islands, about 50 or 60 miles apart, and most of them in sight of each other, formed of precisely the same rocks, placed under a quite similar climate, rising to a nearly equal height, would have been differently tenanted; but we shall soon see that this is the case. It is the fate of most voyagers, no sooner to discover what is most interesting in any locality, than they are hurried from it; but I ought, perhaps, to be thankful that I obtained sufficient materials to establish this most remarkable fact in the distribution of organic beings.
>
> The inhabitants, as I have said, state that they can distinguish the tortoises from the different islands; and that they differ not only in size, but in other characters. Captain Porter has described those from Charles and from the nearest island to it, namely, Hood Island [now called Española], as having their shells in front thick and turned up like a Spanish saddle, whilst the tortoises from James Island are rounder, blacker, and have a

The giant tortoises of the Galápagos (*Geochelone nigra*) differ from island to island. These are from Santa Cruz.

> better taste when cooked. M. Bibron, moreover, informs me that he has seen what he considers two distinct species of tortoise from the Galápagos, but he does not know from which islands. The specimens that I brought from three islands were young ones: and probably owing to this cause neither Mr. Gray nor myself could find in them any specific differences.

How much Darwin actually took from this it is difficult to say. In *The Origin of Species* he does not even mention the tortoises, though he discusses the differences among Galápagos birds at considerable length.

Fifteen distinct forms of tortoise have been described from the Galápagos (only 11 still survive). They have been divided into as many as 14 species, though today most scientists think there is only one. The island tenants that finally caught Darwin's eye are now considered subspecies, one each on six of the seven islands where tortoises occur today, and five on the largest island, Isabela—one on each of its five volcanoes.

Like the Aldabra tortoise, the Galápagos tortoise has been saddled with a plethora of names. With its South American mainland relatives, it has been placed either in *Geochelone* or *Chelonoidis*, and its species name (depending on some obscure rules in the International Code of Zoological Nomenclature) is either *elephantopus* or *nigra*. Two of the consultant editors for this book have disagreed rather vigorously in print on the question. I am not such a fool as to try to solve this problem myself, so I am going to call the Galápagos tortoise *Geochelone nigra* on the grounds that that is what the most recent study of its relationships, the one that tied it to the chaco tortoise of the mainland, called it.

That study, by Adalgisa Caccone and her colleagues, turned up some very interesting information, not just about the Galápagos tortoise's nearest relatives, but about the way the tortoises have evolved within the islands. In particular, the most northerly of the five forms on Isabela turns out to be less closely related to the other four than to the tortoises from another island, San Salvador. Isabela, then, appears to have been colonized twice. Apparently, the tortoises have been able to travel between the islands more frequently than we had thought.

The study also had something to say about the most famous individual tortoise in the Galápagos, if not in the world. The island of Pinta has its own distinctive race of tortoise (or species; some herpetologists treat the various island forms as separate species in their own right), *Geochelone nigra abingdonii*, but it had not been seen since 1906 and was thought to be extinct. In 1971, national park wardens trying to eliminate goats from the island came across a single male. The last of his kind, he now lives at the Charles Darwin Research Station on Santa Cruz. His name is Lonesome George.

At the station, scientists have been trying for years to do something about Lonesome George's love life. Short of cloning (something that has actually been considered), a single male cannot produce offspring; but if Lonesome George could be mated with a close relative, then at least some of his genes could be preserved. Attempts to interest Lonesome George in females from islands near Pinta, though, have met with indifference. The molecular study may show why.

It turns out that Lonesome George's closest relatives are not from close at hand, but from the more distant islands of Española and San Cristóbal. Pinta may have been originally

Last of his line: "Lonesome George," the only Pinto Island giant tortoise (*Geochelone nigra abingdonii*) in the world.

colonized from one of these islands, perhaps by tortoises drifting on the strong current running from San Cristóbal. Like them, Lonesome George is a saddleback, with the front of its carapace, in Darwin's phrase, "turned up like a Spanish saddle." A saddleback shell—something that could only evolve on a predator-free island because it exposes the entire front of the tortoise to attack—may help its owners stretch up to reach higher browse. It certainly assists in the struggle for dominance among rival males, who determine contests by how high they can reach. The scientists may have, it seems, been fixing Lonesome George up with the wrong dates.

The fate of the other Galápagos tortoises remains uncertain. The scientists at the Charles Darwin Research Station have done a heroic job. Using captive breeding techniques, they have brought the Española tortoise (*Geochelone nigra hoodensis*) back from the brink of extinction, raising its numbers from 4 to over 1000. But the Galápagos is no longer an isolated paradise. Fishermen fighting for unlimited rights to the waters around the islands have held station scientists hostage, and have threatened to kill the tortoises as a protest if they do not get their way. The politics of wildlife conservation have begun to rear an ugly head in Darwin's living laboratory.

In this way, the Galápagos is a microcosm of the world, and its tortoises of our planet's whole complement of turtles. Lonesome George, living out his life in captivity on an island not his own, is the only turtle that we can honestly call the last of his kind. In years to come, I fear, he may have many others for company.

CHAPTER FIVE

# *Under the Hood*

MUCH OF THE SECRET of not only turtle success, but turtle variety, lies in their physiology. Turtles have had to cope with just about every kind of environmental stress that nature can throw at them: heat, cold, drought, salt, even lack of oxygen. To survive, they have evolved a range of internal mechanisms that you might not suspect from the seeming uniformity of their body plan.

Much of turtle behavior, too, is a response to physiological stresses. That includes dormancy—not "behaving" at all when it is simply too cold, or too hot, or too dry to carry on safely with normal activities (see Chapter 6). The key to many things that turtles do, from basking on a log to burying themselves in the mud, is temperature.

Turtles need heat to fuel body processes like growth, digestion, egg production, and muscular activity. But, with one remarkable exception that we will consider at the end of this chapter, they cannot generate that heat themselves, or regulate their body temperatures by internal means as mammals do. They must absorb the warmth they need from outside, from the sun or from the surrounding air, earth, or water. At the same time, they must avoid, as best they can, dangerous extremes of heat or cold.

A green sea turtle (*Chelonia mydas*) basks on lava rocks at Kaloko-Honokohau National Historical Park on the island of Hawaii.

## Temperature Inside and Out

Active turtles function at their best within a range of optimum body temperatures. What that range is can vary, depending on the species. Along one stretch of the Murray River, Australia, herpetologist Bruce Chessman found that while broad-shelled turtles (*Chelodina expansa*) stopped taking bait in his traps when the temperature dropped below 18.4°C (65°F), implying that below that temperature the turtles were much less active, he caught Macleay River turtles (*Emydura macquarii*) down to a temperature of 16.3°C (61°F), and snake-necked turtles (*Chelodina longicollis*) entered his traps even at 11.9°C (53.4°F).

North American turtles are generally most active at a temperature range roughly between 20 and 30°C (68°F–86°F). In Minnesota, however, Blanding's turtle (*Emydoidea blandingii*), which has a more northerly distribution than most other North American turtles, is more active at lower outdoor temperatures than are painted turtles (*Chrysemys picta*) or common snapping turtles (*Chelydra serpentina*) living in the same wetlands. Blanding's turtles can be active several weeks earlier in the spring and later in the fall. On the other hand, many tortoises are adapted to high body temperatures. The normal internal temperature for an active gopher tortoise (*Gopherus polyphemus*) is close to 35°C (95°F).

The spotted turtle (*Clemmys guttata*), like Blanding's turtle, prefers cool temperatures, particularly at the northern end of its range. Its activity peaks when the mean monthly air temperature is 15.5°C (60°F), and it will mate in water at only 8.5°C (47.3°F). A male spotted turtle was once found crawling slowly beneath the ice in shallow water; its body temperature (or at least the temperature of its cloaca, the most convenient place herpetologists can find to stick a thermometer) was 3°C (37.4°F), the lowest temperature known for any active turtle.

Optimum temperatures may change with the seasons. Turtles in a laboratory can be acclimated to prefer the higher or lower ends of their normal temperature range, and something very like acclimation may happen in nature as winter gives way to spring. In Chesapeake Bay, juvenile loggerheads (*Caretta caretta*) and Kemp's ridley turtles (*Lepidochelys kempii*) prefer warmer water in summer than in winter, even when cooler waters are available. Temperature preference can even change over the life of an individual. Very young turtles may need to seek out higher temperatures to speed up their growth; the sooner they grow, the less likely many predators are to eat them. In a laboratory experiment, hatchling and yearling Florida red-bellied turtles (*Pseudemys nelsoni*) preferred water temperatures over 30°C (86°F). Young sliders and common snapping turtles also prefer warmer water if they can get it.

Loggerhead sea turtles (*Caretta caretta*) can become acclimated to a range of temperatures. This loggerhead is in the Bahamas.

Common musk turtles (*Sternotherus odoratus*) control their temperature in cool weather by migrating into warmer water.

Behavior can help an active turtle fine-tune its body temperature. Achieving this can be as simple as moving to a warmer spot when the turtle gets too cold, or back again when it gets too hot. Common musk turtles (*Sternotherus odoratus*), whose internal temperatures may remain close to that of the water where they swim, migrate into warmer water during cool weather. Freshwater turtles in warmer latitudes may need to do very little; as long as the water temperature remains within a reasonable range, there is no reason for turtles to devote a great deal of effort either to warm up or cool down.

Yellow-spotted river turtles in Manu National Park, Peru, bask on a log as butterflies seek mineral salts on their skins.

A basking chicken turtle (*Deirochelys reticularia*) stretches its neck and limbs and spreads its toes for maximum exposure.

Over the course of a day, turtles may switch between periods of activity and periods of rest, either basking in the sun or cooling off in the shade, or in the water, depending on the temperature. Many turtles are more active during the middle of the day on cooler days, but on hotter days switch to early-morning and late-afternoon activity, interrupted by a midday siesta. In temperate regions, this sort of switch happens over the course of the seasons. Midday activity is commonest in spring and fall, and morning and evening sessions (a so-called bimodal pattern of activity) the rule during the summer. This can depend on habitat, too: during long, hot summers in Greece, Hermann's tortoise (*Testudo hermanni*) has a bimodal activity pattern in open country, but in woodland, where there is more shade, it is active in the middle of the day.

### A Place in the Sun

After a chilly night, the easiest way for a turtle to restore its optimum body temperature is to bask in the sun. A log crowded with turtles, often of several different species, is a familiar sight in certain parts of the world. The turtles certainly look as though they are doing their best to warm up, often positioning themselves for maximum exposure, their legs and webbed feet outstretched to soak up the sun.

Some turtles will put up with considerable discomfort to continue basking. Australian herpetologist Grahame Webb watched captive Macleay River, Jardine River (*Emydura subglobosa*), and saw-shelled turtles (*Elseya latisternum*) that appeared to be undergoing heat stress, at least on their heads, but continued to bask anyway. The turtles gave off a watery discharge from their eyes, panted, and frothed at the mouth. They dipped their front feet into the water and wiped them over their heads in an apparent attempt to cool off. They did everything, in fact, except go back into their pool. Did they have to wait until the interior of their bodies reached the proper temperature, no matter how uncomfortable their extremities?

What, in fact, are basking turtles doing? Warming themselves up, certainly; but sliders will "bask" in the dark, or in the rain, so warming in the sun cannot always be its purpose. Once a turtle goes back into the water, it may lose heat more rapidly than it warmed in the sun. Australian zoologists Ben Manning and Gordon Grigg found, to their surprise, that although Brisbane River turtles (*Emydura signata*) spend considerable time basking, once in the river their body temperature scarcely differs from that of the water around them. Manning and Grigg wondered if basking had to do with body temperature at all.

However, in a recent study of Blanding's turtles in central Minnesota, herpetologists Todd D. Sajwaj and Jeffrey W. Lang found that basking turtles warmed up quickly to a high body temperature, sometimes greater than 28°C (82.4°F), and kept that temperature for hours each day. Basking time varied from season to season, from individual to individual, and from day to day depending on the weather. On cool overcast days, the turtles kept to the water, and their bodies stayed close to the water temperature. In spring, more than 90 percent of the turtles crawled out of the water to bask in the morning, usually on mats of sedge and in protected areas where they could be very hard to find. By July, that figure had dropped to 40 percent.

Female yellow-bellied sliders (*Pseudemys scripta scripta*) bask longer than males during spring and summer.

Even on sunny spring days, some turtles stayed in the water, perhaps because they needed to feed more than they needed to bask. The basking turtles that Sajwaj and Lang examined all had food in their stomachs. A hungry turtle may need to spend more time in the water, but for one that has fed, climbing out for a bask may be a good idea even if it loses that heat when its sunbath is over. Basking after a meal can speed digestion especially in herbivores; digestive enzymes work better at higher temperatures (though common snapping turtles in Algonquin Provincial Park, Ontario, did not bask after feeding; instead, they buried themselves in sediment under logs and organic debris).

In September and October, female Blanding's turtles basked, but males did not. This sexual difference is known in other turtles,

Basking can inhibit algal growth that may slow or even injure turtles like Purvis's turtle (*Elseya purvisi*) from Australia.

too: female sliders (*Pseudemys scripta*) bask longer than males during spring and summer, but not during autumn and winter. Perhaps female turtles need extra warmth for egg production, while male turtles need to concentrate on finding mates instead of sitting around on a log.

This balance between temperature and a turtle's other needs may be a delicate one. In the former Yugoslavia, in the north of their range, summers are cool and Hermann's tortoises must spend longer basking in the morning to reach their optimum temperature of around 34°C (93°F). This presents females with a problem. Female Hermann's tortoises may be twice the size of males, and so take longer to warm up. However, if they take too long to bask, they may not have time to get the amount of food they need to produce a full clutch of eggs. In the end, they must give up basking time in favor of feeding time,

Cagle's map turtle (*Graptemys caglei*) of central Texas basks extensively on floating logs, rocks, and cypress knees.

especially on cool days, and operate at a lower temperature, trading off the ability to heat up quickly for a large body capable of producing, and holding, as many eggs as possible.

There are other reasons for basking. Soaking in the sun may help turtles synthesize vitamin D. Basking makes a turtle not only warmer but drier. This may help it get rid of leeches and other parasites, and inhibit the growth of algae on its shell. A thick growth of algae may make a turtle a less efficient swimmer, may cause the shell to deteriorate, or even spread disease that can kill the turtle. Keeping algal growth down may be an important reason to bask for, among others, the European pond turtle (*Emys orbicularis*).

Some turtles bask much more than others. In North America, emydid turtles bask regularly and extensively. Map turtles (*Graptemys* spp.) even have traditional basking sites that have been used for generations. Some absolutely require suitable places to bask as a component of their habitats, usually exposed snags in deep water. In Africa, helmeted terrapins (*Pelomedusa subrufa*) and serrated hinged terrapins (*Pelusios sinuatus*) are also frequent baskers. So are many Australian turtles, including the Mary River turtle (*Elusor macrurus*) and the Victoria River snapper (*Elseya dentata*). On the other hand, North American mud and musk turtles and Australian longnecks (*Chelodina* spp.) bask comparatively seldom. Though American softshells bask frequently, the Chinese softshell (*Pelodiscus sinensis*) does so only occasionally, and, at least in Hawaii, the wattle-necked softshell (*Palea steindachneri*) rarely basks.

Many turtles, including softshells, bask by floating on the surface of the water, or just below it, rather than by crawling out of the water. The Central American river turtle

Female green sea turtles (*Chelonia mydas*) in the Galápagos crawl onto the beach to bask or to escape overattentive males.

(*Dermatemys mawii*) only basks in this way; it is sometimes found floating on the surface during the day, apparently asleep. Sliders in Florida preferred to bask out of the water when water temperatures average 28.5°C (83.3°F), but switched in their preference to basking while afloat when the water temperature rose to 31.5°C (88.7°F). Sliders living around a nuclear reactor cooling reservoir in South Carolina, where the water temperature was as much as 9°C (16°F) higher than usual, preferred to bask in the water even though perfectly good basking sites were available out of it. Even then, they were being warmed, to some extent, by the sun; their body temperatures ended up between 1–3°C (3.8–5.4°F) higher than the water they swam.

Peninsula cooters (*Pseudemys peninsularis*) sun themselves along the Wacula River in northwestern Florida.

Sea turtles, to the extent they bask at all (and how much they do is not known), will normally do it afloat. In a very few places in the world, including Hawaii and Australia, green turtles (*Chelonia mydas*), particularly females, bask on shore. Perhaps surprisingly, they prefer cooler beaches to warmer ones. Too warm a beach might overheat a basking sea turtle. Green sea turtles studied by Causey Whittow and George Balazs on the beaches of French Frigate Shoals in the northwestern Hawaiian Islands had carapace temperatures as high as 40–42.8°C (104–109°F). To prevent their surface layers from becoming overheated while their interior rose to an appropriate temperature, the turtles flipped sand onto their flippers and carapace, a behavior that lowered their surface temperature by as much as 10°C (18°F).

For many dry-country tortoises, the problem may be too much heat rather than too little. Their basking sessions may be comparatively short. Even then, a tortoise may have to tuck its head and neck out of the way to avoid overheating them while the rest of its body warms up. Gopher tortoises (*Gopherus polyphemus*) bask at the entrance to their burrows on cool days. The Texas tortoise (*Gopherus berlandieri*) warms up in its pallet instead, and will not emerge until its body temperature rises above 28°C (82.4°F).

Shape has a good deal to do with how well a turtle absorbs and retains heat. A tortoise's domed, rounded shell gives it a low surface-to-volume ratio, which is better for retaining heat than for absorbing it. In general, tortoises are able to maintain their body temperatures above 30°C (86°F) for a long time. Size matters, of course; giant tortoises, whose size gives them an even lower surface-to-volume ratio than their smaller cousins, are particularly good at maintaining a stable body temperature. Color also matters: emydid turtles are not only flatter-shelled than tortoises, but usually have darker shells. The closer to black any object is, the faster it will soak up radiant heat.

A basking turtle does not simply have to wait for the sun's heat to penetrate its body. Blood flowing near its skin, or just under its shell, can carry heat into the turtle's interior. When a basking turtle starts to warm up, the blood vessels near its surface dilate and its heart rate increases, shunting more blood to its skin. When it begins to cool (for instance, when it goes back in the water), the vessels constrict again, conserving heat within its body. The results of this sort of adjustment can be dramatic: small spiny softshells (*Apalone spinifera*) are so efficient at it that they can warm up twice as fast as they cool down. The heating rate in wild desert tortoises (*Gopherus agassizii*) may be as much as 10 times faster than their cooling rate.

This increase, or decrease, in cardiovascular activity is triggered by feedback from temperature sensors in the nervous system. Experiments on common snapping turtles and cooters (*Pseudemys floridana*) have shown that blood flow increases as soon as radiant heat is applied to their carapace, even if the interior of their bodies was already fairly warm. However, studies on painted turtles, box turtles (*Terrapene* spp.), and others have shown that not just surface sensors, but the central nervous system, particularly the area of the brain stem called the hypothalamus, is involved in helping turtles set their internal thermostat.

### Too Darn Hot

Overheating can kill a turtle, sometimes in a few minutes. At high body temperatures, turtles lose some of their ability to function, including a righting reflex that is a normal response to being turned on their backs. Few turtles can survive body temperatures much higher than roughly 40°C (104°F). Blanding's turtle has a somewhat lower tolerance level (39.5°C or 103°F) than most other North American turtles. By contrast, gopher tortoises have survived at body temperatures up to 43.9°C (111°F), the highest figure known for any turtle. The angulate tortoise (*Chersina angulata*) of South Africa can also withstand body temperatures over 40°C (104°F).

Adult Galápagos tortoises (*Geochelone nigra*) are too large to take shelter under rocks or bushes on hot days.

This eastern box turtle (*Terrapene carolina*) may risk death from heatstroke if it cannot right itself.

Eastern box turtles (*Terrapene carolina*) can avoid overheating by sitting in a creek, often for hours at a time.

The obvious way to avoid overheating is to get out of the sun, or into the water. Eastern box turtles (*Terrapene carolina*) avoid overheating by sheltering under rotting logs, in mud, or among decaying leaves. If this is not enough, they may find a shady puddle and sit in it for anywhere from a few hours to a few days until conditions become more comfortable.

But the obvious way may not always be possible or sufficient. Tortoises, particularly

Ornate box turtles (*Terrapene ornata*) survive dangerously high temperatures by urinating on their legs and plastra.

Hatchling ornate box turtles (*Terrapene ornata*) burrow deeply into the soil so they can winter below the frost line.

giant tortoises, may be too large to take shelter under rocks or bushes, or may not have burrows, like those of the desert tortoise, to retreat into. To compensate, tortoises and American box turtles can adjust their blood flow so that they heat up slowly and cool down rapidly, the reverse of the effect we see during basking.

A turtle can cool off by evaporation if it can somehow get its skin wet, as Grahame Webb observed with his basking turtles in Australia. If there is no water available, the turtle will have to provide the liquid itself. When its body temperature rises over 40°C (104°F), an African spurred tortoise (*Geochelone sulcata*) begins to salivate. It uses the saliva to wet its head, neck, and front legs. As the saliva evaporates, it cools the tortoise quite effectively. A desert tortoise that cannot get into its burrow will do this too, and the elongated tortoise (*Indotestudo elongata*) can reportedly withstand outside temperatures up to 48°C (118°F) by using the same technique.

An ornate box turtle (*Terrapene ornata*) can compound the cooling effect by urinating on its legs and plastron. This works surprisingly well. An ornate box turtle kept at 51°C (124°F), a temperature that would surely kill it in minutes otherwise, managed to keep its body temperature at a dangerous, but not lethal, 40.5°C (105°F) for an hour and a half by both salivating and urinating on itself. Obviously, these techniques, even for the turtles that can use them, are last-ditch solutions; too much of them, and the animal will end up trading overheating for dehydration.

To survive the winter, painted turtle hatchlings (*Chrysemys picta*) must keep ice crystals from forming in their bodies.

**Turtles on Ice**

Cold can stop a turtle in its tracks as surely as heat. A temperature that is too low can force a turtle to slow its activity rate, stop it altogether, or, if the drop is too extreme, kill it. No animal can survive extensive freezing within its cells because the expanding ice crystals rupture cell membranes. A number of animals, including at least some turtles, can withstand limited freezing of body fluids outside their cells (blood plasma, for instance, or the liquid in their eyes). However, this sort of tolerance can go only so far. As ice crystals continue to form, the concentration of dissolved substances in the turtle's remaining body fluids goes up. If that concentration becomes higher outside the cells than inside them, the cells will lose water through osmosis and, eventually, collapse.

Turtles may go to considerable lengths to avoid freezing temperatures. The communal burrows desert tortoises retreat into in order to escape the winter cold of southwestern Utah, at the northern edge of their range, may be 10 m (33 ft) long. Such structures are hardly the work of a season. This far north, a burrow needs to be at least 4 m (13.2 ft) long to protect a tortoise from freezing, which may mean that those first 4 m (13.2 ft) had to have been excavated before Utah winters grew as cold as they are today. That could, possibly, make those first excavations at least 5000 years old.

The risk of freezing is a particular problem for hatchling turtles. Most North American turtles hatch in the fall, and some, including painted and box turtles, spend their first winter underground. If the soil freezes too deeply, the hatchlings may die. Some turtles cope with this threat by getting out of the danger zone. When ornate box turtles and yellow mud turtles (*Kinosternon flavescens*) hatch, they dig down through the bottom of their nest and spend the winter deep enough in the soil, if they are lucky, to stay below the frost line. Other hatchlings, like those of Blanding's turtles and common snapping turtles, leave their nest and spend the winter at the bottom of a lake, stream, or marsh where the water does not freeze.

Painted turtles, though, stay in their shallow nest all winter, no more than 8–14 cm (3–5.5 in) below the surface. For them, freezing is a real threat, one that they face not by behavioral but by physiological means. We do not understand exactly how they manage it, but ongoing research by Gary and Mary Packard, of Colorado State University, and their colleagues has given us some interesting ideas.

We used to think that painted turtles got through the winter by an unusual ability to survive freezing. Hatchling painted turtles have, indeed, survived for several days with more than half their body water frozen. However, the Packards found that hatchlings of several other species—even sliders, which live farther south and seldom encounter freezing temperatures in the wild—were just as successful at this as painted turtles. Furthermore, none of their frozen turtles—even painted turtles—survived if the temperature dropped much more than 2°C (3.6°F) below the freezing point, something that must surely happen to painted turtles in the wild.

If hatchling painted turtles are to have a real chance of surviving through northern winters, tolerance to freezing will not be enough. They must prevent ice crystals from forming in the first place. Ice does not form simply because the temperature drops below freezing; it usually needs a nucleus to solidify around, and the most common nucleus is another ice crystal. Unlike some animals, turtles do not have a natural antifreeze in their bodies. If a crystal of ice in its nest can penetrate a baby turtle's skin, the hatchling will freeze quite rapidly.

The Packards have discovered that the skin of hatchling painted turtles has a dense layer of lipids lying at the base of the epidermis, a layer that is lacking in turtles that do not face the same threat of freezing. This layer is particularly well marked on the head and forelimbs, the parts of a hatchling that are most likely to come into contact with ice crystals. Hatchlings in the nest normally withdraw their heads into their shells during very cold periods, so that skin of their necks, which is not as well protected, is hidden within their bodies. Is this lipid barrier a hatchling painted turtle's chief line of defense against invading crystals of ice?

### Nor Any Drop to Drink

Turtles in dry country risk not only heatstroke but desiccation. A thick, tough skin is some protection against the loss of water, the shell does provide armor against drying, and a turtle with a hinged shell can close itself in still farther, shutting away its head, limbs, or tail. But none of these can entirely protect a turtle from losing water through the lining of its mouth or cloaca, or when it voids its bodily wastes.

Of course, the easiest way to avoid desiccation is to drink regularly or, failing that, to eat succulent plants or other foods that contain stored water. Tortoises can be adept at finding water in unlikely places. In South Africa, angulate tortoises and tent tortoises (*Psammobates tentorius*) raise their hind legs, extend their necks, and push their snouts into damp sandy soil, filtering and drinking the water that puddles around the edges of their shell, head, or front limbs. Desert tortoises (*Gopherus agassizii*) sometimes dig small

Tortoises in dry country, like this radiated tortoise (*Geochelone radiata*) from Madagascar, must avoid dehydration.

A desert tortoise (*Gopherus agassizii*) can use stored liquid in its bladder if it cannot find water to drink.

basins to catch water during showers and thunderstorms; the basins can hold water for up to six hours. Young desert tortoises in the Mojave Desert of California are so efficient at finding succulent food during the spring that they took in twice as much water as the scientists studying them had predicted. During the summer, if rainwater was available, they drank it and stayed active. If not, they simply retreated into their burrows and remained inactive until the hot, dry weather passed.

Beating a retreat may be the safest thing for a desert turtle to do during a dry spell. Desert tortoises spend much of their lives in their burrows. The yellow mud turtle reportedly can spend as long as two years in a state of dormancy during an extended drought. The Sonoran mud turtle (*Kinosternon sonoriense*), the most desert-distributed of its family, may be more likely to set out across country if its own pond dries up. Herpetologist Paul Stone has recaptured individuals over a kilometer from where they were first found. They seem well equipped to survive an overland trek: under laboratory conditions, the turtles survived at least 80 days without food or water. Whether they migrate or simply burrow out of sight, Sonoran mud turtles seem remarkably able to withstand harsh dry periods. After major droughts in the Peloncillo Mountains of New Mexico in 1996 and 1999, their population seemed to be completely unaffected.

Their tenacity appears partly based on a physiological trick first found in tortoises: the ability to store water in their urinary bladder and draw on it when necessary, particularly to top up the fluid levels in their blood plasma. This technique will only work as long as the water in the turtle's bladder is more dilute than its body fluids. As the turtle uses up its bladder storage and more waste products

The diamondback terrapin (*Malaclemys terrapin*) is the only turtle that spends its entire life in brackish water.

accumulate in its urine, it finally reaches a point at which the concentration of dissolved substances in its bladder becomes the same as in its blood. At that point, its ability to use the bladder as a physiological "canteen" ends. Nonetheless, this can be a long time in coming: with a full bladder, the turtle's canteen amounts to over 15 percent of its body mass, and Stone and his colleague Charles Peterson showed that Sonoran mud turtles can draw on their bladders for at least 28 days after taking a drink.

Desert tortoises, which are larger animals with relatively larger bladders, can supply themselves with water in this way for perhaps as long as six months. When water becomes scarce, they can switch from producing liquid urine to excreting a semisolid sludge, or not urinating at all for months on end. The water they save, in the form of dilute urine, is stored in their bladder canteen.

Water problems seem obvious for desert animals, but for an animal in need of a drink, the ocean is as much of a desert as the Kalahari. Reptiles, or for that matter birds and mammals, that take in any amount of sea water—if only in the course of swallowing another marine creature—are taking in a fluid whose concentration of salt is almost three times greater than that in their own body tissues. Even if they do not drink sea water, if their skins are at all permeable, they will lose water through osmosis into the saline bath surrounding them.

One of the simplest ways to cope with this problem is to grow. The bigger a turtle is, the lower its surface-to-volume ratio. That means that there is simply less opportunity, proportionately speaking, for water loss through its skin. All adult sea turtles are large. So are most other turtles that enter salt water. The pig-nose turtle (*Carettochelys insculpta*) is a good-size animal. The river terrapin (*Batagur baska*) and painted terrapin (*Callagur borneoensis*) are among the largest batagurds, and the seagoing softshell turtles, the Nile softshell (*Trionyx triunguis*), and giant Asian softshell (*Pelochelys cantorii*), rank with the largest of their family. The diamondback terrapin (*Malaclemys terrapin*) does not fit this pattern—it is a fairly small turtle—but it lives in brackish water that is considerably less salty than the open ocean. Placed in 100 percent sea water, a diamondback may dehydrate. Hatchlings of both diamondback terrapins and painted terrapins, which are, of course, much smaller than adults, do not seem able to tolerate any level of salt water for long and must locate fresh water to drink. Young painted terrapins may have to cross 3 km (1.8 mi) of salty water to find it.

Tolerance to sea water has also been suggested as an explanation for the giant size of some tropical sliders (*Trachemys scripta*). The biggest sliders known live on the Caribbean slope of Costa Rica, where they make long journeys along the coast to nest on sea

The watery "tears" of this nesting olive ridley (*Lepidochelys olivacea*) are actually the discharge from its salt glands.

beaches. Perhaps, though, they, and all seagoing turtles, became large because larger animals were better able to avoid predators and survive ocean waves and currents. Increased salt tolerance may be a serendipitous side effect.

Sea turtles have developed a mechanism to cut down on their salt water intake while they eat. Feeding hawksbill, loggerhead, and green turtles have all been seen exhaling sea water out of their nostrils. What is happening is this: when the turtle swallows, its meal does not go straight to its stomach. The passage from a sea turtle's esophagus into its stomach is guarded by a powerful sphincter muscle, which the turtle closes as it swallows. It then constricts its esophagus, forcing water back out through its mouth or nostrils. It does not lose its food because its esophagus is lined with long, stiff projections, or *papillae*, that point backward towards its stomach. The papillae catch the turtle's meal as the water rushes out. Then the turtle relaxes its sphincter, and its meal continues through the digestive tract accompanied by the least amount of sea water possible.

Some does get in all the same. Even if it did not, much of the food that a sea turtle eats has the same high salt concentration as the ocean. Sea turtles still have to have some way to get rid of that excess salt. We mammals have specialized kidneys that can produce urine with a higher salt concentration than our body tissues. Reptiles and birds, including turtles, normally cannot do that.

Instead, they have repeatedly evolved salt-excreting glands, usually by modifying one or the other of the various pairs of glands in their heads. Within turtles, salt glands appear to have evolved independently at least four times, not just in the true sea turtles, but in other seagoing turtles such as the diamondback terrapin (though the diamondback's salt glands may not be particularly efficient). The salt glands in turtles are modifications of the lachrymal gland, the gland that supplies the tear ducts. This is why a sea turtle on its nesting beach often appears to be crying. Its "tears" are actually viscous streams of highly concentrated salt solution.

The salt glands of a sea turtle can take up quite a bit of space. A leatherback turtle (*Dermochelys coriacea*) has glands that are, together, almost twice as massive as its brain (a statistic that would admittedly be more impressive in a brainier creature). Though adult sea turtles probably do not drink sea water deliberately, hatchlings may need to do so during their initial growth spurt. They have proportionately much larger salt glands than adults, presumably because they require a higher desalination capacity.

When a sea turtle's glands are working, they turn out a solution that is impressively salty—more than twice as salty as sea water, and more than six times as salty as the rest of the turtle's bodily fluids. That means that the salt glands can remove all the salt from a

Softshell turtles (e.g. smooth softshell *Apalone mutica*, above) and the matamata (*Chelus fimbriatus*, below) have independently evolved long necks and snorkel nostrils that allow them to take a breath while remaining almost completely submerged.

gallon of sea water using less than half a gallon of secretion, leaving the rest for the turtle to use. In most sea turtles, the glands only work intermittently, presumably when the animal's salt load gets too high. Leatherbacks eat jellyfish, which have a higher concentration of salt than the food of other sea turtles, and their glands seem to work almost continuously. Perhaps, too, the tunic of the leatherback is not as effective as the armor of its relatives in preventing water loss through its skin.

**Taking a Dive**

Whether a turtle is in salt water or fresh, its most immediate physiological problem may not be temperature regulation or water balance, but respiration. While a turtle is at the surface, it can breathe air. The long necks and snorkel nostrils of a softshell turtle or a matamata (*Chelus fimbriatus*) may allow their owner to breathe from as far below the water as possible, but there are still likely to be lengthy periods in any aquatic turtle's life when it will be cut off from the surface. Turtles that winter at the bottom of ponds may be submerged, in a state of dormancy, for months at a time. They must rely on physiological mechanisms that free them, to a greater or lesser degree, from having to come to the surface at all.

If our body cells cannot get oxygen, they will continue to respire anaerobically, building up lactic acid as a by-product of their metabolism. Lactic acid is a toxin. When it escapes into the blood it can be carried to the brain, which has an extremely low tolerance for it. At the surface, or on land, the fresh supply of oxygen we take in with each breath breaks down the lactic acid in our blood. Under water, though, this cannot happen. It is the buildup of lactic acid in the brain that drives us back to the surface. In almost all vertebrates, the brain will die in a few minutes if it is not continuously supplied with oxygen. The brains of at least some turtles have a remarkable ability to survive without oxygen for many hours, but only if the turtle is dormant and its brain activity is very low.

Diving or bottom-walking may take a turtle away from the surface for only a matter of minutes, but during that time it must be active. Its muscles must operate, its brain must function, and the rest of its body must get along with only the oxygen that it has

When a hawksbill sea turtle (*Eretmochelys imbricata*) dives, its heart rate almost doubles.

been able to gulp before beginning its descent. Whales and seals, which can stay under far longer than we can, do so not by carrying more oxygen with them, but through a series of adaptations to keep lactic acid away from their brains, including the ability to close off the major blood vessels that would carry it there so that it builds up in their muscle tissues instead.

All sea turtles are accomplished divers. Most of them spend as little as 3 to 6 percent of their time at the surface. The different species vary in the amount of time they spend under water. Sea turtle dives usually range from a few minutes to almost an hour, though a subadult Kemp's ridley (*Lepidochelys kempii*) is reported to have stayed under for five hours. Like marine mammals, a green turtle can shut off blood vessels coming from its limbs to prevent lactic acid from reaching the brain. Sea turtles, though, normally deal with the stresses of diving through physiological adaptations that increase the efficiency of the system that delivers oxygen to their tissues. These adaptations begin in the lungs.

The amount of fresh oxygen that an animal can take into its lungs depends not just on lung capacity, but on how thoroughly it can empty its lungs of the air that is already there before it draws a fresh breath. The volume of air that an animal can actually exchange with each breath is called its *tidal volume*. Sea turtles have a much higher tidal volume—anywhere from 27 percent to more than 80 percent of the total volume of the lungs—than other reptiles, including other turtles. For comparison, human beings can only manage about 10 percent. A sea turtle can completely change the air in its lungs in only a few breaths, so it has to spend less time breathing at the surface before a dive.

All but the smallest passageways in a green turtle's lungs, and presumably in other sea turtle lungs as well, are reinforced with cartilage, protecting them against collapse under the pressures of a dive, at least down to depths of 80 m (260 ft) or so. Keeping these passageways open helps prevent excess nitrogen from being forced into the turtle's tissues—the cause of the terrible diving ailment known as the "bends."

Getting oxygen into the lungs, however, is only the beginning. Now it must be carried to the tissues. That is determined by the rate at which oxygen can be taken up through the lung wall, and sea turtles are better at doing that than any other reptile. The tissues of a sea turtle's lungs are extensively subdivided, greatly increasing their surface area. The greater the surface area, the faster the uptake. Also, when a sea turtle dives, its heart rate nearly doubles, so more blood can be pumped to the lungs to pick up oxygen and deliver it to the tissues.

Typical sea turtles store most of the oxygen they carry for a dive in their lungs, transferring it to their blood supply as needed. A loggerhead stores some 72 percent of its oxygen in its lungs. This is a perfectly good strategy for a turtle that does not dive particularly deeply, but storing air in the lungs is not really an option for the deep-diving leatherback. A leatherback may descend to over 1000 m (3300 ft), more than three times the depths other sea turtles reach. At those depths, even reinforced lungs will collapse. Instead, the leatherback transfers as much oxygen as possible out of its lungs, and carries it in its tissues.

A vertebrate stores and transfers oxygen by binding it with hemoglobin (in the blood) or myoglobin (in the muscles). The higher the concentration of hemoglobin and myoglobin, the more oxygen it can store. The leatherback has among the highest concentrations recorded for any reptile, enough to carry twice as much oxygen in its blood as other sea turtles.

Strangely enough, though it descends to greater depths, the leatherback's dives are the shortest of any sea turtle. A typical leatherback dive may last as little as four minutes, and is rarely longer than 40 minutes, compared to routine dives of over 50 minutes by hawksbills and female olive ridleys. That may be, in part, because, the leatherback's oxygen storage system faces a heavy demand from another, and quite different, set of adaptations—ones we will consider at the end of this chapter.

Common map turtles (*Graptemys geographica*) winter in cold, highly oxygenated water, breathing through their skins.

## Turtles with Gills

There are three ways for a turtle to avoid coming to the surface. It can carry air in its lungs, as cheloniid sea turtles do during a dive; this solution is, at best, a short-term one. Freshwater species like the Caspian turtle (*Mauremys caspica*) can carry an increased amount of oxygen both by having large lungs and an increased capacity to bind oxygen to the hemoglobin in its blood. At best, though, these adaptations stretch a turtle's time under water by minutes, not hours, days, or weeks.

An Australian broad-shelled turtle (*Chelodina expansa*) releases bubbles of air from its lungs.

For a long stay beneath the surface, a turtle must either be able to take up oxygen directly from the water, like a fish, or to do without oxygen altogether. These may seem to be highly unlikely solutions for any reptile, but freshwater turtles do both.

Not all freshwater turtles, however, do both equally well. The tendency has been to opt, primarily, for one solution or the other. Kinosternids and, particularly, softshell turtles, have gone in for the first solution (though some kinosternids, such as the Mexican giant musk turtle [*Staurotypus triporcatus*] and the white-lipped mud turtle [*Kinosternon leucostomum*], are only moderately good at it). Painted turtles and sliders have mastered the second. Thus, while a softshell turtle is better than a painted turtle at picking up oxygen from the water, it has a lower tolerance for anoxia (doing without oxygen altogether).

This choice of strategies may depend, at least partly, on where turtles spend the winter. A turtle that winters where the water is high in oxygen would be at an advantage to make use of it. Carlos Crocker and his colleagues found that common map turtles (*Graptemys geographica*) in the lower Lamoille River of Vermont assemble for the winter in areas where the water is cold and high in oxygen, and seem able to pass the winter successfully taking up the gas through their skins. On the other hand, painted turtles and common snapping turtles winter in stagnant water with little or no oxygen, and must be able to get along without it for long periods of time.

We have known for almost 150 years that turtles can take up oxygen from water, since this remarkable ability was first discovered by the famous 19th-century scientist Louis Agassiz. We are just beginning to understand how they manage it.

All a vertebrate needs to breathe under water is a thin, permeable skin well supplied with blood, so that oxygen can diffuse across it into the bloodstream and carbon dioxide can diffuse out again. Maximizing surface area is very important; the concentration of oxygen under water may be less than 1/20 of its level in the air. The more of the right kind of skin the animal has for its size, therefore, the better. The gills of fishes are simply places where the skin has been drawn out into hundreds of tiny filaments, well supplied with capillaries. This greatly increases the available surface area for breathing, allowing a fish to take up enough oxygen to carry on an active life. If the sheer volume of oxygen is not an issue, though, any bit of suitable skin can be a gill, or, at least, act like one.

Every turtle that has been studied can breathe underwater, though not always at a high enough rate to be of much use. Gas exchange, though, is a two-way process: as oxygen goes in, carbon dioxide goes out. Carbon dioxide dissolves much more readily than oxygen in water, so getting carbon dioxide out is much easier than getting oxygen in. Common snapping turtles are rather poor at taking up oxygen from the water, but are much more successful at getting rid of carbon dioxide. Brian Bagatto and his colleagues found that the white-lipped mud turtle, when underwater, gave off carbon dioxide through its skin five times more effectively than it took up oxygen.

The advantage of underwater breathing seems obvious enough for wintering turtles. An active turtle may benefit, too, if it can cut down the number of times it has to come to the surface to breathe. It can save energy, spend more time gathering food, and avoid movements that might attract predators or startle prey. These advantages, though, disappear as soon as the turtle comes on land. Skin

A Fitzroy river turtle (*Rheodytes leukops*) swims with its cloaca wide open, pumping water into its cloacal bursae.

thin and permeable enough allow gas exchange in water is more likely to dry out in air. Depending on how much time it spends out of the water, a turtle may be better off with thick, or scaly, impervious skin.

There are, however, other options. A number of turtles breathe through the linings of their mouth and pharynx, or throat cavity, a technique called *buccopharyngeal breathing*. A smaller number use the other end, exchanging gas through the lining of the *cloaca*, the common exit of the digestive and respiratory tracts. Since these areas are not on the outside, it matters far less if their lining is thin. Some turtles have even developed filamentous structures, rather like the gills of a fish, to increase the surface area they have available for underwater breathing. A number of softshells have them on the inside of the pharynx, a fact noted long ago by Louis Agassiz.

Buccopharyngeal breathing can be an important component of a turtle's underwater breathing apparatus. It can provide 30 percent of the oxygen a Nile softshell (*Trionyx triunguis*) takes up below the surface, and about 35 percent for a common musk turtle. A turtle that breathes in this way may enhance the process by pumping water in and out of its mouth. Hibernating red-bellied turtles (*Pseudemys rubriventris*) open and close their jaws slightly about five or six times a minute, drawing water in through their mouths and pumping it out again through their nostrils.

Cloacal breathing may be particularly important in pleurodires. The Fitzroy River turtle (*Rheodytes leukops*) of Australia has unusually large pouches, or *bursae*, extending from its cloaca. The bursae are lined with fingerlike projections, or *villi*. The turtle seems to use these as gills. It swims with its cloaca wide open, pumping water in and out of its bursae at a rate of 15 to 60 times a minute. The Fitzroy River turtle lives in fast-running waters presumably high in oxygen, and apparently it rarely breathes air. John Cann, who studied them in the wild, never saw one put its head above the surface.

No North American turtle is as thoroughgoing a gill-breather as the Fitzroy River turtle, but spiny softshells and common musk turtles seem to be able to survive indefinitely under water as long as they are not active. The common musk turtle can respire for long periods without oxygen, but depends on underwater breathing to survive its winter hibernation. So, probably, does the common map turtle.

The importance of underwater breathing varies from turtle to turtle. In a series of experiments by Brian Bagatto and R. P. Henry of Auburn University, Alabama, spiny softshells took up almost 22 percent of their oxygen from water, Florida softshells (*Apalone ferox*) almost 12 percent, but sliders only slightly over 5 percent. Further, both species of softshells, but not the slider, were able to double their rate of oxygen uptake when necessary. We are not sure exactly how; possibly they pump water in and out of the pharynx more frequently, or supply it, and the skin, with more blood.

This difference is matched by another. Sliders spent longer at the surface than the softshells, taking several breaths before diving again. The softshells took only a single breath. The number of breaths it takes does not seem to be a matter of choice for the turtle, but a genetically programmed behavior. Taking a single breath, as opposed to several, reduced

The painted turtle (*Chrysemys picta*) can survive without oxygen longer than any other known air-breathing vertebrate.

the time softshells had to spend at the surface (though the Florida softshell, which is not as good at water breathing as the spiny softshell, broke the surface more frequently). One-stroke breathing is also characteristic of some other highly aquatic turtles, including sea turtles (which, apparently, do not breathe underwater); it is also the way lungfish breathe air.

### Breathless

The painted turtle can survive without oxygen longer than any other air-breathing vertebrate we know. Perhaps it is no better at this than some of its cousins, but this is the species that has been studied in the most detail, thanks to the work of Donald Jackson and his colleagues at Brown University in Providence, Rhode Island. Dr. Jackson's experiments have shown that a painted turtle has the capacity to survive underwater without oxygen, at a temperature of 3°C (37.4°F), for as long as five months—far longer than it normally has to do in the wild. To accomplish this remarkable feat, it must avoid allowing lactic acid to build up in its tissues to toxic levels. It does so in two separate ways: by lowering the rate at which lactic acid is produced in the first place, and by neutralizing the acid as much as possible once it is released into its body.

Since lactic acid is a by-product of cellular metabolism, the only way a turtle can slow its production is to dial down its metabolic rate. A painted turtle denied access to oxygen can lower its metabolism to about 10 to 15 percent of its normal rate. Even for a dormant, cold-blooded animal at a low body temperature, getting cellular activity down to this level is no easy task. A metabolic rate of 10 percent of normal for a painted turtle is equivalent to less than 0.01 percent for a rat, even if the rat is resting comfortably at normal temperatures. To achieve a rate this low, a turtle has to shut down so many metabolic pathways that it may have to switch on a few special ones held in reserve just to keep it alive. The heart of a turtle in this state may beat only once every five or 10 minutes.

Even at 10 percent of normal activity, the amount of lactic acid an overwintering painted turtle produces, even over a few days, would cause serious problems if it could not be neutralized. To do this, the lactic acid has to be buffered (in other words, the pH in the turtle's body tissues must be restored to neutral levels). Buffering requires mineral compounds, particularly calcium and magnesium carbonates. Though a turtle's body fluids have some built-in buffering capacity, the greatest source of buffering compounds is bone, and the greatest amount of bone by far in a painted turtle is in its shell. Its skeleton, including the shell, holds 99 percent of its calcium, magnesium, and phosphorus, 95 percent of its carbon dioxide, and more than 60 percent of its sodium supply.

Many vertebrates can release mineral ions from their bones into their bloodstream to buffer lactic acid. Turtles have the additional ability, known in no other animal, to take up, buffer, and store lactate directly in the bone itself. Experiments by Donald Jackson and his colleagues have shown that as much as 75 percent of the lactic acid buffering that goes on in a submerged painted turtle (over three months, at 3°C [37.4°F]) is the result of these two mechanisms; 47 percent of the resulting buffered lactate ends up stored in the turtle's skeleton, all but 3 percent of that in its shell.

The more bone a turtle has in its shell, then, and the higher the mineral content of that bone, the better it should be able to handle lactic acid buildup. It is the sheer mass of bone in its shell that makes the key difference for the painted turtle. About 35 to 40 percent of its body mass is bone, all but about 5 percent of it in the shell. For comparison, in a spiny softshell (*Apalone spinifera*), which has a less mineral-rich shell (with only 20 to 25 percent of the calcium and phosphorus in the painted turtle), the figure is only 12 percent. This is not much more than for a snake or mammal of the same size. The two turtles seem equally good at storing lactate in their shells, but studies suggest that softshell bone does not release as much buffer into its system as does painted turtle bone. The difference may explain why the softshell, though better at taking oxygen from the surrounding water, is far less able to get along without it.

### A Warm-Blooded Turtle

When it comes to physiology, the leatherback is, in some ways, more like a reptilian whale than a turtle. It swims farther into the cold of the northern and southern oceans than any

Volunteers weigh a 550 kg (1200 lb) leatherback (*Dermochelys coriacea*) at a nesting beach on Mexico's Pacific coast.

other sea turtle, and it deals with the chilly waters in a way unique among reptiles.

A warm-blooded turtle may seem to be a contradiction in terms. Nonetheless, an adult leatherback can maintain a body temperature of between 25 and 26°C (77–79°F) in sea water that is only 8°C (46.4°F). Accomplishing this feat requires adaptations both to generate heat in the turtle's body, and to keep it from escaping into the surrounding waters. Leatherbacks apparently do not generate internal heat the way we do, or the way birds do, as a by-product of cellular metabolism. A leatherback may be able to pick up some body heat by basking at the surface; its dark, almost black body color may help it to absorb solar radiation. However, most of its internal heat comes from the action of its muscles.

Leatherbacks keep their body heat in three different ways. The first, and simplest, is size. The bigger an animal is, the lower its surface-to-volume ratio; for every ounce of body mass, there is proportionately less surface through which heat can escape. An adult leatherback is twice the size of the biggest cheloniid sea turtles, and will therefore take longer to cool off. Maintaining a high body temperature through sheer bulk is called *gigantothermy*. It works for elephants, for whales, and, perhaps, it worked for many of the larger dinosaurs. It apparently works, in a smaller way, for some other sea turtles. Large loggerhead and green turtles can maintain their body temperature at a degree or two above that of the surrounding water, and gigantothermy is probably the way they do it. Muscular activity helps, too, and an actively swimming green turtle may be 7°C (12.6°F) warmer than the waters it swims through.

Gigantothermy, though, would not be enough to keep a leatherback warm in cold northern waters. It isn't enough for whales, which supplement it with a thick layer of insulating blubber. Leatherbacks don't have blubber, but they do have a reptilian equivalent: thick, oil-saturated skin, with a layer of fibrous, fatty tissue just beneath it. Insulation protects the leatherback everywhere but on its head and flippers. Because the flippers are comparatively thin and bladelike, they are the one part of the leatherback that is likely to become chilled. There is not much that the turtle can do about this without compromising the aerodynamic shape of the flipper. The problem is that as blood flows through the turtle's flippers, it risks losing enough heat to lower the animal's central body temperature when it returns. The solution is to allow the flippers to cool down without drawing heat from the rest of the turtle's body. The leatherback accomplishes this by arranging the blood vessels in the base of its flipper into a countercurrent exchange system.

In a countercurrent exchange system, the blood vessels carrying cooled blood from the flippers run close enough to the blood vessels carrying warm blood from the body to pick up some body heat and raise their blood tem-

Its adaptations for life in the open sea make the leatherback (*Dermochelys coriacea*) more like a whale than a turtle.

perature as they pass. Thus, the heat is transferred from the outgoing to the ingoing vessels before it reaches the flipper itself. This is the same arrangement found in an old-fashioned steam radiator, in which the coiled pipes pass heat back and forth as water courses through them. The leatherback is certainly not the only animal with such an arrangement; gulls have a countercurrent exchange system in their legs. That is why a gull can stand on an ice floe without freezing.

All this applies, of course, only to an adult leatherback. Hatchlings are simply too small to conserve body heat, even with insulation and countercurrent exchange systems. We do not know how old, or how large, a leatherback has to be before it can switch from a cold-blooded to a warm-blooded mode of life. Leatherbacks reach their immense size in a much shorter time than it takes other sea turtles to grow. Perhaps their rush to adulthood is driven by a simple need to keep warm.

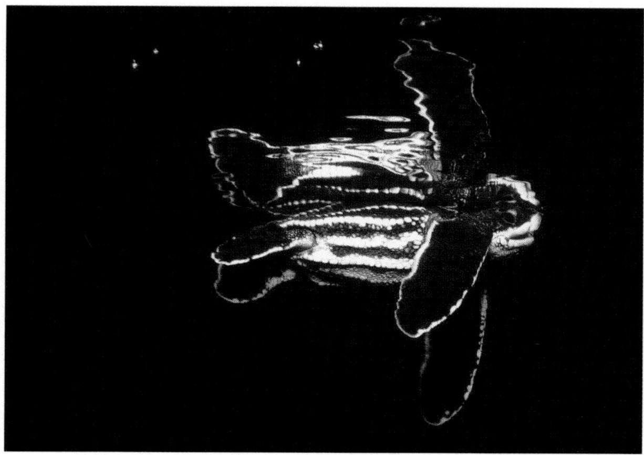

A hatchling leatherback (*Dermochelys coriacea*) swimming in the Gulf Stream, is still too small to control its body temperature.

CHAPTER SIX

# *Life As A Turtle*

WE ARE A LONG WAY FROM knowing the full range of behaviors and lifestyles of land and freshwater turtles. We know little of the lives of the rich variety of turtles in tropical Asia, and considering the plight of most Asian turtles, that is a gap we may never fill. That we do not have that information is not merely unfortunate; it is tragic. Much of what we are learning about the better-known turtles of North America is information we need for intelligent plans for their conservation. If we do not know how much geographic space a turtle needs, how much territory it requires, what types of habitat it uses, or the range of food it needs at different points in its life cycle, we may fail to protect the very elements of its environment that are required if it is to survive.

## How Long a Life?

The lifespan of a tortoise can exceed our own. An Aldabra tortoise (*Geochelone gigantea*), or, just possibly, a native tortoise from the Seychelles, lived on the island of Mauritius for 152 years before its accidental death in 1918. Some Seychellois families living in the granitic islands have kept herds of giant tortoises in captivity in their families for generations. After interviewing the present (now elderly) owners of some of these tortoise herds, Jeanne Mortimer is convinced that certain of the older animals easily exceed 150 years in age.

The oldest turtle ever known may have been the famous Tui Malila (or Tu Imalalia), the tortoise Captain James Cook supposedly presented to the queen of Tonga in either 1773 or 1777. For many years, Tui Malila was thought to be either an Aldabra or a Galápagos tortoise (*Geochelone nigra*). To everyone's surprise, in 1958 he was identified as a radiated tortoise (*Geochelone radiata*) from Madagascar. His characteristic star patterns had faded away during his long and adventurous life.

Tui Malila lived until May 19, 1966. So celebrated was he that, according to Tony Beamish in *Aldabra Alone*:

> . . . he was in later life accorded chieftain's rank and given his own apartments in the royal palace. Not only was he old, but he was tough too. Tu Imalalia [Tui Malila] had survived the loss of an eye . . . being run over and being kicked by a horse which fractured his shell.

If the tortoise that died in 1966 was indeed the one Captain Cook delivered in the 18th century, he may have been over 200

The wood turtle (*Clemmys insculpta*) of eastern North America has a reputation for considerable intelligence.

Eastern box turtles (*Terrapene carolina*) can live at least 80 years, and perhaps more than a century.

years old when he died. Unfortunately, there is some doubt as to whether Tui Malila was ever in Captain Cook's hands at all. Even if Tui Malila was not as old as he seemed, the radiated tortoise can reach a venerable age; one well-documented animal apparently lived more than 137 years.

Eastern box turtles (*Terrapene carolina*) occasionally turn up with initials and dates that seem to be more than a century old carved into their shells. An adult male found in Rockingham County, Virginia, in 1985, bore the date 1874 carved into its plastron. To authenticate records like this can be tricky. There is good evidence, however, that box turtles can live at least 80 years or so, although probably in the wild very few survive past 30 or 40.

Forty may be a ripe age for a wild land or freshwater turtle, but there are longer records in captivity. A common snapping turtle (*Chelydra serpentina*) and a common musk turtle (*Sternotherus odoratus*), both in the Philadelphia Zoo, lived over 70 and 54 years, respectively, and there is a record of 77 years for a Blanding's turtle (*Emydoidea blandingii*). African softshells (*Trionyx triunguis*) have lived over 42 years in captivity, a desert tortoise (*Gopherus agassizii*) over 55, and leopard tortoises (*Geochelone pardalis*) have possibly lived as long as 75 years.

Turtles may no longer be the record holders for vertebrate venerability; we now have evidence that a mammal, the bowhead whale (*Balaena mysticetus*), may live to be over 200, and a number of large ocean fishes may be centenarians. Nonetheless, if they can escape the perils of their youth, it seems that under natural conditions most turtles can look forward to a quite respectable lifespan.

**Are Turtles Intelligent?**

No one who loves turtles will dispute that they have charm and personality. You may not find the personality of, say, a large snapping turtle particularly attractive, but then you should probably have left it alone in the first place. How much intelligence turtles have, though, is another matter. The late Archie Carr dismissed the issue scathingly in his *Handbook of Turtles*:

> Tinklepaugh (1932) averred that the wood turtle equaled the 'expected accomplishment of a rat' in learning the intricacies of an experimental maze, but one must conclude that Tinklepaugh had known only feeble-minded rats.

Carr may have been doing his beloved turtles an injustice. As far back as 1901, famed psychobiologist Robert M. Yerkes put a spotted turtle (*Clemmys guttata*) through a maze that offered it five choices, only one of which led to a nest of moist grass. At first the turtle took 35 minutes to find the nest, but by its 20th try it made the trip in only 45 seconds. By its 5th trial, even O. L. Tinklepaugh's

Japanese turtles (*Mauremys japonica*) have been trained, in laboratory experiments, to distinguish red from blue.

wood turtle (*Clemmys insculpta*), faced with a fairly complex maze, reduced its running time from 15 minutes to less than six.

Wood turtles have something of a reputation for intelligence, as turtles go, though its skill in Tinklepaugh's maze may not reflect true problem-solving abilities. Spotted and wood turtles, which travel back and forth between freshwater areas and woodlands, simply seem to be better navigators than more sedentary turtles.

Turtles are also reasonably good at tests that require them to make choices based on color. Japanese turtles (*Mauremys japonica*), for example, have been trained to tell the difference between red and blue. Similar experiments have shown that turtles not only have color vision, but that they can learn to choose the right color, or pattern, to get a reward. Whatever this behavior says about their intelligence, turtles can engage in quite complicated behavior.

**Patterns of Life**

Outside of the bustle of the mating season, an active turtle's daily round may amount to little more than a cycle of basking, feeding, and resting in the shade. The details of the cycle shift with season and weather. On Egmont Key, Florida box turtles (*Terrapene carolina bauri*) are only active when the temperature rises above 17°C (63°F) and the humidity is above 24 percent (particularly after rainfall). In spring, that may mean a whole day of activity; in summer, only a few hours in the morning. For the rest of the summer day, the turtles burrow beneath leaf litter or bury themselves in shallow, scooped-out depressions, or *forms*, in the cool organic soil.

Many turtles are diurnal, whether they are abroad at midday or only at earlier and later hours (a *bimodal* activity pattern; see Chapter 5). Diamondback terrapins (*Malaclemys terrapin*) spend the night buried in the mud, perhaps to hide from nocturnal predators. Some turtles, though, are active at night, particularly mud and musk turtles (Kinosternidae). In northern Guatemala, white-lipped mud turtles (*Kinosternon leucostomum*) are almost entirely nocturnal; the same may be true of their distant relative, the Central American river turtle (*Dermatemys mawii*). Euphrates softshells (*Rafetus euphraticus*), though mostly diurnal, are often active after dark. About 30 percent of the observations of these turtles by Ertan Taska Yak and Mehmet K. Atatur in southeastern Anatolia, Turkey, were made at night.

How a turtle lives its life may have a great deal to do with its size. There are advantages to being large, and different advantages to being small, even for animals living in the same type of country. One of the smallest tortoises in Africa, the Egyptian tortoise (*Testudo kleinmanni*), lives on the northern edge of the Sahara, and the largest, the spurred tortoise (*Geochelone sulcata*), on the southern. In South Africa, the large leopard tortoise

In the wild, a juvenile leopard tortoise (*Geochelone pardalis*) may need to grow quickly to be safe from predators.

Within its home range, a Blanding's turtle (*Emydoidea blandingii*) may have widely separated "activity centers."

(*Geochelone pardalis*) often lives side by side with the tiny padlopers (*Homopus* spp.).

A large tortoise has fewer potential predators, and may be more likely to survive a fire. A small tortoise, however, can seek shade where a larger animal risks overheating in the sun. It may burrow out of sight or hide in leaf litter, something a large tortoise may find difficult to do. Even large tortoises start off small, but there may be an advantage for them to grow as rapidly as possible. Leopard tortoises in the Serengeti grow more rapidly than in other areas, possibly because faster-growing tortoises are less likely to be eaten by predators or burned in a fire. The faster the young leopard tortoises grow, the more likely they are to escape the risk they run by being small.

The area where an animal lives day after day is its *home range* (which is not the same thing as a *territory*, the area an animal defends). Turtle home ranges (even if we leave out the vast areas covered by sea turtles) vary tremendously in size, even within a single species. The home ranges of juvenile gopher tortoises (*Gopherus polyphemus*) studied by Joan Diemer in northern Florida were smaller than those of adults, and males had, on average, larger home ranges than females. Home range sizes for desert tortoises ranged from a third of a hectare (0.8 acres) to 268 hectares (670 acres) in a single region of Arizona.

If a smaller area can satisfy a turtle's needs, there may be no reason to have a large home range. Red Amazon sidenecks (*Phrynops rufipes*) feed on palm fruits that are available all year and grow commonly along the banks of the streams where they live. They appear to get along with a home range stretching only a kilometer or two, along a stream that may be only 2.5 m (8 ft) wide.

Within its home range, a turtle may have "activity centers" that may be some distance from the places where it hibernates or nests. David Ross and Raymond Anderson studied Blanding's turtles (*Emydoidea blandingii*) in central Wisconsin that had, according to their report in the *Journal of Herpetology*, "well-defined activity centers which were separated by long distances and whose locations changed over time." Like the home ranges themselves, these activity centers may vary in size. The activity centers of Wood turtles (*Clemmys insculpta*) studied by Norman Quinn and Douglas Tate at the northern edge of their range, in Algonquin Provincial Park, Ontario, range from less than a hectare to over 100 hectares (250 acres), usually along streams. Individual turtles tend to use the same centers, large or small, year after year. In mid to late summer, females leave them and travel into the upland portion of their home ranges, presumably on their way to their nesting sites.

This long-necked turtle from New Guinea (*Chelodina* sp.) is aestivating, partly buried in soil and leaf litter.

### The Inactive Life

One advantage of ectothermy is that when food is not available, or the climate is not right, a turtle can simply hole up somewhere and wait for matters to improve. In cool temperate regions, turtles pass the winter in, more or less, a state of dormancy. Many herpetologists call this winter dormancy "hibernation," but it is difficult to tell if turtles really hibernate in the way some mammals and a very few birds do. Some herpetologists do not use the term "hibernation" at all, or use it only when the overwintering turtle seeks out a refuge, or *hibernaculum*, for the winter.

Dormancy at other times of year—for instance, during periods of drought—is usually called *aestivation* instead. It is probably best, for our purposes, to ignore the debate over the definition of these terms (and others, like *brumation*, a sort of reptilian hibernation) mean and simply say that at various times of the year, for one reason or another, turtles may find a sheltered spot, dig a burrow, or dive to the bottom of a pond, and simply stop in their tracks for anywhere from a few days to several months.

A turtle seems to spend much of its life doing nothing at all. Taking into account both winter hibernation and summer aestivation, a Central Asian tortoise (*Testudo horsfieldi*) may be inactive for nine to nine and a half months of the year. The narrow-bridged musk turtle (*Claudius angustatus*) is active only from June to October, when river waters rise and flood the neighboring grassy marshes. The turtles invade the marshes; as the dry season approaches and the waters recede, they nest and disappear. For the rest of the year they probably bury themselves in the mud and simply wait until the rains come again; Jonathan Campbell reported seeing one dug out from beneath about a foot of dried mud at the edge of a temporary pool, or *aguada*.

In East Africa, pancake tortoises (*Malacochersus tornieri*) spend most of their lives in rock crevices, usually only emerging to feed during the early-morning hours. Desert tortoises spend most of their lives in their burrows, driven there both by winter cold and dry summer heat. In October or November the tortoises retire for the winter, emerging any-

A Chinese softshell (*Pelodiscus sinensis*) digs its way into the soil, perhaps in preparation for dormancy.

where from February through April; in summer, they are active only from dawn until about ten o'clock in the morning, or again after seven o'clock in the evening. Even in midsummer, though, the arrival of a thundershower brings the tortoises out to feed and drink.

Spiny softshells (*Apalone spinifera*) spend the winter buried in sediment at the bottom of a stream, just deeply enough to be able to reach the water by extending their necks. This is probably so they can pump water in and out of their throats occasionally for buccopharyngeal respiration (see Chapter 5). If the winter temperature rises above 12°C (53.6°F), the softshells emerge and switch to another burial spot.

There is evidence, strangely enough, that under some circumstances even sea turtles become dormant. Green sea turtles (*Chelonia mydas*) in the Gulf of California have been found buried in the sandy bottom, where they—at least according to the local fishermen—have stayed for up to three months. Loggerhead sea turtles (*Caretta caretta*) off the central coast of Florida have also been caught in circumstances that suggest that they were lying dormant in the mud of the bottom. How the turtles get along without breathing during these periods we do not really know.

In Australia, short-necked members of the family Chelidae rarely become dormant, but some longnecks (*Chelodina* spp.) do quite regularly. If the water holes they inhabit dry up, New Guinea long-necked turtles (*Chelodina novaeguinae*) may try to go elsewhere. If they cannot, they may bury themselves in the mud, their presence revealed only by a slender tunnel to the surface air. They remain there as the mud dries and hardens around them, trapping them in place. Bruce

Volunteers relocated over 900 aestivating turtles from this dry lake in Central Queensland, Australia.

Chessman attached a radio transmitter to the carapace of a snake-necked turtle (*Chelodina longicollis*) and released it into a warm Victorian billabong. The next day, he found it 100 m (328 ft) away from the water, buried 1 cm (0.4 in) below loose soil and leaf litter. Chessman recorded snake-necked turtles remaining in such refuges for up to two and a half months. In colder weather, snake-necked turtles will also enter dormancy underwater; one investigator in northern Victoria found 10 in the end of a hollow log.

The only short-necked Australian chelid known to become dormant is the rare western swamp turtle (*Pseudemydura umbrina*). As summer temperatures rise to 30–40°C (86–104°F) and the swamps where they spend the winter dry out, the turtles find natural tunnels or burrow under leaf litter in the nearby woodlands. Sometimes they bury themselves so that only the top of the carapace and the tip of the snout is showing. There they remain until the rains return in April and May.

The shallow, turbid ponds preferred by the savanna side-necked turtle (*Podocnemis vogli*) of Colombia and Venezuela may disappear completely during the dry season, forcing the turtles to burrow into the mud, sometimes by the thousands. According to Peter Pritchard and Pedro Trebbau in *The Turtles of Venezuela*, the dormant animals "may be buried at almost any angle—head up or down, plastron up or carapace up—as if they had been thrown into the drying mud rather than settled into it voluntarily."

Tortoises of several species aestivate during summer droughts, when temperatures soar and food plants wither. *Gopherus* tortoises in North America, the Egyptian tortoise in North Africa, spurred tortoises south of the

Savanna side-necked turtles (*Podocnemis vogli*) bury themselves by the thousands in the mud of the Venezuelan llanos.

Though striped mud turtles (*Kinosternon baurii*) often winter under water, some burrow into the soil instead.

Sahara, and the speckled padloper (*Homopus signatus*) in the deserts of Namaqualand in Southwest Africa, all become dormant during the hottest and driest times of the year.

The choice of a hibernaculum may depend on the temperature. The underwater hibernacula chosen by many North American and Eurasian turtles often remain above freezing. Because water is most dense at 4°C (39°F), unless a pond or river freezes solid there will be a layer of water above the freezing point near the bottom. Even if that freezes, the turtles may bury themselves safely in the unfrozen mud below. In the Caucasus, European pond turtles (*Emys orbicularis*) bury themselves at least 10–15 cm (4–6 in) deep, and Caspian turtles (*Mauremys caspica*) from 8–30 cm (3–12 in).

Blanding's turtles overwinter underwater, partially buried at the bottom of a marsh, pond, or creek. They may either do the same, or aestivate on land, during the heat of the summer. Turtles that winter on land may shift deeper into the soil as the frost line approaches. Eastern box turtles can reach nearly 0.5 m (1.6 ft) deep in the coldest months, then dig their way slowly towards the surface again as the frost line recedes.

Many North American turtles spend the winter underwater. Ironically, some of the more aquatic turtles winter on land, even though their watery habitats remain available. Eastern, yellow, and striped mud turtles (*Kinosternon subrubrum*, *flavescens*, and *baurii*) winter either on land or in the water. Chicken turtles (*Deirochelys reticularia*), in the northern part of their range, leave their ponds and travel, sometimes for some distance, to bury themselves in mud or moist sand. They do the same in summer, when the temporary ponds they prefer reach their lowest levels; in Virginia, chicken turtles have been recorded burying themselves in wooded areas for as long as two weeks.

In some turtles, the refuges, or hibernacula, chosen for the winter may be quite traditional. Eastern box turtles in southwestern Ohio traveled as much as 1.5 km (1 mi) to reach their hibernaculum site. Snapping turtles may return to the same general area year after year. One used the same site four years in a row. Twelve bog turtles (*Clemmys muhlenbergii*) used the same hibernacula two years running.

Some turtles occasionally winter in groups. Timothy L. Lewis and John Ritzenthaler investigated nine spotted turtle hibernacula buried beneath the ice and mud of a swamp in Ohio. Seven were simple vertical holes in the mat of vegetation, dug and used for only a single season. Five had one turtle each, and two had two. Of the other two hibernacula, though, one contained eight turtles, and the other, at least at the peak of its winter use, held 34. In both group hibernacula, the vertical entrance led to a horizontal passage. Both were used each winter during the three years of the study.

Up to 70 wood turtles have been found wintering together in a beaver pond. They could not have been keeping each other warm because they cannot generate sufficient internal heat and lack effective insulation. Possibly there is something about group hibernacula that attracts numbers of turtles; the wintering sites of common map turtles (*Graptemys geographica*) in the lower Lamoille River of Vermont may be areas where the water is especially high in oxygen (see Chapter 5). Another possible advantage of wintering in a group is that it makes finding a mate in the spring considerably easier. Carl Ernst reported finding 16 spotted turtles in a mating aggregation in a pool in mid-March, and speculated that they may have spent the winter together.

**Shifting with the Seasons**

Freshwater turtles seem to be sedentary creatures, tied to the ponds, lakes, or rivers where they live. Many turtles, though, have more or less regular cycles of migration in and out of their centers of activity. Mud turtles (*Kinosternon* spp.) are confirmed wanderers. Striped mud turtles may bury themselves in the mud when their pond dries, but are equally likely to strike off cross-country in search of another one.

Helmeted terrapins (*Pelomedusa subrufa*) are such voracious consumers of tadpoles that a single hatchling in a temporary savanna pond can eliminate its entire tadpole community in a few days. The terrapins then may switch to other foods, or migrate overland to another pond. Mark-Oliver Rödel found that helmeted terrapins in the Comoe National Park, Ivory Coast, moved regularly from pond to pond even when their ponds were filled with water; so did two other turtles that shared the ponds, the West African mud terrapin (*Pelusios castaneus*) and the Senegal flapshell (*Cyclanorbis senegalensis*).

Wandering may help turtles find mates, as well as the best sources of food and shelter in a patchwork of localized, diverse ecological situations, or microhabitats. Bodies of fresh water, like similar areas on land, can have periods of feast or famine. In some cases they may actually disappear, annually during seasonal droughts or occasionally during unusual dry periods. There are times when it may be advisable for even the most sedentary turtle to move on.

For some turtles, the cycle of the seasons brings travel to and from feeding areas, winter shelters, and summer resting places. At Cedar Swamp in central Massachusetts, spotted turtles spend their winters, from November through March, in permanent red maple–sphagnum swamps. Here, they lie dormant in underwater passageways between the roots of the maples, or crawl slowly through the thick mats of sphagnum that cluster at the bases of the trees.

In late March, melting snows and spring

The annual cycle of the spotted turtle (*Clemmys guttata*) takes it from permanent swamps to temporary upland ponds.

showers fill upland hollows and basins with water, creating temporary pools where invertebrates like caddisfly and damselfly larvae emerge for a frenzy of feeding. The spotted turtles emerge from their winter retreat, and strike out for the newly filled pools, traveling an average of 120 m (400 ft) to reach them (not an unusual distance; in South Carolina a spotted turtle on its spring trek traveled 423 m [¼ mi] in 24 hours). They stay in and around these pools until their waters begin to dry in July and August.

By early August, the turtles begin their journey back into the permanent swamps where they spend the winter. Sometimes—we don't know why—they pause along the way to lie dormant on dry land, resting in shallow forms in grassland or under leaf litter for anywhere from four days to two weeks (there is a Connecticut record of nine weeks) before completing their journey.

As temperature rise in the summer, wood turtles in many parts of their range start to spend more and more time on land, and during late spring and summer they may be almost entirely terrestrial. In Virginia, wood turtles emerge from the deep pools where they spend the winter in late March and early April, and return to the pools in October. Apparently a change in temperature is the chief stimulus that sends them to and from their winter refuges.

Spur-thighed tortoises (*Testudo graeca*) in the western Caucasus live in areas covered by a patchwork of forests, meadows, open glades, and boulder-strewn screes. In the spring, the tortoises prefer open areas, where they dine on dandelions, clover, and other plants; this is the mating season, and glades are favored places for the tortoises to dig their nests. As temperatures rise in the summer, most tortoises move into the cooler woodlands, often shifting back and forth between sunlight and shade depending on the time of day; on really hot days, they may cluster around forest streams or the few lakes and ponds scattered in the woods. By autumn, they scatter through the woodlands looking for winter shelter.

In Madagascar, the local and endangered angonoka (*Geochelone yniphora*) shifts seasonally between microhabitats within its tiny range. In the wet season, when food and water are plentiful, they may spend more time in open areas, feeding or searching for nesting sites. In the cool, but sunny, dry season, their home ranges shrink, and they spend more time sheltering under savanna grasses or among thickets of bamboo. Understanding how the angonoka uses different types of vegetation from season to season is crucial to our efforts to conserve it; the variety within its habitat may be essential to its annual cycle.

Australian snake-necked turtles (*Chelodina longicollis*) can navigate overland, using the sun as a compass.

## Finding Their Way

For river turtles, shifting habitat can simply involve moving up- or downstream, but for turtles in lakes or ponds an overland trek may be the only way to leave home. This presents wandering turtles with a serious challenge above and beyond the need to survive in a less familiar, dry-land environment: how are they to find their way?

American herpetologist Terry Graham found some of the answers in a study of the snake-necked turtle in coastal New South Wales. Here, the turtles spend most of their time in lakes among sand dunes. These lakes have the advantage of being relatively permanent, but they lack an abundant food supply. In some years, at times as rarely as every seven years or so but usually more often, heavy rains fill the dry beds of nearby swamps and temporary ponds. These newly filled pools quickly become hotbeds of invertebrate life, full of rich food. The turtles set off overland to find them. Graham (who has also studied the rather similar, but annual, journeys of spotted turtles at Cedar Swamp, Massachusetts, where the cycle of the seasons is far more regular) followed the turtles by the low-tech method of attaching spools of thread to their carapaces. As the turtles set off, the spools unrolled, leaving a trail of thread for Graham to follow.

Graham's turtles made their way to their bountiful new feeding ground in a remarkably straight line, over hill and dale for almost 0.5 km (0.3 mi). Clearly, they were using some cue to orient themselves. Aboriginal workers suggested to Graham that the turtles used smell to find their way, and indeed Graham found that the animals could find piles of swamp debris by odor alone. However, their ability to orient proved to be much greater on sunny days than on cloudy ones. On cloudy days the turtles interrupted their journeys, burrowed into leaf litter and stayed there until the sun came out again (the opposite of the behavior seen in North American chicken turtles, which tend to wander on rainy days and bury themselves on sunny ones). The animals were obviously using the sun as a compass.

Furthermore, the turtles appeared to have some sort of internal biological clock that allowed them to take the time of day into account while they determined the sun's position. Graham proved this by keeping some turtles under artificial conditions in which "dawn" and "dusk" were shifted ahead by six hours. When the turtles were released, they set off at a right angle to the direction they should have followed, showing that their internal clocks were giving them the wrong information. The turtles seem to rely primarily on their view of the sun for the cues they need, followed by their sense of smell and, where possible, their ability to recognize landmarks on the way. By combining these three sets of clues, they are able to make a highly accurate beeline for the rich swamps awaiting them.

Of course, that isn't the whole story. Graham's study tells us how the turtles steer themselves toward the ponds—but how do they know where the ponds are in the first place? Do they remember from the last rains? If so, how do they make their very first trip? Do they follow other, more experienced turtles? We simply don't know. The turtles are not traveling together, though sometimes that does happen; snake-necked turtles have been seen crossing roads by the hundreds.

Some turtles are capable of remarkable feats of orientation. Sea turtles, of course, are master navigators (see Chapter 8). Leopard tortoises appear to have a well-developed homing instinct. Richard Boycott and Ortwin Bourquin write, in *The Southern African Tortoise Book*:

> Many farmers have mentioned that tortoises taken to remote areas soon find their way home. One report is of several individuals which were marked and taken 8–13 km [5–9 mi] from the area in which they were found, and returned within two weeks, climbing through, under or over 1.2 m [4 ft] wire-mesh fences to achieve this. They were reported as being able to climb such fences, and having reached the top they simply toppled over to the other side.

Boycott and Bourquin report that a leopard tortoise in the Serengeti National Park in Tanzania, taken some 8 km (5 mi) from where it was captured, waited about three and a half months for the rainy season before starting homeward. It took two months to get there, following a roundabout route of about 12 km (7.5 mi).

Hatchling turtles face their own navigational challenges as they make their way from their nests to the water, or search for a safe place to spend their first winter. As no hatchling receives any care or training from its parents, newly-hatched turtles have no opportunity to learn where they must go. Their orienting skills—and they certainly have them—must be entirely innate. Hatchling wood turtles and Blanding's turtles apparently follow the trails of other hatchlings, perhaps using scent to find their way—hatchling Blanding's have a distinctive musklike odor.

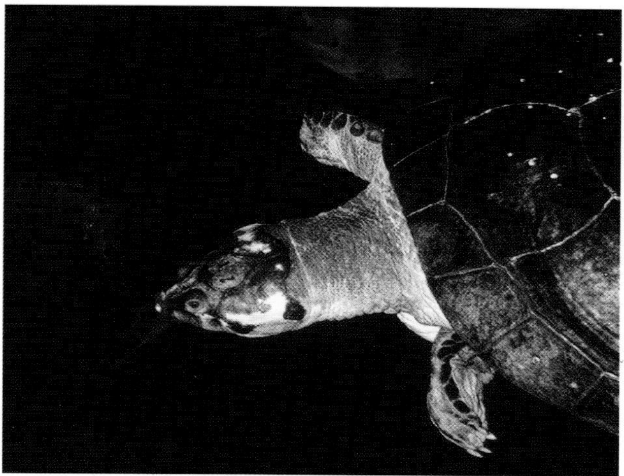

The giant South American arrau (*Podocnemis expansa*) makes the longest migratory journeys of any freshwater turtle.

Other than that, the cues that hatchling turtles appear to use include smell, sight, sound (for example, the noise of rushing water in brooks), and a general tendency to head downhill, usually the best direction to follow to find water.

The longest recorded turtle travels are not on land, but in water. Sea turtles make journeys that no land or freshwater turtle can match, some traveling over 5000 km (3000 mi) (see Chapter 8). In the Orinoco River, female arrau (*Podocnemis expansa*) leaving their nesting beaches have been captured months later a few hundred kilometers downstream, or 100 km (60 mi) upstream. In the Rio Trombetas in Brazil, females tagged with radio transmitters by Richard Vogt traveled up to 45 km (27 mi) in two days. Some softshell turtles (*Apalone* spp.) may travel 5–10 km (3–6 mi) to reach their nesting grounds. Not all river turtles are as energetic: a female alligator snapper (*Macrochelys temminckii*) reportedly traveled only 6.5 km (4 mi) over an entire year.

## Omnivores

Most turtles are, more or less, opportunistic, omnivorous generalists. That is a zoologist's way of saying that most turtles will eat whatever they can get, whenever and wherever they can get it.

Generalists may lack the evolutionary sophistication of food specialists, but if food becomes scarce or hard to find, or if conditions change, they can switch to other kinds of foods and carry on. Thus, Blanding's turtles in some populations eat mostly snails, while elsewhere they dine heavily on crayfish. Local populations of painted turtle (*Chrysemys picta*) may be either entirely carnivorous, entirely herbivorous, or something in between.

Turtle food preferences may take time to develop. In a laboratory study, newly hatched diamondback terrapins ate almost anything they were offered. By the end of their first year, though, they began to refuse items they had accepted only a few months before.

A really aggressive and flexible generalist should be able to shift easily from food to food, from feeding technique to feeding technique, and even from habitat to habitat. An opportunistic generalist will be able to turn a wide range of circumstances to its own best advantage. In a changing world, it may have the best shot at survival.

Australia is a land of frequent droughts, with drastic shifts in the kind and amount of food a hungry turtle can find. In this harsh country, it may be better not to have a dietary preference, but to be able to eat whatever comes along. Australian freshwater turtles do include strict carnivores (for example, the Fitzroy River turtle [*Rheodytes leukops*], the western swamp turtle and most of the long-necked turtles [*Chelodina* spp.]) as well as a few herbivores (for example, the Victoria River snapper [*Elseya dentata*]), but the *Emydura* turtles, the Mary River turtle (*Elusor macrurus*) and the pig-nose turtle (*Carettochelys insculpta*) are omnivorous generalists.

The pig-nose turtle eats a wide range of animal food, but it prefers plants. The plants it eats, though, differ from place to place. In Papua New Guinea, it eats the unripe fruits of crabapple mangrove (*Sonneratia* spp.). In Australia, *Sonneratia* grows mostly along the coast. Farther inland, the turtle feeds mostly on the leaves, flowers, and fruits of riverside trees like fig (*Ficus racemosa*), bush apple (*Syzigium forte*) and screw pine (*Pandanus aquaticus*). This change probably has little to do with what the turtle prefers; instead, it dines on what is available.

In some billabongs in Kakadu National Park in northern Australia, "what is available" to pig-nose turtles includes flying foxes (*Pteropus* spp.). Roosts of these giant fruit bats overhang the waters where the turtles live, and the bats occasionally end up inside the turtles. How they get there we are not sure. Will a turtle actually drown a bat that falls into the water, or will it only scavenge from a carcass? It is hard to say, but perfectly formed hairballs composed entirely of bat fur have been recovered from pig-nose stomachs.

Mathew Allanson and Arthur Georges found that adults of the recently described Purvis's turtle (*Elseya purvisi*) and Georges' turtle (*Elseya georgesi*), both confined to isolated river drainages in New South Wales, eat a wide range of animal and plant foods apparently without being, in any measurable way, selective. This sort of indiscriminate behavior may allow the turtles to thrive in a variety of streams, each with a different supply of foods (or, at least, the same foods in different proportions).

Opportunism may be a good strategy for another turtle that faces a wide range of changing circumstances, the helmeted terrapin of Africa. This is a species of temporary ponds, places that may offer a variety of foods at different times or in different circumstances. Helmeted terrapins will eat almost anything, animal or plant, though they are carnivores by preference. About the only restriction is that they eat all their meals underwater.

Pig-nose turtles (*Carettochelys insculpta*) prefer to eat plants, but will take animal food, including flying foxes.

Helmeted terrapins will scavenge from a rotting carcass, or pull ticks from animals like rhinoceroses that wallow in their pools—a habit they share with the serrated hinged terrapin (*Pelusios sinuatus*). Like North American common snapping turtles, they seize and devour ducklings and other waterbirds. In the Etosha Pan of Namibia, a great flat expanse that only occasionally, after exceptional rains, becomes a shallow lake, helmeted terrapins may act like miniature crocodiles, seizing doves and sandgrouse that come to the pan to drink and holding them under the water until they drown.

Perhaps the ultimate chelonian omnivore is the alligator snapping turtle. Alligator

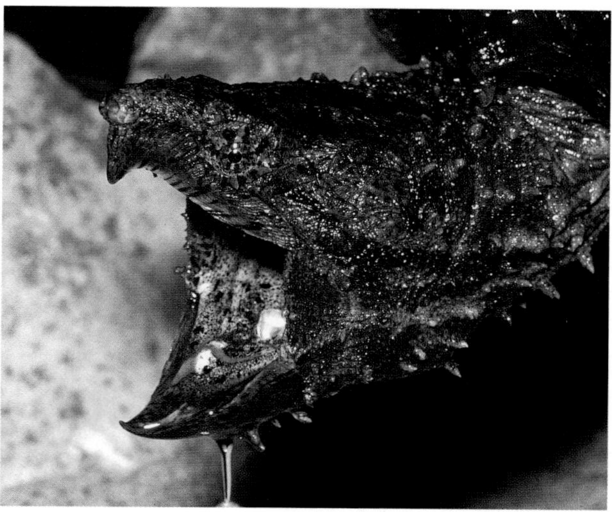

The powerful jaws of the alligator snapping turtle (*Macrochelys temmincki*) can crush mussel shells, and other turtles.

snappers will eat just about anything. Their stomachs have yielded fish like pickerel, gar, and carp, salamanders including the large, eel-like sirens and amphiumas, snakes, turtles including its cousin, the common snapping turtle, and even young members of its own species, small alligators, crayfish, freshwater mussels, snails, wood ducks, and other birds, and mammals, including raccoons and muskrats. As if that weren't enough, the alligator snapper also eats persimmons, wild grapes, acorns (in a study by Kevin Sloan, Kurt Buhlmann, and Jeffrey Lovich, perhaps at the right season, acorns were the most abundant food item by weight recovered from snapper stomachs) and miscellaneous items, including rocks, fishhooks, wood, and cardboard.

These extremely catholic tastes may seem surprising in an animal that appears, on first examination, to be a highly specialized fish trapper. However, even the fish it eats are not necessarily baited in by the wormlike pink lure on the snapper's tongue. Some may be taken as carrion.

Fishing by lure seems to be used mostly by the juveniles. As the turtles grow and their jaws increase in strength, they rely more and more on other foods. The alligator snapper's exceedingly powerful jaws are quite capable of cracking the shells of smaller turtles like the common musk turtle, a dietary staple in some areas. They may actually have evolved, though, for crushing freshwater mussels. Today, however, the huge and diverse fauna of freshwater mussels that once lived in the eastern United States has become, after years of pollution, siltation, and habitat destruction, primarily a catalog of endangered species. Over much of its range, the alligator snapper's first dietary choice may no longer be on the menu.

**Mollusk-Eaters**

The alligator snapper is not the only turtle with mollusk-crushing jaws. The massive crushing surfaces on the upper jaws of the Mexican giant musk turtle (*Staurotypus triporcatus*) allow it to deal with shells of snails and smaller turtles. In Africa, the Zambezi flapshell (*Cyclanorbis frenatum*) apparently specializes in freshwater mollusks. According to Richard Boycott and Ortwin Bourquin's *The Southern African Tortoise Book*, "There are reports from Malawi of this species digging up buried mussels using their forelimbs, which requires considerable effort as the mussels are buried vertically in the mud and sand of the lake bottom. The shells of the mussels are apparently crushed first before the mussels are swallowed."

Not every turtle seeking to penetrate a mollusk shell uses brute force. The Mary River turtle of eastern Australia has no special modifications of the head for crushing shells. Instead, it takes a clam in its mouth, maneuvers it about until it is in line with its neck, and rakes

A macrocephalic female Worrell's short-necked turtle (*Emydura worrelli*) from Australia's Northern Territory.

its foreclaws along the edge of the shell until it manages to slip a claw between the valves and scrape away at the soft body inside. Once it does enough damage, the clam is no longer able to hold its valves together and opens. The turtle breaks up the opened shell as it eats, swallowing some bits and ejecting others.

In Australia, some female turtles of the genus *Emydura* develop, as they age, greatly enlarged heads with bulging, solid jaw muscles—a condition called *macrocephaly*. At the same time, the horny sheath covering the upper jaw grows backward over the roof of the mouth. In northwestern Australian species like the Victoria River and northwest red-faced turtles (*Emydura victoriae* and *E. australis*, respectively), it may reach back far enough to cover the internal openings of the nostrils and form a secondary palate. This extreme development apparently converts the turtle's head into a powerful and effective crushing instrument, but in old females it may proceed to such a degree that the animal is no longer able to tuck its head under its shell and may even have difficulty eating some kinds of foods.

Why does macrocephaly occur in female *Emyduras*, but not males? One suggestion is

The female Barbour's map turtle (*Graptemys barbouri*) is more than twice the size of the male.

that females, which must produce shells for their eggs, may need the calcium that mollusk shells provide. In North America, the same need may drive the tremendous differences that develop between males and females of some species of map turtle (*Graptemys*). In all map turtles, the females are larger than the males. This reaches an extreme in Barbour's map turtle (*Graptemys barbouri*), a species confined to the Apalachicola and Chipola Rivers and their tributaries in Alabama, Georgia, and Florida. A female Barbour's is more than twice the size of a male, with a carapace up to 33 cm (13 in) long and an enormous, broad head. Not surprisingly, the sexes

Yellow-spotted river turtles (*Podocnemis unifilis*) skim food from the water's surface, a technique called *neustophagia*.

have different diets. The males eat mostly insects, while the females crush the shells of aquatic snails and introduced mussels.

Perhaps, though, this difference is less about egg formation than about a simple division of resources. It shows up at the species level, too; in many rivers, a broad-headed, mollusk-eating map turtle lives side by side with a narrow-headed insect-eater.

**Filterers and Drinkers**

The painted turtle and a number of South American pleurodires, especially the yellow-spotted river turtle (*Podocnemis unifilis*) and the savanna side-necked turtle, have a special technique for skimming fine particles of food from the water's surface. The skimming turtle drops its jaw, opens its throat, and slowly adjusts its head until its jaw is just below the waterline. When the cutting edge of the jaw breaks the surface, water, with any bits of food floating on it, pours into the turtle's mouth. After the turtle closes its mouth, it constricts its pharynx and forces the excess water out through its nostrils or its almost-closed jaws.

This rather crude version of the filter-feeding of flamingos and baleen whales has been called *neustophagia* after the neuston, the technical collective name for the bits of stuff that float on the surface of the water. Neustophagia only works if the surface of the water is calm and still. This may be why turtles like the Magdalena River turtle

(*Podocnemis lewyana*), which lives mostly in fast-flowing rivers, do not practice it.

For terrestrial turtles, finding water may be as important as locating food. Box turtles (*Terrapene* spp.)may get much of their liquid requirements from fruit, at least in season. Drinking may be critically important for tortoises (see Chapter 5). Summer thundershowers bring desert tortoises above ground to drink what they can. Aldabran and angulate tortoises (*Chersina angulata*) drink through their noses, allowing them access to very shallow puddles. Standing water may not always be available, and some South African tortoises, including the tent tortoise (*Psammobates tentorius*), have developed a way to drink from desert mists. During a mist or light rain, droplets of water condense on the tortoise's carapace. To get at them, the tortoise straightens its hidlimbs. This tilts the shell forward, and the droplets run forward along gutters in the sides of the carapace until the turtle can use its forelimbs and head to direct the water into its mouth.

**Vegetarians**

Even the least likely turtles can have a streak of vegetarianism. Along the banks of the Tigris near Diyarbakir in Turkey, Euphrates softshells are accused of raiding the local farmers' watermelon crops. A number of turtles, from a variety of families, are more or less committed vegetarians. This is particularly true of tortoises, though some tortoises can be carnivores at times: the South American red-footed tortoise (*Geochelone carbonaria*) is an opportunistic omnivore, and the feces of the endangered geometric tortoise (*Psammobates geometricus*) have revealed snails and even the remains of young common padlopers (*Homopus areolatus*). Many species of tortoise pick up extra nutrients by feeding on dung, animal carcasses, or eggshells. Females may be especially prone to do this, perhaps to build calcium stores to produce the shells of their own eggs.

Does a tortoise simply work its way through its local plant community, eating different species pretty much as it finds them, or is it selective? There is no clear answer. Studies of desert tortoise, gopher tortoise, and angonoka have shown that the commonest plants in their diet were also the commonest plants in their environment, implying that the tortoises were not being selective.

However, the most frequent plants in the diet of leopard tortoises studied by Mervyn Mason, Graham Kerley, and their colleagues in the Eastern Cape of South Africa were fairly uncommon in the tortoise's habitats. The tortoises seem to be selecting them. They preferred succulent plants to the grasses that dominated the area, possibly because succulents provide water as well as nutrition. Leopard tortoises may only be part-time specialists; if desirable plants are easy to come by, a tortoise may concentrate on them, but if its favorites are not available it may eat anything else it can find.

This may also be true for yellow-footed (*Geochelone denticulata*) and red-footed tortoises in South America. The tiny speckled padloper seems to be more selective, with a preference for flowers. The Asian impressed tortoise (*Manouria impressa*) appears to be a true botanical specialist, eating almost nothing but several species of forest mushrooms.

According to Dionysius S. K. Sharma of WWF Malaysia, the painted terrapin (*Callagur borneoensis*) "is herbivorous in the wild, feeding mainly on leaves, fruit, flowers, stems and roots of selected riparian plants." In the tropical rainforests of South America, where annual floods may send the rivers out among the trees, fallen fruits are food for

The impressed tortoise (*Manouria impressa*) of Southeast Asia apparently eats almost nothing but mushrooms.

The meal this Egyptian tortoise (*Testudo kleinmanni*) is eating may take weeks to pass through its digestive system.

many aquatic animals. Fruits and seeds are important items in the diet of several South American sidenecks, from the yellow-spotted river turtle, which eats almost entirely plant material, to the omnivorous big-headed Amazon River turtle (*Peltocephalus dumerilianus*). Palm fruits are particularly important to the big-headed Amazon River turtle, possibly because their high caloric content makes them a rich source of energy.

Adult arrau keep to the main river channels during the dry season, but when the rains come they invade the flooded forests and feed on fallen fruit and other foods. As waters recede, young turtles may stay in the forest pools, but the adults must return to the rivers to nest. The great 19th-century naturalist Henry Walter Bates joined a hunt for turtles that had become stranded in a forest pool about 1.6–2 ha (4–5 acres) in size. In *The Naturalist on the River Amazons*, he noted that:

> . . . younger turtles never migrate with their elders on the sinking of the waters, but remain in the tepid pools, fattening on fallen fruits, and, according to the natives, on the fine nutritious mud. We captured a few full-grown mother turtles, which were known at once by the horny skin of their breast-plates being worn, telling of their having crawled on the sands to lay eggs the previous year. They had evidently made a mistake in not leaving the pool at the proper time, for they were full of eggs, which, we were told, they would, before the season was over, scatter in despair over the swamp.

Bates also commented that the younger turtles ". . . were very fat. Cardozo and I lived almost exclusively on them for several months afterwards. Roasted in the shell they form a most appetising dish."

Since animals cannot digest the cellulose in plants, many plants enlist the help of cellulose-digesting bacteria. Herbivores from termites to cattle carry colonies of such bacteria in their guts; so do green sea turtles (*Chelonia mydas*) (see Chapter 8), and so do some tortoises. The advantages of carrying a bacteria-powered fermentation chamber are considerable. The Aldabra tortoise, which does not have cellulose-digesting bacteria in its gut, digests only about 30 percent of the plant matter (mostly tough grasses) it eats. The red-footed tortoise, which does, manages about 65 percent (but it eats more fruit). One of the reasons tortoises prefer high temperatures (see Chapter 5) is to keep their microflora operating at top efficiency.

Tortoise meals may take weeks to macerate and digest, particularly at cooler temperatures. Peter Holt used barium meals to show that food can take up to a month to pass through the gut of the spur-thighed tortoise. Slow digestion may help desert tortoises deal with periods of feast and famine. They commonly fast for a month even during their active periods. Overall, they may spend only 0.3 percent of their time feeding—a figure amounting to less than 48 hours a year.

Gopher tortoises face a special problem over the winter. Before entering winter dormancy, they empty their digestive systems, and, of course, they may not eat again for some time. This is enough to starve their complement of microflora. To restart their fermentation chambers, the tortoises recycle a starter sample. They do this by defecating in their burrows in the fall, and eating a sample of their dried feces in the spring.

Vegetarianism is a habit most turtles acquire with age. Even strictly vegetarian turtles may be more or less carnivorous as hatchlings or juveniles. Young yellow-spotted river turtles studied in Rondonia, Brazil, catch and eat small fishes, but the amount of fish in their diet shrinks as the turtles grow. By the time they reach adulthood, they are almost entirely plant-eaters; plant material amounts to over 89 percent of their stomach contents.

In Australia, Krefft's river turtle (*Emydura kreffti*) undergoes the same shift from omnivory to vegetarianism. The red-bellied turtle (*Pseudemys rubriventris*) provides a peculiar exception: hatchlings start out as herbivores, then switch to a diet of crayfish.

A few turtles are lifelong vegetarians. Florida red-bellied turtles (*Pseudemys nelsoni*) and Central American river turtles, supported by colonies of microflora in their small intestines, live on plants from the day they hatch. River turtle hatchings, at least incaptivity, will eat their mother's feces to pick up their microflora.

## Hunters and Suckers

Though the turtle body hardly seems designed for active hunting, it is surprising how many turtles are good at it. Turtles can be more agile than they look; creatures as unlikely, and different, as the Nile softshell (*Trionyx triunguis*) and leopard tortoise have scaled wire-mesh fences a meter or more high. A smooth softshell (*Apalone mutica*) has been seen chasing down and capturing a brook trout, one of the fastest freshwater fish in North America. African helmeted terrapins may even hunt in consort. In ponds where their numbers are high, several terrapins may attack a small water bird, drag it under, and tear it apart, though perhaps such behavior is more of a free-for-all than an organized group hunt.

Some wood turtle populations in central Pennsylvania and Michigan draw earthworms to the surface by "stomping," apparently to imitate rain. Carl Ernst, Jeffrey Lovich, and Roger Barbour describe "worm stomping" in *Turtles of the United States and Canada*:

> A stomping turtle typically takes a few steps forward and then stomps several times with one front foot and then the other at a rate of about one stomp per

Earthworms are a common item in the diet of the eastern box turtle (*Terrapene carolina*).

> second. Any worm brought to the surface is eaten. The whole sequence usually lasts 15 minutes or more and may include 2–19 stomps (mean 8.1) from the time stomping begins until a worm is eaten. Sequences start with light stomping with the force of each successive stomp increasing, and some stomps are audible for several meters. The plastron may also be banged against the ground in the process.

The fact that only certain wood turtle populations do this suggests the remarkable possibility that "worm stomping" is a learned behavior—evidence of a sort of turtle culture. How such things could develop in animals with no parental care—even leaving aside the question of the wood turtle's reputed intelligence—we have no idea.

Freshwater turtles like the common musk turtle explore the bottom actively, a method Carl Ernst called a "peer and probe method of food detection." Carnivorous turtles seeking larger prey may depend primarily on stealth, camouflage, and surprise attacks. The alligator snapper, as we have seen, uses its "lure" to bait fish into its jaws.

One of the chief marks of a carnivorous turtle is a long neck, often ending in a streamlined head. With their heads and necks alone, common snapping turtles, most soft-shelled turtles, and the snake-necked turtles of the family Chelidae (*Chelodina* spp. in Australia, *Hydromedusa* spp. in South America) can make a slow, stealthy approach to an unsuspecting fish or small invertebrate, followed by a rapid, snakelike strike and a gulp.

Many softshells, including the American *Apalone* spp., bury themselves in the sand, the better to ambush their prey. Softshells have particularly long necks and narrow heads, culminating in the weirdly proportioned skull of the Asian narrow-headed softshells (*Chitra* spp.; see Chapter 1). The Indian narrow-headed softshell (*Chitra indica*) is, according to an old account, able to "suddenly shoot out its long neck with inconceivable rapidity." Ms. J. Vijaya of the Madras Snake Park Trust sent Peter Pritchard the following account of a captive:

> *Chitra indica* seems very much a fish eater. We have a juvenile at the Crocodile Bank who keeps himself buried in sand most of the time. Just the tip of his snout and his eyes are exposed so it is almost impossible to find him. Just as a fish passes overhead he very quietly puts out his head and gulps it. Dr. [Edward] Moll once observing it said that while grabbing a particularly large fish it almost leapt out of its hiding place, throwing a fountain of sand while grabbing the fish.

Snappers, softshells, Australian snakenecks, and at least one of the South American

An eastern spiny softshell (*Apalone spinifera spinifera*) buries itself in the sand, the better to ambush its prey.

The Indian narrow-headed softshell (*Chitra indica*) normally lies buried in the sand, only the tip of its snout showing.

species, the South American snake-necked turtle (*Hydromedusa tectifera*), have added a further refinement to the final strike—the so-called "gape-and-suck" technique. In effect, they have turned their necks into vacuum cleaners. By using their neck muscles and the *hyoid apparatus* in their throats—the bony and cartilaginous structure that supports the base of the tongue—these turtles can drop the floor of their mouths and expand their necks in a sudden, rapid movement that sucks water, fish, small invertebrates, and anything else in the vicinity into their open jaws. Afterward, they close their mouths loosely and collapse their necks again, forcing out the excess water but retaining their prey.

Maximilian's snake-necked turtle (*Hydromedusa maximiliani*) apparently does not eat fish, but concentrates on invertebrates, including insects, that fall into the rivers where it lives; we do not know whether it uses the same technique as its relatives. Franco Leandro Souza and Augusto Shinya Abe saw crickets and cockroaches enter a river where the turtles lived to escape swarms of army ants, fleeing from one predator only to, possibly, fall into the maw of another.

Softshells aside, the chicken turtle may be North America's closest ecological approach to the snake-necked gape-suckers of the southern hemisphere. It has all the requisite anatomical features: a long neck, narrow jaws, a greatly expanded hyoid apparatus with associated neck-expanding muscles, and eyes placed relatively far forward on its head, the better to draw a binocular bead on its prey. Like Maximilian's snake-neck, though, it does not eat fish. Instead, its diet consists almost entirely of arthropods, mostly aquatic insects and freshwater crustaceans like crayfish. Its eating habits are, in fact, most like those of Blanding's turtle, its distant northern relative.

The most extreme gape-and-suck predator is the largest chelid in South America, the matamata (*Chelus fimbriatus*). The matamata lives in murky "black river" waters where the visibility is very low. It has extremely small eyes, and apparently does not detect its prey by sight. Instead, it relies on the various fringes and flaps of skin on its head. These

An Indian softshell (*Aspideretes gangenticus*) gapes and expands its neck, sucking in water and prey.

The throat of this northern long-necked turtle (*Chelodina rugosa*) is still expanded after an unsuccessful attack.

The matamata turtle (*Chelus fimbriatus*) of South America is the most extreme "gape-and-suck" predator among turtles.

flaps almost certainly provide the matamata with camouflage, increasing its resemblance to a pile of dead leaves, but that is not all the flaps do. They are extensively supplied with nerve endings. They are very sensitive to even slight movements, and apparently allow the turtle to detect vibrations in water.

While its relatives, the *Hydromedusa* turtles, can make an extremely accurate, direct strike at a victim, the matamata is more likely to make a blind, sidewise swipe, guided by its built-in vibration detectors and, perhaps, by sound. Its vacuuming abilities, though, more than make up for its near blindness and relative inaccuracy. Its hyoid bones are enormous, their four branches comparable in size and shape to the bones that make up its lower jaw, and the development of its neck musculature is equally spectacular.

When the matamata detects potential prey, it slowly stretches its long neck out towards

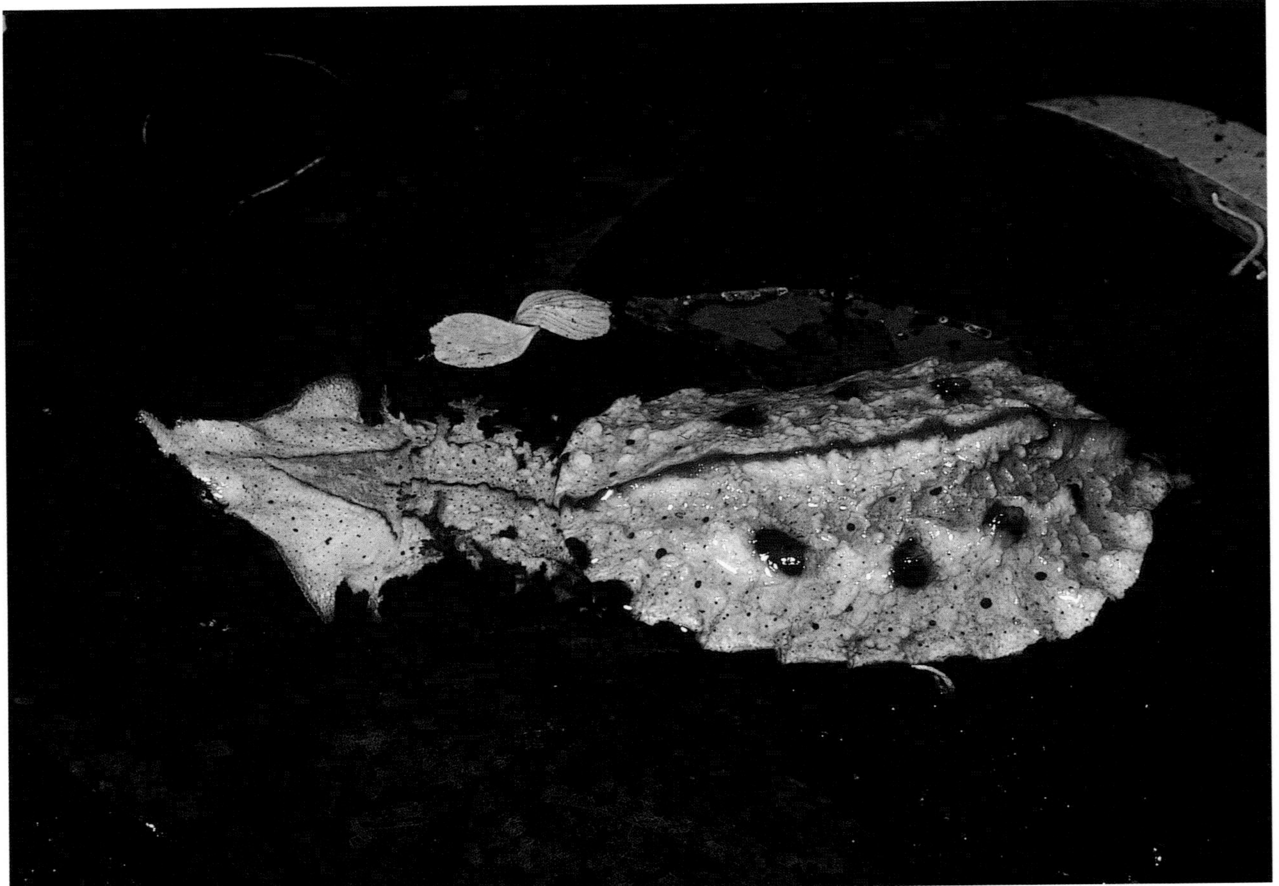

The matamata lives in murky waters, and uses sensitive skin flaps on its head and neck to detect the motion of its prey.

its victim. When it is close enough, it closes off its nostrils, shoots its head forward, opens its cavernous mouth, and contracts its hyoid musculature, greatly expanding its throat. The result is a sudden rush of water that, if the turtle's positioning and timing are right, sweeps its prey into its mouth. All this takes place at lightning speed, literally too fast for the eye to follow.

The matamata may do more than simply wait for fish to come along. Matamatas in the Bronx Zoo, New York, watched by William F. Holmstrom Jr., developed the habit of herding goldfish in their pool by walking towards them, waving one or the other of their front legs. Driving the fish into shallower water made them easier to catch. Matamatas at the Beardsley Zoological Gardens in Connecticut accomplished the same thing by advancing towards the fish with the front part of their bodies pressed against the bottom and the rear part of the shell angled upward. In the wild, matamatas feed close to reeds and other vegetation near shore, and may herd prey towards them, though this has never been recorded. If they do, they may not be alone. Turtle enthusiast John Levell has told me that his captive common snapping turtles herd fish by walking towards them with the rear ends of their shells raised, rather like the matamatas at the Beardsley Zoo. Perhaps this sort of hunting is commoner than we thought.

## Getting Along

Though the species and families involved vary from continent to continent, freshwater turtle communities have a remarkably similar cast of characters:

- long-necked carnivore/omnivores: common snappers, softshells (*Apalone* spp.), Blanding's and chicken turtles in North America, other softshells in Africa and Asia, snake-necked turtles and other longnecks in Australia, and *Hydromedusa* longnecks and matamatas in South America;
- large-headed, strong-jawed mollusk (and, in some cases, turtle) predators: alligator snappers, loggerhead musk turtles (*Sternotherus minor*), and macrocephalic map turtles like the female of Barbour's map turtle in North America, giant musk turtles (*Staurotypus* spp.) in Mexico and Central America, and the Colombian toad-headed sideneck (*Phrynops raniceps*) in South America;
- herbivores: cooters (*Pseudemys* spp.) in North America, the Central American river turtle in Mexico and Central America, the arrau in South America, and the Victoria River snapper in Australia;
- agile omnivores in rapidly flowing rivers, with a preference for aquatic insects and other invertebrates: common map turtles and related species in North America, Geoffroy's sideneck (*Phrynops geoffroanus*) in South America, and the Fitzroy River turtle (a freshwater sponge eater) in Australia; and
- a whole range of opportunistic omnivores.

The parallel members of these varying casts are called *trophic equivalents.* Their similarities, for the most part, have evolved independently under selective forces that have driven evolution along parallel paths in different regions of the world. Family after turtle family has produced candidates to fill the various trophic roles. Not all the equivalencies are exact: the giant narrow-headed softshells (*Chitra* spp.) of Asia, like the alligator snapper, are oversize, bottom-hiding fish ambushers, but lack the snapper's lure and mollusk- and turtle-crushing equipment.

On land, American box turtles partly fill the role of smaller tortoises like the South African padlopers (*Homopus* spp.), though they are not rockpile specialists. In many of the world's forests there are small, more or less land-dwelling turtles that burrow into, and often resemble, piles of leaves on the forest floor: wood turtles in North America, some neotropical wood turtles (*Rhinoclemmys* spp.) and the twist-necked turtle (*Platemys plalycephala*) in the New World tropics, the flat-tailed tortoise (*Pyxis planicauda*) in Madagascar, and the spiny turtle (*Heosemys spinosa*), keeled box turtle (Pyxidea *mouhotii*), and black-breasted leaf turtle (*Geoemyda spengleri*) in tropical Asia.

Similar species of animals that live in the same place and eat the same sorts of food normally must find some way to separate their lifestyles if they are to coexist. This division of living space and resources is called *habitat partitioning.* Sometimes this includes direct competition: larger turtles, for example, may drive smaller ones from favored basking sites.

Map turtles that share the same rivers may have different diets. Conversely, map turtles with similar diets may select different habitats. Common (*Graptemys geographica*), false (*Graptemys pseudogeographica*), and Ouachita

The keeled box turtle (*Pyxidea mouhotii*) of tropical Asia lives among, and resembles, leaf litter on the forest floor.

map turtles (*Graptemys ouachitensis*) have similar diets, at least among males. In eastern Kansas, where the three species occur together, common map turtles studied by Linda Fuselier and David Edds preferred shady streams flowing over rocks and gravel, while false and Ouachita maps were more likely to live in wider, sunnier rivers with more basking sites. False maps were more likely to live over muddy river bottoms, while the Ouachita map preferred sandy areas and warmer water. These differences, though, are neither exact nor universal: farther east, where the other two species do not occur, common map turtles live in the open sunny areas.

Dietary overlap shifts with the seasons for turtles living in a sluggish stream in Belize, studied by Don Moll. During the rainy season, both white-lipped and scorpion (*Kinosternon scorpioides*) mud turtles eat aquatic insects and small snails. As the dry season progresses, white-lipped mud turtles concentrate more and more on snails, while scorpion mud turtles focus on insects. Why this happens we do not know. It may have something to do with an increase in density; as their rainy-season homes dry up, animals living in nearby temporary ponds migrate to the permanent stream, swelling its turtle ranks.

While the mud turtles become more selective, Mexican giant musk turtles in the same permanent stream broaden their food preferences. Unfortunately for their smaller relatives, they do this by shifting from a diet of giant apple snails (*Pomacea*) to one based more and more on the increasing numbers of mud turtle refugees fleeing the temporary ponds.

Not all differences in diet may reflect habitat partitioning. Even within the same species, turtles of different sizes, ages, and sexes often take different foods, either because certain items are easier for a large animal to catch than for a small one, or because smaller items may not be worth a larger turtle's efforts.

Turtles, as we have seen, may change their diet, or even their habitat, as they grow. Even omnivorous turtles like the painted turtle tend to be more carnivorous as hatchlings. Mud and musk turtles tend to take larger prey as they themselves grow bigger. So does Krefft's river turtle in Australia. Juvenile Indian black turtles (*Melanochelys trijuga*) are semiterrestrial animals of hillside streambeds, while the adults are aquatic animals that prefer slow-flowing backwaters and oxbow lakes. In southeastern Brazil, juvenile Maximilian's snake-necked turtles stay in shallow water near the riverbank, where they feed heavily on a tiny amphipod crustacean, *Hyalella pernix*. As they grow, the turtles shift to faster and deeper water, and gradually switch to larger prey, including the larvae of mayflies.

Mature females of the Chinese stripe-necked turtle (*Ocadia sinensis*), like the females of many other bataguridand emydid turtles, are larger than the males (with an average carapace length of 16 cm (6.3 in) as opposed to the males' 12.6 cm (5 in)). In

Where the common map turtle (*Graptemys geographica*) lives with other map turtles, they may differ in diet and habitat.

northern Taiwan, adult females tend to be herbivores, feeding chiefly on marsh dewflower (*Murdannia keisak*). Males, though, are primarily carnivorous, eating mostly the larvae and pupae of simuliid flies. So do young females; but, unlike the males, they shift to vegetarianism as they grow.

The males of all three American softshells (*Apalone* spp.) are little more than half the size of the females. Smooth softshell preferences are the reverse of the stripe-necked turtle: in Kansas, females concentrated on aquatic insect larvae while males ate large amounts of such things as mulberries and cottonwood seeds.

Some turtles show little difference between the sexes in either size or diet. Luca Luiselli found no such differences in the West African mud terrapin in southeastern Nigeria. Augusto Fachin Teran and his colleagues, in a study of five species of turtles in the Guapore river system in Rondonia, Brazil, found differences between the foods eaten by male and female yellow-spotted river turtles but none between the sexes of Geoffroy's sideneck.

Turtles may not need to divide up resources if there is enough to go around. In Florida, food resources are often abundant, and in constant, year-round supply. Three quite similar turtles—the river cooter (*Pseudemys concinna*), the cooter (*Pseudemys floridana*), and the Florida red-bellied turtle (*Pseudemys nelsoni*)—may live comfortably together, although all three are herbivores, eating plants like eelgrass (*Vallisneria americana*) and coontail (*Ceratophyllum demersum*). Karen Bjorndal, Alan Bolten, and their colleagues found the level of overlap in their diet to be very high: from 56 percent between

Painted turtles (*Chrysemys picta*) threaten each other with openmouthed gaping displays.

the two cooters to as high as 94 percent between the cooter and the Florida red-bellied turtle. They differ in nesting habits, but with respect to diet the three have found, in one of the most turtle-rich areas of the world, a chelonian Eden.

## Enemies

Human beings are certainly the greatest devourers of adult turtles on the planet today, but they are not the only enemies turtles have. Alligator snappers and Mexican giant musk turtles are major predators on some of the smaller mud and musk turtles. The alligator snapper poses such a serious threat to other turtles that several have evolved special means of keeping out of its way. Loggerhead musk turtles can detect, and avoid, an alligator snapper in murky river waters using chemical cues alone.

Crocodilians are serious predators of even the largest freshwater turtles. Alligators (*Alligator mississippiensis*) take a number of different species of turtles in the southeastern United States. Sliders (*Trachemys scripta*) in Guatemala may fall prey to Morelet's crocodile (*Crocodylus moreletii*). Mugger crocodiles (*Crocodylus palustris*) have been seen eating Indian softshell turtles (*Aspideretes gangeticus*) in the Chitwan National Park, Nepal. The largest of the crocodiles, the saltwater or estuarine crocodile (*Crocodylus porosus*), will even attack adults of the enormous Asian and New Guinea giant softshells (*Pelochelys cantorii* and *P. bibroni*); a crocodile captured by villagers in New Guinea in 1981 had an adult *Pelochelys* in its stomach.

Otters prey on turtles in a number of parts of the world. Otters in Africa will take Zambezi flapshells, though turtles are not a major part of their diet. In southern Brazil, a giant otter (*Pteronura brasiliensis*) has been seen eating a South American snake-necked turtle. River otters (*Lutra longicaudis*) are apparently among the few animals other than humans that regularly eat the Central American river turtle. Otters are now so rare in its range that they may be more endangered than the turtle.

In North America, birds and mammals, from bald eagles to black bears, will take adults of smaller turtles like the spotted turtle. The most thoroughgoing predator American turtles face, though, is probably the raccoon (*Procyon lotor*). Besides being adept nest robbers and hunters of juvenile turtles, raccoons will kill and eat adult spotted turtles, bog turtles, and wood turtles, attack nesting diamondback terrapins, and are the number one predator on all life stages of painted turtles. Raccoons are adept at digging hatchling gopher tortoises out of their burrows, and their depredations can devastate local tortoise populations. The impact raccoons have on turtles in general is probably greater now than it used to be because raccoons themselves have benefited from human activities and have become much more numerous and widespread.

Smaller tortoises in Africa and Eurasia face predators including jackals, mongooses, monitor lizards, and birds ranging from crows and ravens to eagles and the unique African secretary bird (*Sagittarius serpentarius*). Even ostriches (*Struthio camelus*) have been known to attack tent tortoises. Crows and birds of prey break open small tortoises such as the Karoo padloper (*Homopus boulengeri*) by dropping them on rocks. Legend has it that the ancient Greek dramatist Aeschylus was killed by an eagle that mistook his bald pate for a rock and dropped a tortoise on it.

Turtles, like other animals, attract their share of parasites. In Southeast Asia, the tick *Amblyomma geoemydae* is, as its specific name implies, a turtle specialist. Though it occasionally turns up on birds, mammals, and lizards, it is most likely to be found attached to tortoises like the Asian brown tortoise (*Manouria emys*) or to land-living batagurids like the spiny turtle. We know very little about the effect *Amblyomma* has on the health of its hosts, but the lesions from tick bites can become infected.

Freshwater turtles are often heavily infested with leeches, and loggerhead sea turtles (*Caretta caretta*) may carry the marine turtle leech (*Ozobranchus margoi*). Turtles that do not hibernate buried in sand or mud, like wood turtles and common map turtles, may carry their leeches with them throughout the winter. Over half of the wood turtles in a study area in West Virginia carried leeches of the species *Placobdella parasitica*, usually attached along the front edge of the plastron or the base of the tail near the cloaca.

One of the functions of basking may be to help turtles get rid of their load of leeches. Some turtles look to other animals for help: common grackles (*Quiscalus quiscula*) will pick leeches off several species of basking map turtles, and blacknose dace (*Rhinichthys atratulus*), a species of minnow, have been seen apparently cleaning leeches from wood turtles. Snapping turtles have been found buried in mounds of carpenter ants (*Formica obscuriventris*); we do not know if the ants cleaned the turtles of leeches, but one turtle found by Vincent Burke and his colleagues in an ant mound on the George Reserve in Michigan was the only female, of 39 examined in the area, to be free of them. Dean McCurdy and Thomas Herman recorded wood turtles in Nova Scotia visiting ant mounds. Whether this was helping them get rid of leeches, the investigators could not tell.

Turtles have a number of ways of defending themselves against predators. The most obvious is escape: either getting out of the predator's way in a hurry, or avoiding being noticed in the first place. If an ornate box turtle (*Terrapene ornata*) sees a predator, its first

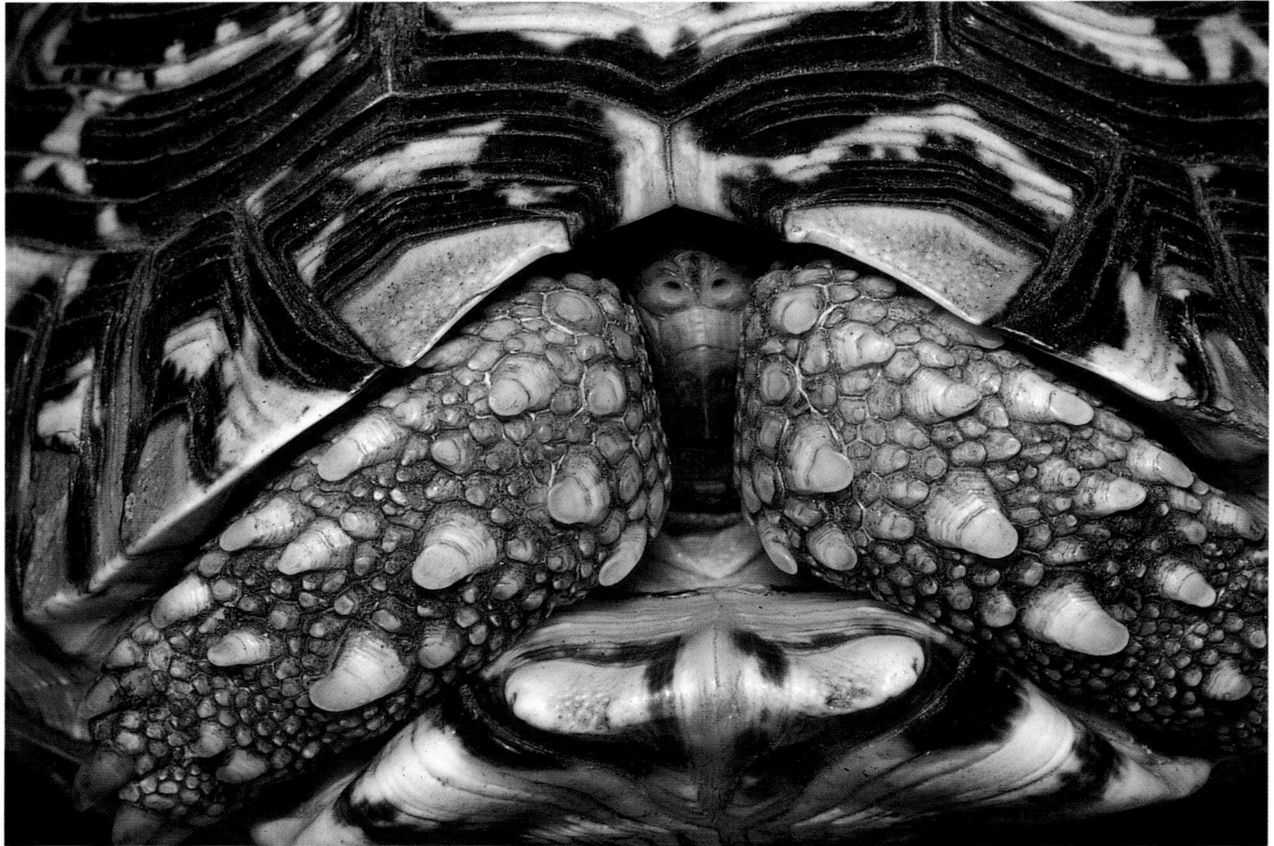

The leopard tortoise (*Geochelone pardalis*) of Africa may have little to fear from predators once it grows to adult size.

reaction is to freeze; only if the predator approaches will it either scuttle away (which it can do surprisingly quickly) or seal itself in its shell. The ability of turtles to sense low-level vibrations can give them advance warning of an enemy approach, and an ample opportunity to flee or hide—anyone who has ever tried to get close to basking turtles will know how difficult it can be to catch them unawares. Many turtles can be remarkably cryptic, especially when they are inactive, even if they are not buried beneath mud or leaf litter. The smaller tortoises, in particular, can be extremely difficult to find.

If withdrawing into their shells is not enough to discourage attackers, many turtles will bite or discharge the contents of the cloaca. Common snapping turtles, musk turtles (Kinosternidae), helmeted terrapins, the South American gibba turtle (*Phrynops gibbus*), and others can release foul-smelling chemicals from their musk glands.

Some turtles have distinct threat displays, including the gaping and hissing threats of giant musk turtles and narrow-bridged musk turtles (see Chapter 3). If you corner a common snapping turtle on land, it will raise the posterior part of its body as it opens its mouth and hisses at you; if you move, the turtle spins around, sometimes using the tail to help pivot, to keep its head facing its attacker. At closer range, it will strike, and if its bite connects, it will hang on doggedly. Bell's hinged tortoise (*Kinixys belliana*) has a less aggressive technique; if threatened, it may play dead.

A common snapping turtle (*Chelydra serpentina*) responds to a threat by opening its mouth widely and hissing.

Despite the range of predators they face, many turtles, particularly the larger species, may be relatively immune to non-human predation once they reach adulthood. Even in predator-rich Africa, a full-grown leopard tortoise may have little to fear. The problem is to reach adulthood in the first place. Nests, eggs, and hatchling turtles are vulnerable to a far greater range of predators than are adults (see Chapter 7).

### A Place in the System

In the Galápagos Islands, a giant tortoise may carry a tiny, brilliantly colored passenger. The vermilion flycatcher (*Pyrocephalus rubinus*) uses the tortoise's carapace as a mobile perch, and the tortoise itself as a beater. As the tortoise plods through the underbrush, it startles insects into the air, where the flycatcher, sallying from its perch, snatches them up.

On Frigate Island in the granitic Seychelles, introduced Aldabra tortoises perform the same function for one of the rarest birds in the world, the Seychelles magpie robin (*Copsychus seychellarum*). Elsewhere, the interactions between turtles and the other animals and plants in their ecosystems may be equally close, even more crucial, and far more subtle.

In Bangladesh, the Indian roofed turtle (*Kachuga tecta*) feeds on introduced plant pests like water hyacinth, scavenges dead animals, and sometimes devours human waste. In doing so, it helps to reduce environmental pollution and may slow the spread of infectious human diseases.

On Santa Cruz in the Galápagos, a vermilion flycatcher (*Pyrocephalus rubinus*) hitches a ride on the back of a giant tortoise.

The composition of oak forests in the floodplains of the lower Mississippi may depend, surprisingly enough, on alligator snapping turtles. The turtles eat a wide range of seeds, from acorns to pecans. The trees that bear them depend on the river to carry their seeds downstream, but they may need the turtles if their seeds are to go upstream as well.

Kevin Sloan, Kurt Buhlmann, and Jeffrey Lovich concluded, after a study of the stomach contents of alligator snappers caught by commercial fishermen, that "little attention has been focused on the overall role of the species as a scavenger, predator, and possible plant disperser. Our data suggest that [*Macrochelys*] has an important function in the trophic structure and dispersal mechanisms of riparian systems."

Dionysius S. K. Sharma has proposed a similar role for the painted terrapin in Malaysia. The terrapins have been tracked swimming upstream, and seeds of crabapple mangrove (*Sonneratia*) and madang (Family Lauraceae) from terrapin droppings germinate well. Though tidal movements may carry seeds some distance upstream in low-lying rivers, the terrapins may be important agents dispersing them farther, past the point of tidal influence.

In South America, the herbivorous arrau may be an important agent for dispersing the seeds of rainforest plants. Other turtles may be seed dispersal agents, too; such a role has been suggested for the big-headed Amazon

More than 360 species of animals make use of the burrows of the Gopher tortoise (*Gopherus polyphemus*).

River turtle, the Mexican giant musk turtle, and a number of tortoises.

As the gopher tortoise grazes, it, too, affects the distribution of plants in its pineland ecosystem, and its droppings may help disperse the seeds of understory plants. But it is in its role as a burrower that the gopher tortoise may have the greatest impact on its community.

More than 360 species, both vertebrate and invertebrate, may make use of gopher tortoise burrows. The Florida mouse (*Podomys floridanus*), an increasingly rare species restricted to the same longleaf pine woodlands as the tortoise, is particularly dependent on them. The mice build their dens in little side tunnels off the main burrow. They may fall prey there to another declining species, the eastern indigo snake (*Drymarchon corais couperi*). The snake, though, is more likely to prey upon frogs and other snakes—perhaps including yet another threatened animal, the gopher frog (*Rana capito*), that finds refuge in the cool, moist tortoise burrows.

A study by Karen Lips showed that other animals made more use of active tortoise burrows than of abandoned ones. An active tortoise burrow is the basis for a whole community of animals, partly because tortoises keep active burrows open and accessible. Probably more important, though, is their habit of defecating in their burrows to recycle their gut microflora. For other burrow animals, tortoise feces form the basis of a food web. An active burrow hosts a constant supply of insects that eat the feces. The insects are food for frogs, lizards, and mice that are preyed on, in their turn, by snakes, spotted skunks (*Spilogale putorius*), and long-tailed weasels (*Mustela frenata*).

We are only beginning to understand how important turtles are to the ecosystems where they live.

CHAPTER SEVEN

# 'Twixt Plated Decks

The turtle lives 'twixt plated decks
Which practically conceal its sex.
I think it clever of the turtle
In such a fix to be so fertile.
—Ogden Nash, *The Turtle*

THE ULTIMATE GOAL of any living creature is to reproduce its kind. Turtles have extracted themselves from Ogden Nash's "fix" not by cleverness, but through a range of life history strategies. These strategies raise the chances that turtles will not only mate, but produce young that will survive to reproduce in their turn.

No turtle gives birth to living young or cares for its offspring. Males have nothing to do with the reproductive process after they have finished mating. Except in the Asian brown tortoise (*Manouria emys*), which guards its nest mound, females pay not the slightest attention to their eggs or young after laying, though there is an old report of a captive Inaguan turtle (*Pseudemys malonei*) that supposedly helped release her newly hatched young by digging away the hard soil over her nest. Everything that a turtle contributes to help her offspring survive she normally provides before her eggs begin to incubate.

The ultimate product of turtle reproduction: a spurred tortoise (*Geochelone sulcata*) struggles out of its shell.

Animals that lack parental care must compensate for its absence. Their eggs must be well suited to survive their incubation period unaided. They may construct carefully concealed and optimally situated nests, hidden from predators and parasites and placed to provide the best available environment for embryological development. To increase the likelihood that at least some young may survive, and reproduce, most turtles lay a large number of eggs over their reproductive lifetimes. Turtles, to greater or lesser degrees, have done all of these; indeed, sea turtles (see Chapter 8) lay the largest clutches of eggs of any reptile.

## Cycles and Seasons

The reproductive life of land and freshwater turtles is governed by the seasons, whether they are the four seasons of the temperate zone or the wet and dry seasons of the tropics. Males and females go through annual cycles of, respectively, sperm and egg production that do not necessarily peak at the same time. In temperate-zone turtles, males produce sperm over the summer that they will need the following spring. Male spotted turtles (*Clemmys guttata*) in southeastern Pennsylvania produce most of their sperm in June

Where baby tortoises come from: a mating pair of spurred tortoises (*Geochelone sulcata*).

and July, concluding in August; their peak mating season is from March through May. This pattern may ensure that the male has viable sperm available when the female ovulates. Most female turtles ovulate in spring, so sperm formation ought, ideally, to take place during the winter. Winter cold prevents this from happening, so males must have their sperm supply ready by the end of the previous fall. They store the sperm in their genital tract until they need it.

Female turtles, too, may store sperm, in tubules in their oviducts. Desert tortoises (*Gopherus agassizii*) do most of their mating in the fall when male sperm production is at its peak, but the females store sperm, and delay laying their eggs, until they emerge from their burrows the following spring. Sperm storage lets the female time fertilization and egg-laying so that her hatchlings develop and emerge under the best possible conditions. It also allows her to lay multiple clutches of eggs from a single mating—a useful insurance strategy for many turtles. Female Chinese softshells (*Pelodiscus sinensis*) can store viable sperm in their oviducts for almost a year. Eastern box turtles (*Terrapene carolina*) and diamondback terrapins (*Malaclemys terrapin*) have produced fertile eggs four years after their last mating.

In North America, most temperate-zone turtles lay their eggs in the spring. In Florida, striped mud turtles (*Kinosternon baurii*) reach their nesting peak, unusually, from September to November, but may nest in any month of the year. Chicken turtles (*Deirochelys reticularia*) in South Carolina

Galápagos tortoises (*Geochelone nigra vandenburghi*) mate on the slopes of Volcan Alcedo, on the island of Isabela.

also nest in the fall, from August to November, and again in the late winter and early spring from mid-February to May. In Mexico, the nesting of the Mexican mud turtle (*Kinosternon integrum*) and the Mexican rough-footed mud turtle (*Kinosternon hirtipes*) may be timed so the hatchlings emerge during the summer rainy season.

The mating season of Galápagos tortoises (*Geochelone nigra*) peaks from January to June, the hottest time of the year. Nesting begins in May, and ends in October, when temperatures are cooler; the young emerge between November and April, once again during the rainy season.

In Southern Africa, temperate-zone terrapins and tortoises tend to nest in the southern spring. Farther north, nesting usually starts, or peaks, in summer after the first rains. Some species, like the leopard tortoise (*Geochelone pardalis*), nest all year. In tropical Asia, the Indian roofed turtle (*Kachuga tecta*) has two distinct nesting periods, one from the beginning of December to the middle of January, and another from mid-February to the end of March. Painted terrapins (*Callagur borneoensis*) nest from June to August on the east coast of peninsular Malaysia, and from October to January on the west coast; the difference may reflect rainfall patterns.

Most chelid turtles in Australia lay their eggs in spring, or late in the dry season in the tropical north. In the southwest, the western swamp turtle (*Pseudemydura umbrina*) nests before the summer drought. Broad-shelled turtles (*Chelodina expansa*) lay their eggs in late fall, with hatching following about a year later. This timing corresponds to the dry-season nesting of tropical Australian turtles. The broad-shelled turtle may be a recent arrival in temperate Australia, its reproductive cycle still tuned to the tropics.

In South America, the nesting of Amazon River turtles (*Podocnemis* spp.) may be governed by cycles of flooding. Six-tubercled Amazon River turtles (*Podocnemis sextuberculata*) probably start nesting as soon as the water levels fall enough to expose the highest beaches, while the arrau (*Podocnemis expansa*), which needs vast expanses of beach for its communal nesting grounds, waits until the river reaches its lowest ebb. In Malaysia, another large riverine turtle that lays its eggs on sand islands and sandbars, the river terrapin (*Batagur baska*), also times its nesting to begin as the storms and floods of the northeast monsoons abate and water levels fall. So does the pig-nose turtle (*Carettochelys insculpta*), which nests late in the Australian dry season, when low water exposes the river sandbanks.

We are still not certain what environmental cue sets turtle reproductive, and hormonal, cycles in motion for more than a few species. Temperature appears to be a key factor. In some turtles, changes in day length trigger various points in the reproductive cycle. Such changes appear to be major cues for female painted turtles (*Chrysemys picta*) and male sliders (*Trachemys scripta*). Blanding's turtles (*Emydoidea blandingii*) begin nesting earlier in years with more warm days in early spring; in this case, temperature is apparently more important than day length. Common snapping turtles (*Chelydra serpentina*) and female common musk turtles (*Sternotherus odoratus*) begin their reproductive cycles in spring as water temperatures rise. They do not seem to be affected by the lengthening days, but decreasing day length is an important trigger for sexual activity in the male common musk turtle.

### Courtship

The first step in turtle mating is for the males and females to find one another, or, more specifically, for the males to find females, as that appears to be the way these things usually happen. Turtles that hibernate in groups, like some spotted turtles, may already be in each other's company when spring comes (see Chapter 6). Most turtles appear to find each other by sight, though female tortoises, musk turtles, and pleurodires release scents that may attract males.

Before a male succeeds in impregnating a female, he may need to deal with his rivals. As many as six male loggerhead musk turtles (*Sternotherus minor*) may attempt to mate with a single female, pushing each other aside in a struggle to mount her. Male common and alligator snapping turtles (*Macrochelys temmincki*) may behave aggressively towards other males when a female is present. Male Geoffroy's turtles (*Phrynops geoffroanus*) have been seen attacking other males that tried to horn in during mating; the victims did not fight back, suggesting that these turtles have a dominance hierarchy, with the attacking turtle being higher in the peck order. Wood turtles (*Clemmys insculpta*) also appear to have dominance relationships; one male was seen to open his mouth and pursue another that moved away without contesting the issue.

Tortoises are particularly aggressive fighters, and equally aggressive in pursuing their prospective mates. Leopard tortoises do not so much court as batter their prospective mates into submission, barging continuously into their backs and sides. In common padlopers (*Homopus areolatus*), battles between males, and the pursuit of females, can result in bleeding wounds. Males watched in captivity used their powerful beaks to bite and hold on to the front edge of each other's carapace, and engaged in pushing matches that could go on for an hour. Even after one of the tortoises surrendered and beat a retreat, the victor usually pursued him, biting at his hind legs and shell until his victim could retreat to safety. Jack Hailman and Rosemary Knapp watched an intermittent half-hour battle between two male gopher tortoises (*Gopherus polyphemus*) that involved shoving, attempting to overturn each other with the gular extensions of their plastra, and kicking sand in each other's faces with their hind feet. At times, tortoises seem remarkably human.

An angonoka or ploughshare tortoise (*Geochelone yniphora*) apparently needs to engage a rival male as a stimulus to mating. The inimitable Gerald Durrell, in *The Aye-Aye and I*, described the joust that ensues as the males deploy their immense ploughshares (*ampondo* in Malagasy):

> The two males, rotund as Tweedledum
> and Tweedledee dressed for battle,
> approach each other at what, for a
> tortoise, is a smart trot. The shells clash
> together and then the Ploughshare's

Courtship in Purvis's turtle (*Elseya purvisi*): the male sniffs at the female's cloaca (left); the pair touch each other with their sensitive chain barbels (middle); and finally, mating takes place.

*ampondo* comes into use. Each male struggles to get this projection beneath his opponent and overturn him to win a victory in this bloodless duel. They stagger to and fro like scaly Sumo wrestlers, the dust kicked into little clouds around them, while the subject of their adoration gazes at their passionate endeavors, showing about as much excitement and enthusiasm as a plum pudding. Finally one or other of the suitors gets his weapon in the right position and skidding along and heaving madly he at last overturns his opponent. Then, he turns and lumbers over to gain his just reward from the female, while the vanquished tortoise, with much leg-waving and effort, rights himself and wanders dispiritedly away.

Competition among males may have affected the evolution of turtle growth. Many freshwater turtle females are larger than males, but in species in which males fight over prospective mates, the males may be the larger sex. The Mary River turtle (*Elusor macrurus*) is one of only two Australian chelids in which the males are larger than the females, and it is the only one in which the males have been seen to fight.

Struggles for dominance may have driven the evolution of shell shape in the saddle-backed races of Galápagos tortoise. Though the peculiar raised front of a saddleback's carapace may have originally evolved to assist in high browsing, it soon provided opportunities for dominance contests among rival males, determined by which male could reach the highest. Once such contests developed, they might have driven, in part, more and more extreme saddleback modifications, permitting an even higher reach, and with it greater social, and sexual, success.

A male painted terrapin (*Callagur borneoensis*) in non-breeding color, left, and another coming into full breeding colors, right. A breeding male in high color has the head almost entirely white, set off by the red diamond on his crown.

After the prospective partners come within range of each other, and after (or while) any rivals have been dealt with, the next step is courtship. As in humans, turtle courtship is not a calm or reasoned process. Captive male Namaqualand speckled padlopers (*Homopus signatus signatus*) pursued females constantly until, according to Nicolas Bayoff, both sexes seemingly became so stressed that their feeding rates declined and they broke out in infestations of nematodes.

When a male Jardine River turtle (*Emydura subglobosa*), a chelid from New Guinea and northern Australia, spots a prospective mate, he stretches his neck to its full extent and swims after her, inspecting and sometimes touching her cloacal region. She may twitch her tail in response to his attentions. He tries to swim around in front of her, bringing them face-to-face. If she stops, or slows down, the male escalates his attentions, stroking her face with his forelimb, blinking his nictitating membranes, and bobbing his head so rapidly that even a slow-motion video camera cannot measure the rate. If all is

Male wood turtles (*Clemmys insculpta*) use "posing" displays to show off the bright orange skin on their necks and forelimbs.

Male false map turtles (*Graptemys pseudogeographica*, left) and yellow-bellied sliders (*Trachemys scripta scripta*, right) vibrate their long foreclaws along the sides of the female's head during courtship.

still well, the female rises to the surface and the male follows; the couple float nose to nose, touching briefly. He backs off and blows a stream of water droplets through his nostrils into her face. He may repeat this as many as four times, sinking and resurfacing each time. What happens next, alas, we do not know, but we can assume that, at some point, at least some of the time, copulation will follow.

When a Jardine River turtle strokes his female's face with his forelimb, or bobs his head up and down in front of her, he is using wooing techniques common to a wide range of turtle families. Some of his postures show up in so many different lines of turtles—including both pleurodires and cryptodires—that it seems at least possible that they go back to the earliest days of turtle evolution.

Head-bobbing is a common maneuver in tortoises, though tortoises court on land instead of in the water. A male impressed tortoise (*Manouria impressa*) approaches the female from the front and bobs his head up and down while simultaneously opening and closing his mouth. If the female approves, she raises her body high, making it easier for him to mount her. Sometimes, though, she lowers herself again before the male can scuttle around behind her and get into position, and her poor suitor must shift back to the front and start all over again. Unlike most tortoises, the male impressed tortoise does not seem to press his suit by butting or biting at his mate.

In many animals—birds and butterflies provide obvious examples—females are the arbiters of the courtship ritual. They do the selecting, while males outdo each other with brilliant colors and extravagant displays. Some turtles appear to do the same. Male wood turtles have "posing" displays during which the male extends his head fully, holds this position for thirty seconds to a minute, and then turns it to one side or the other, exposing the bright orange skin on his neck.

Male European pond turtles (*Emys orbicularis*) display themselves to females by rising to the surface of the water every three to four minutes before retreating once more to deeper waters. Scent may also play a role for some species; secretions from the chin gland of a male Texas tortoise (*Gopherus berlandieri*) seem to stimulate both courtship and combat.

Male river and painted terrapins, unusually in turtles, don bright colors in the breeding season. Few scientists have watched these rare turtles in the wild, but Faith Kostel-Hughes was able to observe their courtship rituals in the Bronx Zoo's Jungle World. Kostel-Hughes found that each species had special displays that seem designed to show off their particular breeding colors. A courting river terrapin faces his female, touches her nose to nose, raises his head, opens his mouth, and pulsates his throat and lower jaw towards her, displaying white throat stripes that stand out sharply against the black of his head. A painted terrapin, on the other hand, sways his head rhythmically back and forth, showing off the brilliant red diamond on his crown. Males of both species use the same displays while facing down other males; perhaps their colors are intended to intimidate as well as to attract.

In addition to displaying their bright colors, wood turtles pursue females aggressively and grab at their shells with all four feet. In many turtles, indeed—particularly in mud, musk, and snapping turtles—courtship may appear, to our eyes at least, to amount to little better than rape. Sometimes the females fight back; basking female smooth softshells (*Apalone muticus*) often snap at males trying

The plastron of the male desert tortoise (*Gopherus agassizii*), is curved, the better to fit the female's carapace during mating.

Tortoise mating can be a particularly noisy affair. These are Travancore tortoises (*Indotestudo travancorica*).

The male northern long-necked turtle (*Chelodina rugosa*) hooks his hind feet under the female's carapace while mating.

7-16. A male slider (*Trachemys scripta*) may have difficulty holding on to his considerably larger mate.

The shell of this female Steindachner's turtle (*Chelodina steindachneri*) is scarred with the claw marks left by her mates.

Common snapping turtles (*Chelydra serpentina*), like most other freshwater turtles, mate in the water.

to mount them, and it is not uncommon, during the spring mating season, to find a male suffering from semicircular bleeding wounds on the posterior edge of his carapace.

Some earlier authors doubted that turtles like the alligator snapper courted at all, beyond males simply overpowering females by brute force. Alligator snapper males, in fact, do court their mates, though their method of doing so may appear, to us, unsubtle. It seems to involve little more than the male pursuing the female and sniffing her, starting at her nose and working his way down, paying special attention to her bridge and cloaca. This behavior may help the male make sure he is dealing with a female. During mating itself, which lasts about six minutes, the male forces bubbles of air out his nose. Whether or not this is part of courtship it is difficult to say, but in common snappers both sexes gulp air and bubble it out their nostrils while facing each other.

Tactile stimulation—turtle foreplay, if you will—is apparently part of the courtship of some pleurodires, who touch each other with the sensitive barbels on their chins. Male and female Macquarie turtles (*Emydura macquarii*) tilt their heads upward, and press their barbels together. A male helmeted terrapin (*Pelomedusa subrufa*) seeking to mate will pursue the female, touching her hindquarters with his snout; if she does not cooperate, he may snap at her tail and hindlimbs. If she is receptive, though, he mounts her, gripping the edges of her shell with all four feet, and rubs his chin barbels on the back of her head. As he mates, he stretches out his neck, sways his head from side to side, and expels a stream of water from his nostrils over his ladylove's face.

Face-stroking is widespread in turtles, but only male American emydid turtles enhance the effect by growing elongated front claws. A male painted turtle uses the backs of his long, curved foreclaws to stroke the female's head and neck; if she is receptive, she responds by stroking his forelimbs with her own claws. Sliders, false map turtles (*Graptemys pseudogeographica*), and chicken turtles do not stroke the female, or even touch her. Instead, they vibrate their foreclaws along the side of her head. Male cooters (*Pseudemys* spp.), which sport extremely long, straight foreclaws like a handful of knitting needles, are also vibrators rather than strokers. A courting river cooter (*Pseudemys concinna*) swims above the female, positions his foreclaws beside her face, and uses a quite uniform pattern of vibratory titillation, each burst lasting an average of 506 milliseconds. Perhaps, for the female, the sensation is somewhat akin to the pulsations of a whirlpool bath.

Turtle fertilization, like our own, is internal. Males have a penis of spongy connective tissue. It normally lies on the floor of the cloaca, in the base of the tail, but at the right moment, it uses blood hydraulics to enlarge and grow erect, and emerges from the cloacal opening. The opening itself is conveniently located some way down the tail, so that the male has at least some flexibility as he brings it into position. Tortoise copulation can be a particularly noisy affair. A mating male may give vent to sounds ranging from the clucking noises made by the red-footed tortoise (*Geochelone carbonaria*) to what Peter Pritchard, in his *Encyclopedia of Turtles*, describes as the "deep bellowing groans which can be heard a great distance" emitted by the giants of the Galápagos.

As Ogden Nash pointed out in verse, simply achieving the physical act is fraught with difficulty. Turtles are not built to intertwine. A male may have to use his claws to grip the female shell if he is to stay in position, particularly if the female is larger than he is. A female Steindachner's turtle (*Chelodina steindachneri*) often carries permanent gouges in her carapace left by the claws of her smaller mates. Many turtles make matters somewhat easier by mating in water, where at least the male does not have to contend, to the same degree, with gravity.

American box turtles, which mate on land and have particularly high, rounded shells, seem to find copulation particularly challenging. A male eastern box turtle must actually lean over backward to get himself into position. To avoid tumbling over on his back, he hooks his toes behind the rear edge of the female's plastron; she helps by clamping her plastral valve on his claws, locking him in place. Once he has anchored himself as firmly as possible, he rocks backward until his shell rests on the ground, shifts his hind feet slightly to improve his purchase, and tilts forward to the vertical before achieving insertion. This maneuver is not only tricky but dangerous. Male box turtles sometimes fall onto their backs after copulating, and, if they cannot right themselves, may die.

### Investing in Motherhood

Unlike other long-lived reptiles such as crocodiles, turtles may take many years before they are ready to begin their reproductive lives. Even without parental care, the investment that reproduction requires can be costly. Painted turtles in Michigan devote approximately 14 percent of their annual energy budget to reproduction. Mexican rough-footed mud turtles may lay only 12 eggs a year, in four separate clutches, but these eggs equal about 28.4 percent of the female's rather small body mass.

The physical, and physiological, task of developing sperm and, particularly, eggs, takes up much of a turtle's energy and resources. This may explain why turtle growth rates slow drastically once the animals become sexually mature. The physiological cost to males, though, may be considerably less than to females, and male turtles often mature at an earlier age.

Even after a female reaches maturity, she may not nest every year if the resources she needs are not available in sufficient quantity. Only an average of 58 percent of mature female spotted turtles in Georgian Bay, Ontario, examined by Jacqueline Litzgus and Ronald Brooks, produced eggs in any given year; only 37 percent of individual females did so three years in a row. In nearby Algonquin Provincial Park, other northern species—common snappers, painted turtles, and wood turtles—are somewhat more likely to nest every year, but it is not unusual for a female, particularly a smaller animal in her early reproductive years, to take an occasional sabbatical.

A female turtle stands a good chance of living a long reproductive life, through good nesting seasons and bad. It may, therefore, be more important, in terms of her lifetime output of young, to survive to nest again than to risk everything for any one season's brood. For an animal that faces unpredictable and difficult conditions from year to year, like a desert tortoise, this may mean hedging her bets by holding her resources in reserve for her own needs rather than committing them to her young. This includes reabsorbing ovarian follicles that fail to ovulate and reclaiming the nutrients they contain, a process called *atresia*.

Unlike mammals such as ourselves, turtles do not have a preset maturation time. Maturity seems to have more to do with size than with age. For sea turtles, reaching sexual maturity may take decades (see Chapter 8). Land and freshwater turtles mature more rapidly—usually in three to six years for common musk turtles and painted turtles. How long maturation takes can vary from place to place. Female common snapping turtles from Cootes Paradise in southern Ontario, studied by Gregory Brown and his colleagues, grew almost four times faster, and produced clutches roughly 30 percent larger, than snappers 300 km (186 mi) farther north in Algonquin Provincial Park, near the northern limit of the species' range—even for hatchlings incubated and reared under identical conditions in the laboratory.

Even among individuals in a single population, time to maturity can range from 9 to 15 years for yellow mud turtles (*Kinosternon flavescens*), from 11 to 16 years for common snapping turtles, or from 12 to 21 years for Blanding's turtles. This variation probably has something to do with how fast individual turtles grow, and that, in turn, may be governed by such things as the amount and quality of food they can find and eat. In northern Australia, the carnivorous northern long-necked turtle (*Chelodina rugosa*) matures relatively quickly, in less than four years for males and just over six years for females, but the Victoria River snapper (*Elseya dentata*), a herbivore with a relatively protein-poor diet, may take more than twice as long to reach maturity.

**Eggs**

In general, the bigger a female turtle is the more space there is for eggs within her body, the more eggs she can lay, and the larger those eggs may be. Smaller female desert tortoises not only lay smaller clutches than larger females, but they lay them later in the spring—perhaps because they have fewer reserves in store. First, they feast on early spring annual plants to get the nutrients they need to produce the yolks for their eggs.

The larger a mother painted turtle or chicken turtle, the more massive, roughly speaking, are her eggs and the bigger her hatchlings. The relationship, though, is not exact, and often is not true for every population or species. In other turtles, including the spotted turtle and the Florida softshell, larger females do not lay bigger eggs.

There may be a minimum size for the egg of even the smallest turtle, or the hatchling it produces will not survive. Small turtles, like the bog turtle (*Clemmys muhlenbergii*) and some of the smaller musk turtles, lay only one, or a few, eggs. Small tortoises like the smaller padlopers (*Homopus* spp.), the flat-tailed tortoise (*Pyxis planicauda*), and the pancake tortoise (*Malacochersus tornieri*) typically lay only a single egg at a time. Malagasy big-headed turtles (*Erymnochelys madagascariensis*) with a carapace length of less than 32 cm (12.6 in) produce a small clutch of eggs, but once they reach 40 cm (15.7 in) or more, they commonly lay more than 60 eggs a season, in up to three separate nestings.

At the other end of the scale, a captive Kanburi narrow-headed softshell (*Chitra chitra*) weighing 108 kg (238 lb) laid a clutch of 107 eggs, each one 34 mm (1.3 in) across and 20 g (0.7 oz) in weight. The turtle died soon after, and its body was found to contain a further 450 developing eggs and ovarian follicles.

This does not mean that a bigger turtle always lays more eggs than a smaller one. Among the batagurids, the large river and painted terrapins lay clutches averaging about 20 eggs each, while the smaller three-striped roofed turtle (*Kachuga dhongoka*) lays from 30 to 35. Most tortoises lay rather small clutches; even the giants of the Galápagos and Aldabra rarely lay more than 20 eggs to a clutch. Alligator snappers grow much larger than common snappers, but do not lay as many more eggs as their size difference might suggest (20 to 30 per clutch is normal for the common snapper, 20 to 50 for the alligator snapper). In Guatemala, the white-lipped mud turtle (*Kinosternon leucostomum*) may lay from one to five eggs in a clutch, while its larger relative and sometime predator, the Mexican giant musk turtle (*Staurotypus triporcatus*) lays from 4 to 18.

Broadly speaking, the bigger the egg, the bigger the hatchling—though within a species, many other factors, like incubation temperature and the availability of moisture, can affect development. Bigger hatchlings often have a better chance of surviving.

Since a female turtle does not help her newly hatched offspring find food, the only way she can give them a nutritional head start is by laying a well-stocked egg. The yolks of common snapping turtle and painted turtle eggs contain more than half again as much food, in the form of stored lipids, as the

An eastern spiny softshell (*Apalone spinifera spinifera*) hatches from a hard-shelled egg (above), while a hatchling wood turtle (*Clemmys insculpta*) breaks through its leathery shell (below).

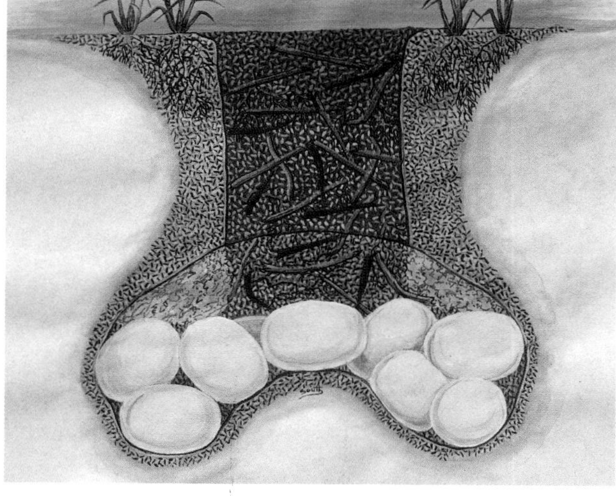

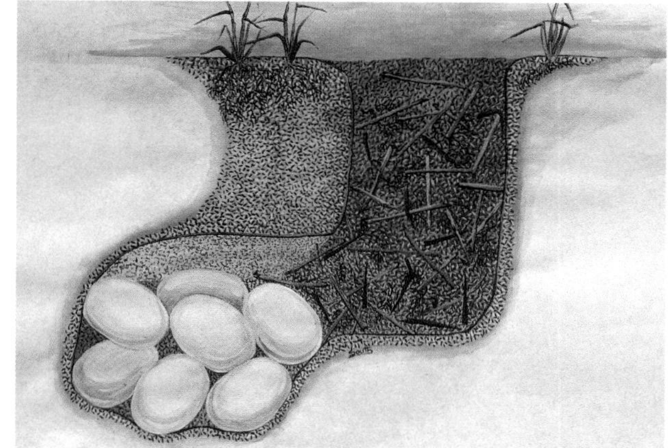

Two cross-sectional views of the nest of an Australian broad-shelled turtle (*Chelodina expansa*).

embryo needs to develop, hatch, and emerge from its nest.

Turtle eggs may be hard-shelled or soft-shelled, spherical, oval, or elongate. All turtle eggshells consist of a shell membrane—thick and tough in the flexible-shelled eggs of sea turtles, quite thin in hard-shelled eggs—under a layer of hard shell units made of, usually, crystals of aragonite, a form of calcium carbonate. Hard-shelled eggs are hard because their crystalline shell units are larger and are packed closely together, except for occasional pores to allow for the exchange of water and gases.

Soft, leathery shells are frequent in turtles, but chelids, some pelomedusids and podocnemids, kinosternids, softshells, the pig-nose turtle, the Central American river turtle, neotropical wood turtles (*Rhinoclemmys* spp.) and many tortoises lay hard-shelled eggs. Eggs with flexible shells are more likely to take up water from their environment than hard-shelled eggs; the amount of water they absorb critically affects the size of the hatchling.

Closely related turtles can lay very different eggs. While the eggs of the yellow-spotted river turtle (*Podocnemis unifilis*) are hard-shelled and elongate, those of the arrau are soft-shelled and spherical. Egg shape may even change as a turtle grows; though the eggs laid by larger, older common snapping turtles are spherical, younger and smaller animals lay oblong eggs, possibly because the passage through their pelvic girdle is too narrow to pass a sphere of the appropriate size. The Texas tortoise lays a hard-shelled egg far larger than the opening between its carapace and plastron. When the female lays her eggs, the bones of her shell apparently bend out of their way, the carapace moving slightly upward as the plastron flexes downward.

When turtle eggs are laid, their yolk is surrounded by a thick layer of albumin. This is the substance we call "egg white," but that name is a misnomer for turtle eggs. I remember, as a child in Jamaica, being greatly amused by the story of an acquaintance who spent about 45 minutes attempting to fry a sea turtle egg without realizing that turtle albumin, unlike that of a bird, does not coagulate. The albumin carries most of the water supply for the newly laid egg. Within the first week or two, most of that water flows inward to the yolk, through the vitelline membrane (see Chapter 5). As the embryo develops, it is this expanded yolk that supplies it with both food and water.

### The All-Important Nest

Some turtles seemingly pay little attention to where they deposit their eggs. Neotropical wood turtles (*Rhinoclemmys* spp.) drop their enormous eggs (see Chapter 1) in crevices or among leaf litter in the forest, as may the gibba turtle (*Phrynops gibbus*). Narrow-bridged musk turtles (*Claudius angustatus*) lay their eggs among tangles of vegetation. Yellow-footed tortoises (*Geochelone denticulata*) may bury their eggs completely, cover them with a thin layer of soil or dead leaves, or simply drop them in leaf litter on the forest floor. The white-lipped mud turtle scrapes out a shallow depression in the ground and covers her eggs with a thin layer of soil or leaf litter. Loggerhead musk turtles leave their eggs in a barely concealed scratch in the ground, often against a tree stump or rock, and yellow mud turtles may leave their eggs in burrows they dig for their own use in the sandy soil. Karoo (*Homopus boulengeri*) and speckled padlopers (*Homopus signatus*), tiny tortoises that live in stony places where digging a nest may be impossible, apparently hide their single egg under an overhanging rock.

Most turtles, though, lay their eggs in a specially prepared nest. In some species, the labor involved in preparing it seems minimal. The impressed tortoise apparently digs only a very shallow nest, and covers the eggs with leaves that she scrapes into place with her hind legs. A more typical nest is usually a flask-shaped chamber in the soil, excavated by the female with alternating strokes of her hind legs. The shape and depth of the nest depend on how far she can reach from the surface. Many land and freshwater turtles use their hind feet to arrange the eggs as they are laid, sometimes into two or three layers separated by thin partitions of dirt.

Some turtles, including the alligator snapper, dig a body pit first, clearing away vegetation and loose soil that might otherwise fall into the nest chamber. A female eastern mud turtle (*Kinosternon subrubrum*) starts excavating with her front feet, throwing the dirt aside until she is almost out of sight, before

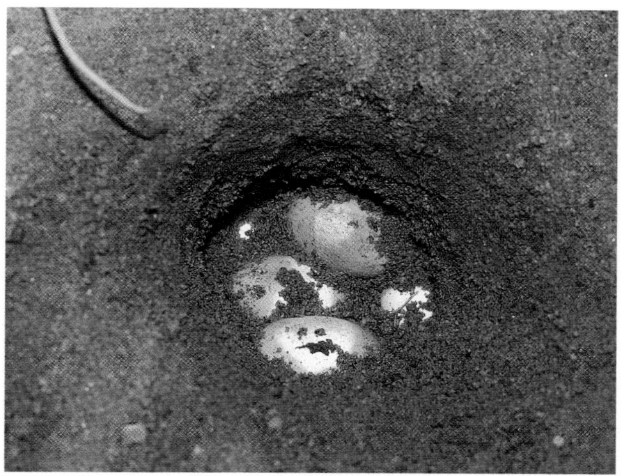

A Mary river turtle (*Elusor macrurus*) probably floated on the surface for a week before coming ashore to dig this nest.

she turns around and uses her hind feet to finish digging the nest chamber.

Gopher tortoises start their nests by swinging their bodies in a circle and scraping out a shallow bowl with their front feet. Peculiarly, these tortoises, which have specially modified front limbs capable of digging deep and extensive burrow systems, use their hind feet to dig out the nest cavity. Digging a nest hole, though, is physically quite a different task from excavating a burrow.

Many turtles, even diurnal ones, dig their nests at night, though they sometimes start before dusk. On cloudy days, Blanding's turtles and false map turtles may nest in daylight hours as well. In Guatemala, sliders will only nest on particularly dark, often moonless nights. Common snapping turtles lay their eggs during the day; common snappers (*Chelydra serpentina*) usually nest in the early morning or late afternoon.

Excavation may take a long time. The columnar hind feet of tortoises make digging especially slow. Captive Bell's hinged tortoises (*Kinixys belliana*) in Tanzania usually took up to six hours to dig their nests, starting just before dusk. The whole process, from digging, through laying, to covering the finished nest, can take more than two and a half hours for river terrapins, and up to eight hours for spotted turtles in Ontario. This last is exceptional: other turtles in the same area usually take less than 90 minutes, and even spotted turtles farther south in Pennsylvania took no more than two hours. At the western limit of their range, and in Illinois, however, spotted turtle females needed up to 12 hours to complete their task.

### The Right Spot

Turtles may devote considerable time and effort to selecting a suitable place to nest. Spotted turtles in Georgian Bay, Ontario, dug their nests in soil and lichen that collected in crevices in the outcrops of Precambrian stone, probably the sunniest and warmest places available so far north. Females often spent days wandering around among the rocks before digging their nest, presumably searching for just the right spot.

Mary River turtles drift around on the surface for a week or so before coming ashore to nest, possibly deciding on the best place to emerge. An eastern mud turtle may try several sites before she finds one that satisfies her. Some Blanding's turtles on the Edwin S. George Reserve in Michigan start digging the evening they leave their marshes, but others may stay on land for several days first.

Diamondback terrapins appear to select their nest sites using such physical features as the slope and height of sand dunes and the extent of plant cover, and choose their timing on the basis of the tidal phase and the weather. Black-knobbed map turtles (*Graptemys nigrinoda*) sniff the sand on their chosen river beach, make test scratches with their front feet or, occasionally, dig a test nest. Apparently they are gathering information that helps them find the best place to dig when they nest in earnest.

A properly constructed and situated turtle nest is an incubator that must expose the eggs to the proper range of temperature and humidity. Too little heat, and the eggs do not develop; too much, and the embryos die. The right spot may depend on the female size. In the temperate zone, medium to large turtles often nest in open sunny areas, placing their eggs deeply enough in the soil to avoid overheating. Striped mud turtles are too small to dig a deep nest, and in central Florida, where Dawn Wilson studied their nesting ecology, they avoid sunny nest sites. Instead, they place their nests near clumps of grass or other vegetation. Because their eggs incubate fairly close to the surface, they apparently need the shade that the plants provide.

Tradition may be important in nest site selection. Texas tortoises may use the same sites for years. Arrau studied by Richard Vogt on the Rio Trombetas in Brazil nested in large numbers on the same beach in 1978, 1979, and 1980, completely ignoring three apparently similar, and equally suitable, beaches a short distance away. Most Blanding's turtles return to nesting areas they have used before, and individual striped mud turtles in Florida repeatedly returned to the same general area to nest, year after year.

The soil or sand where a female proposes to nest must be the right consistency for digging. Gopher tortoises may excavate their nests in the apron of sand at the mouth of their burrow. Sometimes, the best spot may be where another animal has already done some of the work. Nests of the New Guinea giant softshell (*Pelochelys bibroni*) have been found buried in the nest mound of a freshwater crocodile (*Crocodylus novaeguinae*), and Florida red-bellied turtles (*Pseudemys nelsoni*) may carry out their excavations in the nest of an alligator (*Alligator mississippiensis*). Since crocodilians, unlike turtles, guard their nests, a turtle nest in an alligator or crocodile mound may gain an extra measure of protection.

A common map turtle (*Graptemys geographica*) deposits an egg in its nest.

Several Australian freshwater turtles, including the snake-necked turtle (*Chelodina longicollis*), prefer to nest after heavy rains, perhaps because water-soaked soil is softer, easier to dig, and less likely than dry soil to break up and fall into the nesting hole. Perhaps for the same reason, many turtles void copious amounts of highly dilute urine

After laying its eggs, a broad-shelled turtle (*Chelodina expansa*) sealed its nest with this plug of soil and urinary fluid.

into the nesting cavity. The water that a painted turtle carries in her bladder may weigh more than the eggs themselves.

Water desposition may be an essential part of the nesting process. At the E. S. George Reserve in Michigan, Owen Kinney and his colleagues found that if they interrupted a female painted, common snapping, or Blanding's turtle on her way to nest, and she urinated in response (as turtles often do when handled), on release she headed back to the water to take on another supply before setting off once more to lay her eggs. John Cann watched a snake-necked turtle that made several trips to the nearby lagoon, over an hour and a half, to take on water, returning each time to release it into the nest cavity.

Turtles may spend some time finishing off their nests. Instead of simply burying their clutches, painted and other turtles will, if possible, work the moistened soil into a plug that seals the opening, leaving an airspace between the roof of the nest and the eggs. Many turtles drag their plastrons over the completed nest, smoothing the soil above it. In Galápagos tortoises this can go on for days. A female red-bellied turtle (*Pseudemys rubriventris*) flattens the soil over her eggs by raising herself as high as possible and dropping heavily on top of her nest. A leopard tortoise or angulate tortoise (*Chersina angulata*) may do the same by lifting and dropping her shell repeatedly on the spot.

A female Suwannee River cooter (*Pseudemys suwanniensis*) or Florida cooter (*Pseudemys floridana*) has a unique trick to fool nest robbers. Instead of building one nest, she digs three: a principal nest where she lays most of her eggs, and two satellite nests, one on each side, with only a few eggs in each. After she lays her eggs, the female covers the principal nest as thoroughly and carefully as possible, doing her best to obscure all traces of its existence. The satellite nests, though, she covers only haphazardly, if at all. A predator like a raccoon may have no trouble finding the satellite nests, but may miss the principal nest altogether. The satellite nests, in short, are decoys, and the eggs laid in them sacrifices—but not always successful ones—to the safety of the rest of her brood.

**On the Beach**

Sandbars and sandy beaches are the preferred nesting places of a number of river turtles in different parts of the world, including Amazon River turtles, many softshells, river and painted terrapins, map turtles, and the pig-nose turtle. Nesting on a sandbar has its advantages: less distance to travel from the water, and, if the sandbar is on an island, possibly some protection from predators. However, open sand can get very hot and dry in the sun; it may not be a coincidence that many of the turtles that nest on sandbars are large, and can dig relatively deep, well-insulated nests.

Sea turtles are not the only turtles to nest on ocean beaches. In North America, diamondback terrapins nest on barrier beaches and dunes. In Asia, river terrapins nest en masse (or once did) on beaches on the Perak and other rivers in Malaysia, the Irrawady River delta of Myanmar, or on islands in the Sunderbans of India and Bangladesh. Its relative, the painted terrapin, nests on sea beaches in West Malaysia, and, on the west coast, on river sandbanks. The giant Asian softshell (*Pelochelys cantorii*) and, and the Mediterranean population of the Nile softshell (*Trionyx triunguis*), may nest on sea beaches. In southern New Guinea, pig-nose turtles do the same.

Turtles that use open sand may have physiological adaptations to nesting there. Eggs of most sandbar-nesting turtles can tolerate remarkably high incubation temperatures. Smooth softshells require relatively higher incubation temperatures than many other North American freshwater turtles—no less than 25°C (77°F), and preferably more; 30°C (86°F) may be optimal. Temperatures in the nest of an arrau may climb to over 36°C (96.8°F) for more than half the incubation period, a temperature that kills even smooth softshell eggs if it is maintained for too long. Eggs of diamondback terrapins and pig-nose turtles can tolerate—briefly—temperatures of 40°C (104°F) or more.

Perhaps the most unexpected beach nesters are a population of oversize sliders in Tortuguero National Park on the Caribbean coast of Costa Rica, a place much more famous for its green sea turtles (*Chelonia mydas*). According to Don Moll, females emerge from the rainforest and actually swim a considerable distance along the sea coast to reach their nesting beaches. The journey is a risky one. Sliders are not particularly tolerant of saltwater, and not every female survives her passage through the sea. The risk may be worth it; if the turtles traveled directly through the forest to the beach, they might be tracked, or leave a scent trail that predators could follow.

Before human beings began to demolish the rainforests, these beaches may have been the only place the sliders could go to lay their eggs in a reasonably open area exposed to the right amount of sun. The turtles do not nest entirely in the open, where temperatures of the surface of the sand may approach 73°C (163.4°F); here, it would take a much larger turtle, like a green, to dig to a safe depth. Instead, they prefer the shade of cocoplum trees (*Chrysobalanus icaco*) growing on the upper beach. The trees provide not just shade, but camouflage, both for females as they nest and for the young as they eventually emerge.

**Off the Beach**

How far should a freshwater turtle's nest be from the water? Some species travel a considerable distance, possibly because the farther a nest is from an obvious marker like a shoreline, the harder it may be for a predator to find. Though some Blanding's turtles nest only a short distance from the water's edge, others travel more than a kilometer before laying their eggs—an unusually long distance for a North American freshwater turtle. Even yellow-spotted river turtles, which nest on sandbanks, may travel more than 100 m (328 ft) from the river to find a suitable patch of sand. Savanna side-necked turtles (*Podoenemis vogli*) walk for kilometers.

Painted turtles usually nest no more than 50 m (164 ft) from the water, while sliders normally travel more than 10 times farther away. John Tucker has suggested that the difference has to do with the dangers that emerging hatchlings face on the journey back. Painted turtle hatchlings are smaller and have more permeable skins than hatchling sliders. They may need to make the shortest, and fastest, dash possible to the water's edge to avoid drying out on their journey.

Female striped mud turtles in central Florida may journey over 240 m (787 ft) to their nest site. A trip of 60–180 m (197–262 ft) is, though, more usual. After the exertions of their overland trek and egg-laying activities, Dawn Wilson found that females did not head back to the water but buried themselves nearby, below the soil and leaf litter, for anywhere from one to 35 days. This type of behavior, which is typical of a number of other mud turtles and is also known in the western swamp turtle, may not simply be the result of exhaustion. The females may be waiting for rain, either to avoid dehydration themselves, or so that, when they do head back to the water, they will not leave a scent trail that could guide a predator to their nest. Yellow mud turtles bury themselves directly on top of their eggs, possibly to protect them from predators—perhaps as close as any turtle gets to parental care.

If the best nesting grounds are far from their feeding grounds, turtles may need to migrate to reach them. Such migration is common for turtles that nest on sandy beaches; the journeys of the arrau are famous (see Chapter 6), and of course the sea turtles make the longest migratory journeys of any reptile (see Chapter 8). Painted terrapins migrate, usually downriver, to find nesting sites unlikely to be flooded by river and sea tides. Even land turtles may migrate. On Santa Cruz (Indefatigable) Island in the Galápagos, nesting sites are particularly hard to come by. The giant tortoises (*Geochelone nigra porteri*) must travel from their upland feeding grounds to the southwestern lowlands, where the silty soil of special nesting areas called "campos" is suitable for digging. To reach them, the tortoises follow a system of trails they have trodden for generations.

### Taken at the Flood

A nest that is too close to the water risks being flooded. Flooding can have serious consequences. If the eggs of smooth softshells are flooded for even a few days during the early stages of development, their chance of survival is slim; six days underwater is enough to kill the whole brood. In some years, six-tubercled Amazon River turtles in the Samiria River in Peru may lose all their nests when the river floods. Early yellow-spotted river turtle nesting beaches are often washed away. Perhaps as an evolutionary response, their eggs develop more rapidly than is usual in turtles, increasing the chances that the young will hatch, emerge, and be safely in the river before the floods arrive.

Central American river turtles are so

The northern long-necked turtle (*Chelodina rugosa*) of tropical Australia is the only turtle known to nest under water.

Northern long-necked turtles build their nests in shallow lagoons; the eggs do not develop until the lagoon dries.

helpless on land that they must nest very near the water—usually less than 1.5 m (5 ft) from the waterline. Nest flooding is regular. Their eggs show a remarkable ability to tolerate a prolonged spell underwater, at least during early development. *Dermatemys* eggs have survived 28 continuous days beneath the surface. Flooding, as long as it does not last too long, may even be an advantage for these turtles if it hides the nest from predators.

We do not know if Central American river turtles deliberately lay their eggs where they are likely to be flooded, or even if they occasionally nest in shallow water. A tropical Australian turtle, though, lays its eggs underwater as a matter of course—the only reptile in the world known for certain to do so.

Rod Kennett discovered the remarkable nesting habits of the northern long-necked turtle almost by accident. These turtles are commonest on tropical floodplains in Australia's Northern Territory. Here, from December through April, monsoonal rains inundate thousands of square kilometers, creating great shallow marshlands that dry almost completely in the near-rainless months outside the monsoon. As the waters recede, turtles retreat, if they can, to the few permanent water sources, or bury themselves in the drying mud to await the next coming of the rains.

Kennett could not understand why, no matter how common these turtles were in the billabongs he was studying, he was unable to find a single nest despite two years of searching. The Aboriginal people knew the answer: they said that the turtles buried their eggs in the mud beneath the shallows as the rainy season came to an end, before the waters themselves dried.

In 1991, Kennett and his co-workers tried an ingenious test of the Aboriginal story. They implanted tiny egg-shaped radio transmitters in the oviducts of gravid turtles at a lagoon near Darwin. When the turtles laid their eggs, the transmitters became part of the clutch, broadcasting the nests' location. The trick worked twice. The transmitters' signals proved that the Aboriginals were right: both nests were under water, buried beneath clumps of aquatic vegetation 17 m (56 ft) and 14.5 m (47.5 ft), respectively, from shore.

The transmitters solved one mystery but left the scientists with even greater ones. How do the eggs survive such treatment? And why do the turtles nest underwater in the first place?

Flooding kills most reptile eggs because it

The plains of Australia's Northern Territory, home to the northern long-necked turtle, dry completely outside the monsoon.

denies them the oxygen they need to develop, and because water entering by osmosis may cause the egg to swell and the shell to crack. In some turtles, water entry is not necessarily a problem, and may even be a normal part of development, as long as the underlying membranes surrounding the embryo remain intact. If these rupture, though, the embryo will likely die.

Northern long-necked turtle embryos solve these problems by having a particularly water-resistant vitelline membrane surrounding the yolk and the embryo, and by simply not developing until the waters above them recede and air can reach the eggs. They can survive for at least 12 weeks in a state of arrested development—a remarkable feat, considering that embryos of their close relative, the snake-necked turtle, will die after only about two weeks under water.

Why do these turtles have such a bizarre adaptation? The answer probably has to do with the temporary and unpredictable nature of their watery habitat. Other tropical Australian turtles live in more or less permanent waters, and nest on land in the traditional manner, even, apparently, including the newly described Arnhem Land long-necked turtle (*Chelodina burrungandjii*), which may be the northern long-necked turtle's closest relative. Because they live in temporary floodwaters, though, northern longnecks may not be able to find a safe dry-land site at the right time of year.

If the turtles lay on land during the rainy season, or towards the end of the dry season, their eggs may be flooded when the waters rise again, and flooding during the later stages of development, as opposed to immediately after laying, will kill even their embryos. Waiting until the waters recede at the end of the rainy season could delay nesting by several months, and cost them the chance to lay more than one clutch.

By laying when they do, and where they do, northern long-necked turtles put their young in the best position to survive to hatching. By having their eggs develop during the dry season, they ensure that their young will only emerge when times are right. If the hatchling turtles finish development before the dry season ends, the hard-baked ground over their heads will probably be impossible to break through. Only when the rains come again will the hatchlings be able to leave the nest, and only then will there be safe habitat for them to prosper in.

### Multiple Clutches

One of the best ways for a female to increase the selective odds in her favor is to, literally, avoid placing all her eggs in one basket. The more nesting seasons she survives, and the more clutches she can lay per season, the greater her chance of winning through against the very high levels of predation and loss her nests and hatchlings face.

Laying multiple clutches may be a particularly good strategy for turtles that live in uncertain environments. In tropical Australia, northern long-necked turtles face unpredictable monsoons that may vary considerably from year to year; they lay several clutches at different times during the season, increasing the chances that at least some will emerge when the rains come again. By contrast, Victoria River snappers, which live in permanent rivers, lay only one clutch per year.

Snapping turtles (Chelydridae) and the arrau also nest only once a year. Wild spotted turtles and Blanding's turtles lay only once a year, though captive female spotted turtles may produce a second clutch; this probably reflects the shorter nesting season, or smaller resource supply, available to these northern species.

Other turtles may lay only two clutches. Desert tortoises in the Mojave Desert of California lay one clutch in late April or early May, and, about 70 percent of the time, a second clutch in late May or early June. The Mary River turtle of eastern Australia nests in late October and again about a month later. Central American river turtles may lay anywhere from one to four clutches in a season, though two or three are more likely.

Some turtles lay several clutches per season, probably using sperm they have stored up from the spring's (or even an earlier year's) mating. Several North American turtles, including the Escambia map turtle (*Graptemys ernsti*), the Florida cooter, and the Florida red-bellied turtle lay as many as six clutches a year. Florida softshells may also lay six clutches annually (three to five is more likely), but some females appear not to nest every year. Western pond turtles in the Los Angeles basin may nest only every other year, but may lay double clutches in the years they do nest.

Multiple clutching is not universal for all populations of sliders, nor even for all individual females within a population. Florida box turtles on Egmont Key have the capacity to nest up to three times a year, but only a small percentage of the females on the island actually do so. Not every female nests every year, and not every female that does nest lays multiple clutches; perhaps it is better to lay several clutches in a good year, and none the next, rather than laying every year no matter what the conditions.

Pancake tortoises lay only one egg at a time, but do so two or three times a year. Even larger tortoises, which can produce more eggs per clutch, may nest several times a year if they can. Leopard tortoises usually lay from 10 to 12 eggs per clutch. Over the course of the nesting season, a leopard tortoise may lay four to five clutches, with 22 to 31 days elapsing between each nesting.

Multiple clutches may be particularly important for small turtles like the striped mud turtle. A small turtle simply cannot carry within her body as many fully developed eggs at one time as a large one; mud turtle clutches normally contain only two or three eggs. The only way for a female mud turtle to increase her annual reproductive output is to lay several clutches over the course of the season. Some tropical mud turtles may lay clutches throughout the year; the Mexican giant mud turtle, which is capable of producing a much larger clutch, may still nest as many as nine times a season.

**Putting Development on Hold**

In the temperate zone, the number of eggs a female turtle can develop, and the number of clutches she can lay, may be restricted by the number of warm days available. Her ability to nest over and over again through the year may be cut off by the approach of winter. A hatchling that emerges in, say, mid-December might not have a very long life expectancy.

Many turtles in the warmer parts of the temperate zone, however, have developmental tricks that may allow them to extend the nesting season. A female chicken turtle can carry fully formed, shelled eggs in her oviduct for more than eight months. That ability allows her to generate a clutch in the fall but not lay it until the following spring.

In other turtles, clutches laid late in the year simply stop developing when winter comes, the embryos overwinter in the eggs, and development is completed in the spring. Sonoran mud turtles (*Kinosternon sonoriense*) are active during a brief period of summer rain, and eggs laid one summer may not deliver their hatchlings to the surface until the following summer, more than 11 months later. Apparently they pass the winter as embryos.

If development is put on hold early in incubation, but after the eggs are laid, it is called *embryonic diapause*. An egg in diapause will not continue its development even if the temperatures are right. Many of the eggs of broad-shelled turtles, which nest in late fall, are laid in a state of diapause, and will not develop for several weeks, no matter how high the temperature. In some turtles, such as the pig-nose turtle, the pause occurs late in development, with the embryo, almost ready to hatch, dialing down its metabolism, rather like an adult turtle entering dormancy; this is called either delayed hatching or *embryonic aestivation*.

Diapause usually occurs among certain turtles with a number of things in common: they tend, for example, to lay hard-shelled eggs, though the chicken turtle is an exception. Turtles with diapause, including a number of mud turtles such as the striped mud turtle, tend to live in the warmer parts of the temperate zone. No turtle living north of 35°N is known to exhibit diapause. This makes some sense if the advantage of diapause is to permit turtle embryos to overwinter in the egg; north of a certain point, the risk of freezing may be too great for any overwintering egg to survive. The painted turtle, a northern species that overwinters in its nest, does so not as an egg but as a hatchling (see Chapter 5). Some tropical turtles exhibit embryonic diapause too, including some chelids, kinosternids, the Indian softshell (*Aspideretes gangenticus*), at least two batagurids, and the pancake tortoise.

Once diapause is under way, how does the

The sex of this juvenile yellow-bellied slider (*Trachemys scripta scripta*) was determined by its incubation temperature.

embryo in the egg "know" when to start developing again? Clearly it needs some sort of environmental cue. A fall in temperature appears to do the trick for Geoffroy's turtle, at least in the laboratory. The cue for the pig-nose turtle, which nests on land but whose hatchlings must reach water to survive, is a cutoff of the egg's oxygen supply—something that presumably happens in the wild when the eggs are flooded by the first heavy rains of the season, or by the rising waters where the hatchlings will spend their lives.

### Heat, Sex, and Turtle Eggs

In the mid-1960s, scientists made the bizarre discovery that, for many reptiles, incubation temperature does not merely determine an embryo's development rate, but its sex. Reptiles with temperature-dependent sex determination (TSD, or TDSD) do not all react to temperature in the same way. In some crocodilians, eggs incubated at higher temperatures develop into males, while those at lower temperatures hatch out as females.

No turtle shows this pattern, termed *Pattern Ib* by J. J. Bull. As far as we know, temperature produces the opposite result in podocnemidids, sea turtles, most emydids, tortoises, and some batagurids. If the average temperature within the egg is sufficiently below a certain level during a critical period when the embryo is starting to generate its reproductive organs, hatchlings will be a male; above this *threshold* or *pivotal temperature*, they will be female. Bull called this *Pattern Ia*. A clutch of eggs incubated at or near the pivotal temperature will produce a mixture of males and females, or even some individuals with the characteristics of both; at the pivotal temperature, males and females should be produced in equal numbers.

Pattern Ia is known only in turtles. Pelomedusids (at least the few we know about), snapping turtles, some kinosternids, and at least one batagurid, the Indian black turtle (*Melanochelys trijuga*), have still another pattern, shared with some lizards and crocodilians. In this type of TSD, known as *Pattern II*, there are two threshold temperatures, a lower and higher. Females are produced both below the lower threshold and above the higher one. Males result from eggs incubated at an intermediate range, between the two thresholds. In nature, Pattern Ia and Pattern II may have the same practical effect; Pattern II turtles are unlikely to place their eggs where incubation temperatures will be cool enough, for long enough, to turn out low-temperature females.

The sex of a number of turtles has nothing to do with incubation temperature. So far as we know, all chelids, softshell turtles, and giant and narrow-bridged musk turtles (Staurotypinae), plus an emydid, the wood turtle, and a batagurid, the black marsh turtle (*Siebenrockiella crassicollis*), have genetically based sex determination (GSD). In chelids, staurotypines, and a few others, gender is probably determined by the presence or absence of a sex chromosome, as our own gender is. Turtles with TSD do not have sex chromosomes, as far as we can tell; perhaps such chromosomes prevent TSD from operating.

However, the difference between TSD and GSD may not be as great as it seems. Within a single genus, *Clemmys*, spotted turtles have TSD while wood turtles have GSD. To make matters even more confusing, the distinctions between the three turtle patterns—TSD Pattern Ia, TSD Pattern II, and GSD—are not always clear-cut. Mud and musk turtles include clear Pattern Ia species like the Mexican mud turtle, Pattern II species like the loggerhead musk turtle, and a number of others that seem to be intermediate, including one, the striped mud turtle, that may have both TSD and GSD. In the alligator snapping turtle there appears to be no temperature that will cause all the eggs in a clutch to develop as males. This suggests the possibility that at least some of its eggs—perhaps as many as a third—may be destined to be female no matter what their incubation temperature.

In some turtles, sex determination may depend on both temperature and moisture. In one experiment, painted turtle eggs were incubated above and below the pivotal temperature. Those incubated above the pivotal temperature hatched, as expected, as females, but below, eggs incubated in damp conditions produced males, while those kept in a dry environment hatched out a significant, and surprising, number of females.

The pivotal temperature in Pattern Ia turtles varies from species to species. It ranges from as low as 27.5°C (81.5°F) in the painted turtle to as high as 32.5–34°C (90.5–93.2°F) in the arrau. It is highest in podocnemidids and tortoises. For the gopher tortoise, it lies somewhere between 29°C (84.2°F) and 32°C (89.6°F), while in Hermann's tortoise (*Testudo hermanni*) it is 31.5°C (88.7°F). In most emydid turtles, the pivotal temperature ranges from 28–30°C (82.4–86.0°F). In Pattern II turtles, which have two pivotal temperatures, the lower threshold is usually less than 27°C (80.6°F). It reaches below 22°C (71.6°F) in northern populations of the common snapping turtle, but may be as high as 29°C (84.2°F) in the helmeted terrapin. The upper threshold ranges from 25.5°C (77.9°F) in the white-lipped mud turtle, to 32.1°C (89.8°F) in the helmeted terrapin.

The pivotal temperature may have a great deal to do with where female turtles choose to nest. Conversely, the kinds of nesting sites available may have affected the way pivotal temperatures have evolved. In North America, northern turtles tend to have lower pivotal temperatures. In the hotter Southeast, some species deliberately choose nest sites in the shade. In the dry Southwest, where cool nest sites may be hard to find, turtles with TSD tend to have higher pivotal temperatures. The South American arrau, which has the highest-known pivotal temperature, selects sites on river sandbars where the incubation temperature may be very high indeed. This increases the chances that its young will develop quickly enough to escape rising floodwaters, and its pivotal temperature appears to have shifted upward in consequence.

In turtles with TSD, gender may depend on where an egg lies in its nest. The nests of common snapping turtles may be warmer at the top, near the surface, and cooler at the bottom. Eggs from the upper reaches of the nest chamber are more likely to hatch as females, while hatchlings emerging from its depths tend to be males.

We are not exactly sure how TSD works. Only the temperature during the middle third of the incubation period, though, seems to be important. During this critical period, temperature does not have to be constant; but laboratory experiments have shown that, in order to turn out females, eggs have to be exposed to temperatures above the threshold level for at least four hours a day.

What are the chances that, during this crucial period, nest temperatures will be close enough to pivotal to produce both males and females? In many cases, the answer seems to be "not very good." Studies on nests of map turtles (*Graptemys* spp.) and painted turtles have found that most of the young coming out of any given nest were of the same sex. Colin Limpus and his colleagues found the same for sea turtle nests in Australia. A female may need to ensure, not that each nest produces a gender mix, but that at least some of her nests produce males and others females. According to Dionysius S. K. Sharma and his colleagues, painted terrapins at Rhu Kudung, Malaysia, "lay their eggs at different sites on the beach platform, exposing their eggs to different temperature regimes. This will produce a natural male-female hatchling sex ratio."

TSD would seem to expose turtle reproductive success to the mercy of the environment. It is not that turtles have no evolutionary choice; GSD has apparently evolved at least five times in different turtle families, so an adaptive alternative is available. TSD must, then, have its advantages.

Nonetheless, scientists do not yet know exactly what these advantages are. The tendency of TSD nests to produce same-sex hatchlings may somehow prevent interbreeding years later, when the turtles mature, but it is difficult to see how, especially if the same female lays several clutches in different places, some producing males and others females. All her hatchlings would be just as closely related to each other as if they had been laid the same nest.

Incubation temperature not only affects a turtle embryo's sex, but how it grows. As far as we know, the sex that ends up larger at maturity also grows faster. Michael Ewert and Craig Nelson, among others, have noticed that in Pattern Ia turtles, females are usually larger than males. They speculate that there may be some advantage in having turtles that are going to become larger as adults start out as eggs that are incubated at higher temperatures. This may even hold true for Pattern II species. Common snapping turtle males are larger than females, and tend to grow faster at the intermediate temperatures that produce males than at temperatures either above the higher threshold or below the lower one—though experiments by Shyril O'Steen suggest that other factors, such as the water temperatures juveniles prefer after hatching, also affect how fast juvenile snappers grow.

Ewart and Nelson's explanation makes sense for a species like the diamondback terrapin (*Malaclemys terrapin*) in which the female is the larger sex, is incubated at higher temperature, and grows more rapidly. It does not, however, seem to fit the desert tortoise (*Gopherus agassizii*). The larger and more rapidly growing tortoise gender is the male, which is incubated at a lower temperature.

Another possible explanation for TSD has to do with sex ratios. Under GSD, males and females ought to be produced in roughly equal numbers. However, TSD turtles tend to produce, overall, more females than males—possibly a better outcome, as each male can fertilize several females. In the helmeted terrapin, a Pattern II turtle, the span of temperatures that can produce males covers only 3.1°C (5.6°F).

It may be, though, that the sex ratios we

are seeing today are the result, not of a long period of evolution, but of the fact that, as a result of human forest-clearing activities, there are fewer cool, shady, male-producing nest sites available than there used to be. The sex ratio of yellow-spotted river turtles in the Guapore River in Rondonia, Brazil, is strongly skewed the other way, at six males to one female. Perhaps the nests that produce females are laid out in the open, where they may be easier for predators, including humans, to find; or it may be because humans are catching the adult females at their nesting sites.

Whatever the evolutionary advantages of TSD in the past, global climate change could turn it into a serious liability. Turtles could have responded to changing temperature regimes by nesting earlier or later in the season, evolving a different threshold temperature, or shifting their geographic range. Today, turtles—and other reptiles with TSD—may not have time to adapt. Fredric Janzen suggested, in 1994, that a rise of about 4°C (7.2°F) in the mean July temperature could "effectively eliminate the production of male offspring" in the population of painted turtles he studied in central Illinois. This is a rise that may well happen as the 21st century progresses.

A common map turtle (*Graptemys geographica*) emerges from its shell.

### Hatching

The amount of time it takes for turtle eggs to incubate varies tremendously, both among different species and within a single species. Some of this variation is simply a matter of temperature; all things being equal, eggs at cooler temperatures take longer to develop. In a laboratory experiment, eggs of the narrow-bridged musk turtle (*Claudius angustatus*) incubated at a range of temperatures took anywhere from 95 to 229 days to hatch. The shortest incubation time known for any turtle, 28 days, was recorded for the Chinese softshell, though 40 to 80 days is more normal for that species.

By the time a baby turtle is ready to hatch, its eggshell has been greatly weakened. The eggshell is a major source of calcium for the embryo. By hatching time much of its calcium has been extracted, and what remains is not a serious barrier to escape. Late in incubation, the brittle outer layer of the furrowed wood turtle's eggshell (*Rhinoclemmys areolata*) expands and cracks. When the hatchling finally emerges, its carapace is less than 40 mm (1.6 in) wide. Within a day or two, though, the carapace expands to almost 50 mm (2 in)—considerably wider than the egg from which it emerged.

Escape, though, can still be a slow process. Archie Carr described the hatching of a com-

This hatchling northern long-necked turtle (*Chelodina rugosa*) uses the egg-tooth on its snout to cut its way out of its shell.

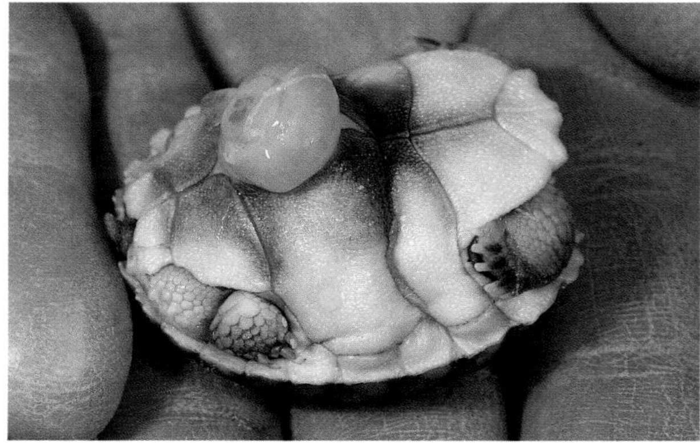

This hatchling desert tortoise (*Gopherus agassizii*) has yet to absorb all of the food reserves in its yolk sac.

mon musk turtle that took almost three days to break out of its shell. A hatchling turtle slices its way out with a sharp little bump of keratin on the tip of its snout, the *egg-tooth* or egg-carbuncle, that falls off within a few weeks. Sometimes it makes a number of holes and cracks that join together until a piece of the eggshell falls away.

For turtles that hatch in an underground nest, their first task after emerging from the shell is getting to the surface. Hatchling turtles are good diggers. Yellow mud turtles and ornate box turtles (*Terrapene ornata*) are quite capable of digging their way through the bottom of the nest to escape the advancing frost line, to depths of 60 cm (24 in) or more (see Chapter 5).

If the ground above them is frozen, or baked hard by the sun, digging out may, nonetheless, be impossible; but under such conditions, emergence is something that it may be well to delay. Hatchling spur-thighed tortoises (*Testudo graeca*) in the Caucasus normally take only about 80 days to develop, and may appear on the surface as early as late July. However, some hatchlings, presumably those resulting from late nestings, may winter underground and emerge in late May the following year. There is no point in hatchlings digging out into a world that is too cold, too hot, or too dry to sustain them. It is better to wait until conditions are right.

Extra yolk, which is often not fully absorbed on hatching, gives buried young the food reserve they need to wait it out. Turtles whose hatchlings remain in the nest have a proportionately greater supply of lipids in their egg yolks than those whose young emerge at once or, like speckled padlopers, are on the surface already. Newly hatched padlopers have already absorbed their yolk sac and begin to feed almost immediately. Hatchling gopher tortoises, though, still have enough yolk, even after they hatch, to sustain them for some time. They are not interested in food, but instead soon turn their attention to the activity that will occupy them through much of their adult life—burrowing. Often, they dig their first burrow within a week of emerging from the nest.

When hatchlings do emerge, it may take the efforts of the entire brood to breach the plug of soil over their heads and escape into the open air. It has been suggested that the rather large brood of common snapping turtles, some 30 to a clutch, may be needed to break out of their nest before winter sets in. However, hatchling turtles do not always emerge from the nest together. Though all hatchlings emerged on the same day in 70 percent of 59 Blanding's turtle nests monitored by Justin Congdon and his colleagues on the E. S. George Reserve in Michigan, individual turtles from other nests took up to four days to make their escape. Joseph Butler and Todd Hull reported that hatchlings from the same gopher tortoise nests in Florida emerged as much as 16 days apart.

Hatchling helmeted terrapins and leopard tortoises do not emerge until the ground above their heads has been softened by rain. In the Mediterranean climate near the Cape of Good Hope, it is the first winter rains that trigger the emergence of young geometric tortoises (*Psammobates geometricus*). Rain not only acts as a "hatchling releaser" for the Indian softshell; it may help the hatchlings make their first journey. If the nest is a long way from the water's edge, hatchling softshells hide under bushes or in cracks in the soil. It takes the next heavy rain to sweep them into the river.

Red-eared sliders (*Trachemys scripta elegans*), which overwinter in the nest, are apparently cued to emerge by changes in temperature. As the sun warms the surface, the soil over their heads becomes warmer than the soil at the bottom of the nest, and emergence may simply be the result of the hatchlings trying to find the warmest spot possible.

Once the hatchlings emerge from their nest, they have to find their way to suitable habitat (assuming they are not there already). For freshwater turtles, this can involve a substantial, and dangerous, journey. Some of the hatchling Blanding's turtles studied by Brian Butler and Terry Graham in Massachusetts made the trip—an average straight line distance of just under 35 m (115.5 ft)—in less than 12 hours, but others took up to nine days. On the way, they kept under cover when they could, burrowing under sphagnum moss and rotted logs, spending the night in the burrows of short-tailed shrews (*Blarina brevicauda*) or among the roots of royal ferns (*Osmunda regalis*). Both Blanding's and wood turtle hatchlings will follow the trails of others, presumably picking up their scent (see Chapter 6).

Hatchling Mary River turtles (*Elusor macrurus*) dig their way out of their nest.

Though hatchling Blanding's turtles normally spend their first winter under water, a few may overwinter on land if the circumstances are right. Their earliest instincts may not be to head for water at all. Though Massachusetts hatchlings traveled more or less directly to water, newly emerged hatchlings studied by Natalie McMaster and Thomas Herman in Nova Scotia generally did not. Some avoided it altogether, even if the investigators released them right next to the water's edge. They seemed to prefer to seek cover instead. Newly hatched diamondback terrapins, perhaps also seeking cover, may head for the nearest clump of vegetation, regardless of whether it is uphill, downhill, or in the same direction as the water they will eventually reach.

## Staying Alive

Turtles stand only a modest chance of hatching at all, and an even smaller one of surviving to the end of their first year. During a seven-year study of snapping turtles in Michigan, nest losses averaged 70 percent (with a range of 30–100 percent). Environmental stresses account for some losses; winter kill may be substantial for northern painted turtle hatchlings overwintering in the nest, despite their ability to withstand freezing (see Chapter 5).

Predators dispose of a high proportion of turtle nests—often well over 90 percent. All 20 unprotected nests of striped mud turtles monitored in Florida by Dawn Wilson and her colleagues were destroyed by predators, including raccoons (*Procyon lotor*) and snakes. Nests protected by the researchers did much better, with 23 out of 26 nests producing at least one young.

Of the unprotected Blanding's turtle nests studied by Butler and Graham in 1990, 33 of 35 were robbed by predators. Predators, primarily raccoons, destroyed 78.2 percent of 182 Blanding's turtle nests monitored over the 23 years of the George Reserve study in Michigan. In nine out of the ten years between 1985 and 1994, they destroyed every one. Raccoons patrol the nesting grounds, probably watching for female turtles at work. Most nests were raided the night they were dug. Sometimes raccoons stole eggs from the nest chamber while the female was still laying.

The nest of this Australian broad-shelled turtle (*Chelodina expansa*) has been raided, perhaps by an introduced fox.

Mongooses raid many nests of the angulate tortoise in South Africa. In Australia, introduced foxes find and destroy almost 90 percent of snake-necked, broad-shelled, and Macquarie turtle (*Emydura macquarii*) nests along the lower Murray River. Jackals and monitor lizards are the chief predators on the nests of the Indian softshell. Water monitors (*Varanus salvator*), wild pigs, feral dogs, crabs, and ants attack the nests of painted terrapins in Malaysia. Tegu lizards (*Tupinambis* spp.) in the Pacaya-Samfria National Reserve, Peru, were the most important predators on the nests of South American river turtles (*Podocnemis* spp.). The crested caracara (*Caracara cheriway*), a carrion-eating relative of falcons, and the tegu lizard *Tupinambis nigropunctatus* are the chief raiders (other than human beings) of the nests of yellow-spotted river turtles on the Capanaparo River in Venezuela. Ants can be a serious problem for nesting turtles; introduced fire ants attack the nests of chicken turtles and gopher tortoises in the southeastern United States.

The bright colors on this hatchling Florida red-bellied turtle (*Pseudemys nelsoni*) may be a warning to predatory fish.

Once they emerge, hatchlings run a gauntlet of predators from alligators to crabs and from bullfrogs to eagles (see Chapter 6). In South Africa, fiscal shrikes (*Lanius collaris*) may skewer hatchling angulate tortoises on thorns. Gopher tortoises stand a 94 percent chance of being killed before they are a year old, and in their early years over 50 percent of young gopher tortoises and Bolson tortoises (*Gopherus flavomarginatus*) die each year.

Hatchling turtles are food for so many animals that you might not expect them to draw attention to themselves. The hatchlings of a surprising number of freshwater turtles, though, from several different families, are nonetheless brilliantly colored. Many bear startling patterns of red, yellow, or orange, particularly on their undersides. Usually, when otherwise defenseless (or tasty) creatures are decked out in this way, it is as a warning to predators that their prospective prey is toxic, venomous, or in some other way an item to avoid. No hatchling turtles, though, carry dangerous poisons. What, if anything, are the turtles warning of, and what kind of predator is the target of the warning?

Carol Britson and William Gutzke suggested a possible answer in 1993. One of North America's most voracious aquatic predators, the largemouth bass (*Micropterus salmoides*), avoids hatchling turtles. Britson and Gutzke tried a number of experiments to see why. They found that bass had no problem eating hatchling meat, or whole hatchlings of painted turtles, as long as turtles were dead or anesthetized. Taste, then, was not the problem. Nonetheless, if a bass attacked a live hatchling, it usually spat it out after three to five seconds. It appears that the hatchlings put up such a

Some Australian turtle hatchlings, like this *Chelodina* sp, may also warn fish away with their bright plastral colors.

struggle, clawing and biting at the inside of the fish's mouth, that they were not worth eating. The bright colors may warn fish that have had a bad experience with a hatchling turtle to pass on the next opportunity.

In Australia, C. Dorrian and H. Ehmann have tested hatchlings of the snake-necked turtle on predatory fishes like the Murray cod (*Maccullochella peeli*) and long-finned eel (*Anguilla reinhardtii*). One hatchling faced down an eel by flipping on its back, exposing the brilliant orange-red blotches on its plastron to its pursuer. After being denied food for nine days, an eel did try to swallow a hatchling but spat it out at once. Dorrian and Ehmann were only able to persuade a Murray cod to eat a dead hatchling after they blackened out its patches of color. The problem, for the fishes, may not be the hatchling's claws but its stink; snake-necked turtles, both young and old, can emit a particularly pungent liquid containing six different acids, and few predators will attack them.

The idea that hatchling colors are there to warn off fishes gets some support from the behavior of the turtles themselves. Hatchlings with bright colors—at least in North America—tend to be active swimmers; those without, like the young of common snapping turtles, usually burrow in the mud. Keeping the colorful areas largely on the underside may alert fishes, which are likely to attack from below, without signaling the hatchling's presence to other predators, like birds, that strike from above. Sea turtle hatchlings are active swimmers that lack bright colors, but they also lack the equipment to rake the inside of a fish's mouth. They are devoured by a host of fishes.

The juvenile razorback musk turtle (*Sternotherus carinatus*) lacks the bright colors of juvenile emydid turtles.

Hatchlings of the big-headed turtle (*Platysternon megacephalum*) may have another way to use their colorful undersides. Hatchlings, but not adults, can make a squealing noise. The combination of the squeal with a flash of bright color, suddenly revealed, might just startle a predator enough to give the hatchling a chance to escape.

The best defense young turtles have, though, is simply to reach adult size. By the time gopher and Bolson tortoises reach adulthood, their annual mortality has fallen from 50 percent to about 2 percent. Other adult turtles may not do as well—annual survivorship for common snapping turtles, painted turtles, and sliders in eastern North America runs at about 81 to 85 percent—but nonetheless, if a hatchling can escape the gauntlet it runs in its early years, its chance of living to a ripe old age is vastly improved.

Since turtles suffer their greatest mortality as small animals, it usually makes adaptive sense for them to grow as quickly as possible. Most turtles grow rapidly in their first years of life. Rapid growth may be essential to weather not only predation but also environmental stresses like drought. If western swamp turtles do not reach a minimum size in their first year, they are unlikely to survive their first period of aestivation.

For large tortoises, reaching adult size can involve an enormous change of magnitude: by the end of its journey from egg to adulthood, an adult male spurred tortoise (*Geochelone sulcata*) with a carapace length of 83 cm (33 in) and a weight of 105.5 kg (233 lb) has increased its length by 18 times, and its weight by 3822 times.

CHAPTER EIGHT

# *The Endless Journey*

THE CYCLE OF LIFE that sends freshwater turtles out onto the land to nest becomes, in most sea turtles, a grand sweeping arc carrying them across thousands of kilometers of open ocean. It may return them, decades later, to the general area, if not the precise spot, of their birth. It is a cycle that, for some sea turtles, follows the circular paths of the great current systems, or gyres, that swirl about whole ocean basins.

The life cycle of a sea turtle is much the same for all seven living species. It begins when a hatchling, aided by its nest mates, struggles out of the sand and scrambles down the beach to the sea. If it escapes the predators that wait for it on the beach, in the surf, or in the sky overhead, the hatchling will leave the coast behind and become a denizen of the upper layers of the open sea (except for the flatback [*Natator depressus*], which apparently spends its whole life on the Australian continental shelf). Here it floats, perhaps carried vast distances by the currents, feeding on tiny creatures that share the surface waters, and growing, probably quite rapidly.

Once it reaches a certain size, its habits, and its diet, change, though the leatherback (*Dermochelys coriacea*) remains, by and large, a creature of the high seas. Perhaps thousands of kilometers away from the waters where the hatchling first entered the sea, it takes up residence on a small patch of the continental shelf, hunting or grazing near the bottom. Here it may live for years, even, as we are now beginning to learn, for decades, until it reaches sexual maturity; or, perhaps, it settles in a series of such habitats one after the other, as it grows and develops; or even spends some of those years wandering the ocean or roaming the coasts. We do not know for sure.

When a sea turtle matures, it enters a more predictable cycle of migration, to and from

A hatchling leatherback (*Dermochelys coriacea*) scrambles to the sea to begin its lifelong journey.

A mature green sea turtle (*Chelonia mydas*) spends much of its life migrating to and from its nesting grounds.

the nesting grounds, alternating reproduction with a return to the feeding grounds to build up their resources for another breeding season. That journey is not always a long one—green sea turtles (*Chelonia mydas*) nesting in the Red Sea may spend their whole lives within its waters—but for others, it extends for thousands of kilometers.

Female sea turtles do not make their nesting journey every year, except for the two ridley species. Kemp's ridley (*Lepidochelys kempii*), and two-thirds of the females in a Surinam population of olive ridleys (*Lepidochelys olivacea*), nest annually. For the others, it may take from two to five years, or even more, before a female is ready to commit to another round of migration and nesting. This period, from nesting year to nesting year, is called the *remigration interval.*

In the years she does nest, a female sea turtle commonly lays multiple clutches, sometimes six or more per season depending on the species. Between nestings, she returns to the sea while her next clutch ripens in her ovaries. This *internesting interval* takes anywhere from 9 to 30 days. Only after she lays her final clutch will she return to her feeding grounds, perhaps by a different route from the one she followed to her nesting beaches.

We once believed that male sea turtles remained on the feeding grounds, or only traveled occasionally to the breeding areas. We thought that females stored their sperm for years, as some land turtles do (see Chapter 7). We now know that males may migrate, too, and may hover in the seas off the nesting beaches, fertilizing the eggs a female carries roughly a month before she digs her first nest. There may be a good reason for males to follow their mates across the ocean. Over the huge range of most sea turtles, nesting seasons vary with climate and latitude. The same species of sea turtle may nest in December in one area, and in June in another.

The lifelong journeys of sea turtles are governed and regulated by the physical parameters of their ocean environment, the richness and availability of their food supply, which may in turn be driven by changing patterns of climate and ocean temperature, and the cycling of hormones and apportionment of resources within their own bodies. Their ability to traverse the seas depends on their internal mechanisms for finding their way out of their nest, down to the sea, across it and, eventually, back again.

In some ways, we know far more about sea turtles than we do about their land-bound relatives; in others, we know far less. Nesting sea turtles have been minutely studied, prodded, and examined. So have their eggs and their new-hatched babies. But when sea turtles return to the sea, they vanish, and the mystery of what they do with most of their lives is only beginning, over the last few decades, to yield to science.

**Down to the Sea**

A sea turtle begins its life buried about 60 cm (24 in) deep in its nesting beach. Though sea turtles have been hatched for years in incubators, we know little of what goes on when they hatch in the wild. Some observations have been made with artificial nests, mostly on greens and loggerheads (*Caretta caretta*), and these give us an idea of what must take place under the sand.

A fully developed hatchling lies curled in its egg around the remnants of its yolk sac. It uses its egg-tooth to cut its way through its parchment-like shell, but before it finally escapes it may rest for anywhere from one to three days while its plastron straightens and the remainder of its yolk is absorbed into its

A dead green sea turtle hatchling from Ascension Island in the Atlantic, still curled around the remnants of its yolk sac.

A hatchling green sea turtle (*Chelonia mydas*) may take up to three days to emerge from its shell.

body. The liquid left in the eggshells drains into the sand, and the shells themselves collapse, leaving room for the hatchlings to struggle free.

Normally, the eggs in a clutch hatch within a few hours of each other, and emergence is a project for the whole group (though sometimes, instead of emerging together, hatchlings may leave the nest in small groups over 24 to 74 hours). Over several hours, the brood alternates rest periods with bouts of seemingly coordinated digging, each hatchling stimulated into activity by the struggles of those around it. The hatchlings at the top of the writhing mass of newborn turtles scrape away at the sand over their heads. As the sand falls into the nest cavity, their siblings at the bottom of the heap trample it into the floor beneath them. The hatchlings are not really coordinating their efforts, but these individual struggles slowly carry the whole brood upward.

The process of hatching and escape can take from three to seven days—long enough for a hatchling to use up its reserves of yolk. Loggerheads studied by Elaine Christens in a turtle hatchery in Georgia averaged 5.4 days from the first pip of the shell to final emergence. Most of this time was spent resting; the hatchlings did not start to climb in earnest for the surface until the last day or two. By the time the hatchlings do break through to the surface, they will have lost weight. They will gain it again by drinking seawater once they enter the ocean (see Chapter 5). Olive ridleys gain 0.2 g (0.07 oz) per day once they begin drinking, according to recent research by Susan Clusella Trullas and Frank Paladino.

If the temperature of the nest, and particularly of the sand over their heads, becomes too warm, the hatchlings will stop—a response that makes it more likely that they will break through to the surface at night, when the sands are relatively cool. Emerging by day can be very dangerous; the sand can be extremely hot, and many a (staged) natural history film has shown hatchling sea turtles scrambling down to the sea by day, only to be snapped up by hovering flocks of predatory frigatebirds (*Fregata* spp).

Once they are exposed to the dangers of the surface, it is vital for the hatchlings to get to the sea as soon as possible. They use a number of cues to tell them where the sea is.

A cluster of hatchling green sea turtles emerge from their nest at dawn on Sipadan Island, Borneo, Malaysia.

The most important seems to be light. The sky is usually brightest over the sea. Cover a hatchling's eyes, and it cannot find the sea even if there is other information available, such as a downward slope of the sand towards the water's edge. The hatchlings respond to light cues in a "cone of acceptance" that takes in a horizontal sweep of roughly 180°, but covers a vertical range of only about 30° above the horizon or, depending on the species, even less. Because hatchlings only react to lights that are close to the horizon, they are less likely to be confused by moonlight. They seem less attracted to yellow light than to other colors—loggerheads show a aversion to yellow light—and this preference may keep them from becoming disoriented by the rising sun.

It is usually safest to have more than one internal compass, and hatchlings seem to be guided by more than light alone. They will steer away from the high, dark silhouettes of sand dunes and vegetation. This may merely be because they block out the light, but sensitivity to shapes around them does apparently help hatchlings find the sea on a bright moonlit night, when the strongest light might not be coming from the horizon above it.

All these reinforcing cues, however, are not enough to guide hatchlings away from the artificial lights that now burn on many a beach environment. Artificial lighting is often strong enough to completely overcome the signals a hatchling sea turtle is programmed to recognize. Artificial light, if it is bright

A green sea turtle hatchling plunges into the ocean in the Turtle Islands National Park, Malaysia.

enough, becomes a stimulus so powerful that the hatchlings respond to nothing else, crawling towards it from hundreds of meters away. The result is often fatal. Lured away from the sea, the stranded hatchlings fall to predators, desiccate in the hot sun the following day, or simply starve. In highly developed areas like the Florida coast, artificial light may be the greatest threat hatchlings face.

If all goes well, and the hatchlings scramble over the sand in the right direction, avoid their enemies, and reach the surf, a new set of orienting mechanisms takes over. As soon as they are afloat, the hatchlings begin to swim, at something over 1.5 km/hr (0.9 mph); Osamu Abe and his co-workers measured hatchling green sea turtles that swam away from their nesting beaches in southwestern Japan at 1.62 km/hr (1 mph). They dive into the path of the wave undertow, where the receding waters sweep them outward, away from the beach. When they surface again, they head for open sea.

This time, they are guided not by sight but, apparently exclusively, by the direction of the incoming waves. Experiments with loggerheads, greens, and leatherbacks have shown that hatchlings swim towards approaching waves, but if the sea is calm they swim randomly or in circles. Hatchlings will swim into the waves even if, under experimental conditions, doing so sends them back to the beach again.

The farther a hatchling gets from shore, the less reliable wave direction becomes as a pointer to the open sea. Nicolas Pilcher and his colleagues have shown that hatchling green sea turtles released from a hatchery in Sabah, Borneo (East Malaysia), are able to navigate around small islands, and keep swimming offshore, even when there are few waves to guide them. They may be relying on yet another internal compass, this time oriented to the earth's magnetic field.

Recent experiments by Jeanette Schnars and her colleagues suggest that leatherback and olive ridley hatchlings "switch on" their geomagnetic compasses almost as soon as they are out of the nest. Though the hatchlings get a geomagnetic "fix" as soon as they leave the nest, and appear to be able to use it as a reference point, they will not follow it blindly if other cues, such as light and sound, are available.

Their geomagnetic compass will only be of use if hatchlings can determine their heading in the first place. A simple directional compass—one that always sent the turtles westward, for instance—would be useless if the open sea lay in some other direction. Therefore, a magnetic compass does not so much tell a hatchling turtle which way to go as keep it on course once it has determined its heading from some other cue.

Hatchling loggerheads and leatherbacks tested in an experiment, once they had

A hatchling olive ridley (*Lepidochelys olivacea*) swims in the seas off the Pacific coast of Costa Rica.

established a heading towards a distant light, successfully used the magnetic field of the earth to keep them on course even in total darkness. Another set of experiments, using a wave tank instead of a distant light, showed that once the hatchlings had set themselves on course by heading into the waves, their magnetic compass was able to keep them on track after the wave action was turned off. If the hatchlings did not have the chance to see the light first, or sense the wave direction, their magnetic compass was useless to them.

Once in the open sea, as they enter the gyres that will carry them to the habitats where they will develop into subadults, their geomagnetic compass is possibly their most important navigational guide. It provides them with two kinds of information: one, called the *magnetic inclination angle*, that tells them whether they are heading towards the pole or the equator; and another, the *magnetic field strength*, that tells them whether they are heading, roughly, eastward or westward. By combining these two indicators—how, we don't exactly know—a hatchling is able to determine its position within the gyre, and stay in its waters as it is swept across the ocean.

**Lost and (Partially) Found**

The hatchlings struggling seaward from their natal beaches are about to enter a period Archie Carr termed the "lost year." This mysterious interval stretches from the moment they disappear into the surf until the day they return, as much larger juveniles, to coastal waters. Today we know that the "lost year" is often far longer than a year—it may last one to three years for Atlantic hawksbills (*Eretmochelys imbricata*), or a decade or more for loggerheads. The hatchlings, thanks to a number of recent discoveries, are no longer quite so "lost" to turtle biologists.

Sea turtle biologists call the instinctive rush that hatchlings make, first for the surf and then for the open sea, the frenzy. Like most frenzies, it doesn't last long—about 24 hours in loggerheads, greens, and leatherbacks leaving the nesting beaches of eastern Florida. According to Jeannette Wyneken and M. Salmon, by their second night on the water hatchlings have begun to enter a more relaxed phase, the *postfrenzy*. They still swim seaward by day, but rest during at least part of the night. Loggerheads swim mostly by day. Greens spend about 10 percent of each night swimming as well, while leatherbacks swim as much as 30 percent of the night.

The frenzy is not a rush to get to their pelagic feeding grounds, a journey that, after all, may take months. It is an adaption to avoid predators. The longer a hatchling turtle stays in inshore waters, the greater its chance of ending up inside a predator's stomach. Predators took 40 to 60 percent of green sea turtle

A hatchling Atlantic loggerhead (*Caretta caretta*) conceals itself in a clump of floating *Sargassum* weed.

hatchlings leaving hatcheries in Sabah, Borneo, during their first two hours at sea. This may be partly because the hatchery caretakers released the hatchlings en masse, usually at the same time and place each day—thus creating "feeding stations" for the local fish. Farther out at sea, though, the hatchlings are likely safer.

Not every hatchling faces an equal risk of predation on its first swim; Emma Gyuris found that larger green sea turtle hatchlings had a better chance of surviving their passage over the outer reef surrounding Heron Island, Australia, than smaller ones. Just as on land (see Chapter 7), size provides a measure of protection. This selective advantage is probably why the flatback, which stays in predator-rich inshore waters, produces hatchlings twice as large as those of other sea turtles.

Florida hatchlings do not have to swim far offshore before they encounter the brown, knotted mats of the alga *Sargassum*. Hatchling leatherbacks, already searching for the gelatinous creatures that are their only food, avoid any kind of floating debris. Greens hover around its edge, hunting free-floating prey in open water. For loggerheads, though, floating debris—especially *Sargassum*—is home. Their brown and tan colors match the *Sargassum* itself, concealing the young turtles among the fronds. When researchers led by R. G. Mellgren added clumps of *Sargassum* to tanks holding captive loggerhead hatchlings still in their frenzy, the turtles climbed into, or on top of, the floating weed, and stopped swimming.

In the waters around the Azores, 4000 km (2500 mi) eastward from Florida, rich upwellings of plankton from the Mid-Atlantic Ridge support a floating ecosystem. Hosts of jellyfish and other marine organisms float here among clumps of *Sargassum*. Here, too, are young loggerhead turtles, ranging from tiny animals not much bigger than hatchlings

A green sea turtle (*Chelonia mydas*) feeds on a moon jellyfish (*Aurelia aurita*) in the seas off Hawaii.

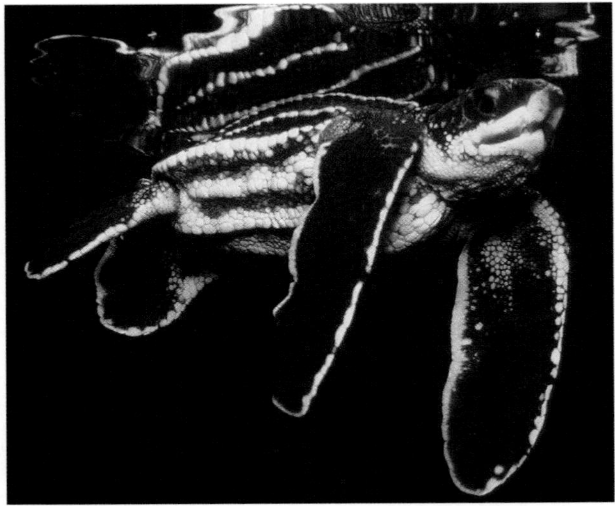

It may be years before hatchling leatherbacks, like this one in the Gulf Stream off Florida, are seen again.

to medium-size juveniles weighing some 23–27 kg (50–60 lb). Thanks to tagging and DNA-fingerprinting studies by Alan Bolten and his colleagues, we now know that these are turtles that hatched on the southeastern coasts of the United States and in the Caribbean, about 80 percent of them on the beaches of Florida. They are carried to the Azores by the clockwise current of the North Atlantic gyre. After 10 or 12 years, they follow its southward loop back to the Caribbean.

Though we are far from having all the details, it appears that an eastward passage around the circuit of a gyre is typical for hatchling loggerheads, whether they hatch in Florida, eastern Australia, Japan, or on the east coast of South Africa. George Hughes has tracked South African loggerheads into the Indian Ocean gyre. Mitochondrial DNA markers and, as we shall see, satellite tracking have proven that loggerheads from Japan and, probably, Australia aggregate off the coast of Baja California, where biologists have recorded a concentration of some 10,000 juvenile turtles. Wherever they are, they spend their early years feeding on floating animals in the upper layers of the sea, concentrating on the stinging jellyfish *Pelagia noctiluca* in the Atlantic and on the delicate violet snails (*Janthina* spp.) and a colonial polyp, the by-the-wind sailor (*Velella velella*), in the Pacific.

We still have very little idea of what other sea turtles do during their "lost year." The one thing we can say for certain is that, if they survive, they grow, but we are not even precisely sure how that happens. Milani Chaloupka and George Zug have suggested that the growth of at least one species, Kemp's ridley, happens in two spurts. The first peaks at about 12 months of age; the second at around seven years.

The fate of hatchling leatherbacks in particular is, according to John Musick and Colin Limpus, "one of the great mysteries of sea turtle biology." Once they leave their nesting beaches, young leatherbacks simply disappear. It probably takes about four years for growing juveniles to reach over 1 m (3.2 ft) in carapace length, the size of the smallest post-hatchlings

Its pelagic years over, a loggerhead (*Caretta caretta*) swims past an artificial reef near Key West, Florida.

researchers see. Where they go during that time we have no idea. All we can say is that by the time they are adults, leatherbacks have increased their length about thirtyfold and weigh some 6000 times more than they did when they first emerged from the egg.

## Miners and Gardeners

At the end of its "lost year," only the leatherback remains a pelagic animal; the projecting, toothlike cusps on its upper jaw may help it hold on when it bites into the huge, gelatinous bells of open-ocean jellyfish like the lion's mane (*Cyanea capillata*). The flatback never leaves the continental shelf in the first place. The other sea turtles leave the open ocean for feeding grounds on the continental shelf, often far from the beaches where they were born. Mitochondrial DNA evidence collected by Anna Bass and Wayne Witzell shows that green sea turtles hatched in Costa Rica, Mexico, Venezuela, and Surinam may assemble on feeding grounds on the east coast of Florida, while juveniles in the nearby Bahamas may even include a few individuals hatched as far away as Ascension Island or Guinea Bissau.

By the time they arrive in their new homes, young sea turtles are big enough to ignore most of the predators that threatened them on their first journey through continental waters. Once a sea turtle reaches more than about 20–30 cm (8–12 in) in carapace length, it has little to fear from anything but a large shark or a killer whale (*Orcinus orca*); it is simply too big and too hard-shelled for most other predators.

Leatherback turtles have actually been seen harassing sharks. One full-grown leatherback that successfully chased off a large gray reef shark (*Carcharinus* sp.) then turned its attention to the observers' boat, ramming it half a dozen times with the leading edge of its carapace before swimming off, only to return to the attack after two members of the crew decided to swim after it.

Immature sea turtles may resort to places that adults rarely visit. Kemp's ridleys, which nest almost exclusively in Mexico, may wander throughout the Gulf of Mexico as adults; juveniles and subadults range farther, north along the Atlantic coast as far as maritime Canada or, in small numbers, to northwestern Europe. Long Island Sound appear to be an important feeding ground for Kemp's ridleys at roughly three to six years of age. Of the 5000 to 10,000 loggerheads, and perhaps over a thousand Kemp's ridleys, that spend the summer in Chesapeake Bay off the eastern coast of the United States, some 95 percent of the loggerheads and probably all of the ridleys are juveniles. From September onward, the turtles move south to waters off Georgia and the Carolinas, returning again the following May and June.

In the Pacific, green sea turtles may be less

A loggerhead (*Caretta caretta*) dines on a spiny lobster (*Paniluris argus*) in the Bahamas.

likely to wander. Turtles from Hawaii and the Galápagos tend to stay around their island reefs, though tagged Galápagos females have been recovered from Costa Rica to Peru. In the southern Great Barrier Reef, some juvenile turtles have been under observation in the same localized feeding areas for over 20 years.

Once they reach about 30–50 cm (12–20 in) or so in carapace length (in the western Atlantic), or 70 cm (28 in) or more (in Australia), loggerhead sea turtles move to the continental shelf and shift their foraging attentions from the surface to the bottom. In the Gulf of Mexico, these immature loggerheads dine on animals like sea pen (*Virgularia presbytes*) that live on, or beneath, the sea floor. Off the coast of Virginia, crabs are the most common loggerhead prey items. The diet of loggerheads feeding on the continental shelf may shift with the seasons; off the coast of Texas, the turtles switch from sea pens to crabs in the summer and fall, when crab numbers go up.

Loggerheads in Moreton Bay, Queensland, trench-mine the bottom sediments for prey, particularly clams and polychaete worms. According to Anthony Preen, the turtles "dug body pits in the sediments, and used their foreflippers to erode the sand wall in front of their faces. The turtles fed from this eroding wall, picking at individual organisms as they were exposed." The turtles' mining activities leave hundreds of trenches 40 cm (16 in), deep scarring the floor of the bay.

Adult green sea turtles are, usually, vegetarians, though the populations in the eastern Pacific are somewhat more carnivorous. The plants they feed on may vary from place to

A green sea turtle (*Chelonia mydas*) grazes on a sea grass bed off Sipadan Island, Sabah, Borneo, Malaysia.

place. Their preferred foods, when they can get them, are various sorts of seagrass—Zosteraceae, one of the few flowering plant families to grow on the sea floor. In Hawaii and on some parts of the Great Barrier Reef, where there are no seagrass beds, the turtles graze on algae growing on the rocks, often within a few feet of shore. The gut microflora seagrass-eaters rely on to digest their meals cannot break down algae, and vice versa. Most turtles carry only one type or the other. In some areas of Australia, though, green sea turtles eat both seagrass and algae, apparently digesting both with equal ease.

In northern Australia, green sea turtles eat the leaves and shoots of mangrove trees, even reaching out of the water to snip off a tasty shoot. Seagrass, though, remains their food of choice. The turtles can be picky about precisely which seagrass shoots they will graze, and may travel long distances to find the plants they prefer. Green sea turtles along the Queensland coast tend to prefer the seagrasses *Halodule uninervis* and *Halophila ovata* over a third species, *Zostera capricorni. Zostera* has lower nitrogen levels and higher fiber content than the others, and is therefore less nutritious. The only other large seagrass grazer in the area, the dugong (*Dugong dugon*), a mammal related to manatees (and elephants), has the same preference.

Young shoots of seagrass are more nutritious than older plants. Green sea turtles in the Caribbean take advantage of this by continuously regrazing the same beds. This regular pruning ensures a continuous crop of nutritious new shoots; in effect, the turtles are

A hawksbill (*Eretmochelys imbricata*) digs through coral rubble off Sipadan Island, Borneo, Malaysia, while reef fish wait for scraps.

gardeners, tending their plots to produce the highest-quality yield.

Stomach contents from different parts of its range give wildly different pictures of what an olive ridley eats. A stomach from Papua New Guinea was almost entirely filled with snail shells. Stomachs from Sri Lanka were often full of algae. One from Surinam contained, among other things, two fresh catfish, while another taken off Japan was full of what appeared to be the liquefied remains of either jellyfish or the floating, gelatinous, colonial sea squirts known as salps.

Adult hawksbills spend their lives around coral reefs, where they live mostly on a diet of sponges.

The only large-scale study of the olive ridley's diet examined the stomachs of 24 males and 115 females taken off Oaxaca on the Pacific coast of Mexico. The Mexican ridleys seemed to be dining largely on fish and salps, but although these items accounted for the bulk of the material, they all came from a small proportion of the turtle stomachs: 14 percent for salps, and 5 percent for fish. The rest of the stomachs may have been largely empty. These turtles were on their nesting grounds, and many breeding female sea turtles eat little or nothing during the nesting period.

Hawksbill sea turtles resort to coral reefs, where they become not just carnivores but, in some cases, extreme specialists. Over most of their range, hawksbills live primarily on sponges. In the Caribbean, they eat little else.

Hawksbills are particular about the sponges they eat. They avoid sponges (like the once-ubiquitous bath sponge) with a skeleton of tough, organic fibers or *spongin*. They have no problem, though, eating sponges whose bodies are studded with sharp *spicules*—tiny, often ornate spikes of calcium carbonate or silica—and, in many species, laced with a variety of toxic compounds. The turtles store the toxins in their own bodies. Hawksbill meat is occasionally poisonous. The medical literature has documented hundreds of cases of human mortality from eating hawksbills, especially in the Indo-Pacific.

Juvenile hawksbills floating in the open sea are apparently herbivores, reportedly living mostly on algae like *Sargassum*. Before they become adults and sponge specialists, juvenile

While a surgeonfish cleans a Hawaiian green sea turtle, the white-spotted toby nearby is more likely to nip its skin.

A Hawaiian green sea turtle being cleaned by surgeonfish (*Ctenochaetus strigosus* and *Zebrasoma flavescens*).

hawksbills may pass through an intermediate, omnivorous stage, foraging on the sea bottom for a variety of invertebrates and algae.

### Fellow Travellers

A sea turtle, like many a large fish, may provide shelter for pilot fish and other small fry, or a hitching post for a remora (*Remora remora*) or its close cousin, the sharksucker (*Echeneis naucrates*). Generations of turtle hunters from East Africa to the West Indies have taken advantage of the remora's predilection for clamping itself, by its suckerlike dorsal fin, to the shell of a sea

A juvenile loggerhead sea turtle (*Caretta caretta*), from the Azores in eastern Atlantic, carries a load of gooseneck barnacles.

turtle. In 1494, on his second voyage, Christopher Columbus watched the Taino Indians of Cuba fish for turtles by tying a rope to a remora's tail, lowering it into the water near a turtle, waiting for the fish to attach itself, and slowly reeling in their catch.

On coral reefs, turtles may take advantage of the attentions of the local cleanup crew—fishes that clean others of their parasites. Julie Booth, who made some of the first regular underwater observations of green sea turtles on the Great Barrier Reef in Australia, watched a large female being cleaned by two species of fish:

> About a dozen [sergeant-majors] *Abudefduf sexfasciatus*... were plucking algal growths from her head and plastron, while several moon wrasses, *Thalassoma lunare*... were picking at small barnacles attached to the skin of her neck. The female would arch her neck first to the right, then to the left, apparently to stretch and tighten the skin and provide access to the barnacles for the fish.

Besides these camp followers and hangers-on, some sea turtles carry their own miniature ecosystems around with them. At least 100 different species of animals and plants have been found attached to loggerhead turtles, though the greatest variety seems to be on sick animals. These range from the algae and barnacles that you might expect (and that also attach to other turtles like the diamondback terrapin) to colonies of star coral, snails, sea anemones, and crabs.

Organisms that live on others are called *epibionts*. All sea turtles have them, but loggerheads, for some reason, have the largest and most diverse epibiont communities; a Kemp's ridley may have no more than a few barnacles. Some loggerhead epibionts grow on other epibionts; all the known examples of star coral (*Astrangia danae*) on loggerheads were actually growing on a variety of barnacles. Others need a protected place to attach themselves to avoid being swept off as the turtle swims; one loggerhead sported some 200 sea anemones of the species *Diadumene leucolena*, all growing in a series of deep scars left from a collision with a boat propeller.

Supporting this little community can be, quite literally, a drag. A load of epibionts on a turtle's shell may cut down considerably on its streamlining, making swimming much harder work. This drain on energy may be particularly serious for younger turtles, which will have to work harder to compensate for the added drag on their shells. At least some turtles, though, may carry around their own cleanup crew.

Loggerheads, green turtles, and hawksbills in the central Atlantic have all been found carrying specimens of the Columbus crab (*Planes minutus*), usually clinging to the tail, hindlimbs, or around the cloaca. It appears that the Columbus crabs feed on the other epibionts, both plant and animal, and may be quite important in keeping the turtles relatively clean. The turtle, in return, provides the crab with its most important breeding ground. Pairs of adult crabs, and females bearing eggs, are much commoner on turtles near Madeira than on the floating mats of *Sargassum* weed surrounding them. Conversely, juvenile crabs are commoner in the weed, which may mean that the *Sargassum* weed, by providing a nursery for the crabs, is essential to the health of young sea turtles. In 2000, the U.S. National Marine Fisheries Service decided to allow commercial harvest of *Sargassum*. Are the turtles about to be robbed of their cleaners?

A green sea turtle (*Chelonia mydas agassizii*) from the Galápagos feeds on green algae (*Caulerpa* sp.).

This Kemp's ridley (*Lepidochelys kempii*), digging its nest at Rancho Nuevo, Mexico, carries a flipper tag.

## Turtles in Space!

An enormous span of years may elapse between the time a female sea turtle hatches to the day when she drags herself ponderously out of the ocean to dig her first nest. How long that span may be varies from species to species, from population to population, and from individual to individual. Not all turtles mature at the same age, or at the same size. Estimates for time to maturity for loggerhead sea turtles off the coast of Georgia, based on studies of bone growth (a technique called *skeletochronology*) by James Parham and George Zug, range from 20 to 63 years.

Green sea turtles, whose diet of plants may not support rapid growth, have possibly the longest time to maturity of any vertebrate. Colin Limpus found that greens on the coast of Queensland may take from 30 to the astonishing age of 70 years to reach their first breeding season. Derek Green has suggested that maturity takes even longer in the Galápagos; by one calculation, reaching the size of the smallest nesting females seen in the islands could take 92 years! An alternate calculation suggests that Galápagos greens may "only" need 63 years to reach this size, but the figure, if correct, is still remarkable.

Once a sea turtle reaches maturity—however long that takes—it begins a migratory cycle to and from its breeding grounds that will last for the rest of its life. The simplest way researchers have to tell where a sea turtle travels on its migratory journeys involves attaching a metal or plastic tag to its flipper. Tagging is cheap and at least partly effective, though information is obviously lost when tags fall off at sea. If the turtle is found again, the information on the tag can tell us where, and how long ago, it was last seen. In 2000, newspapers in Hawaii carried the story of green turtle 5690, tagged as a juvenile by George Balazs near Hilo on the Big Island in 1981 and next seen nesting on Maui 19 years later.

Sea turtle biologists have used flipper tags for many years, and still use them widely today. For leatherbacks in particular, they now are being partly replaced by the more sophisticated PIT (Passive Integrated Transponder) tags, glass microchips about the size of a grain of rice injected into a turtle's shoulder muscle and read with a handheld scanner.

Tagging was not enough to satisfy pioneer sea turtle biologist and conservationist Archie Carr. In the 1960s, during his investigations of the turtles of Tortuguero National Park in Costa Rica, he fitted a number of green sea turtles with radio transmitters, carried above the waves by bright yellow helium balloons on 15–18 m (50–60 ft) cords. As the turtles put out to sea, Carr and his co-workers were able to follow them with binoculars and directional antennas.

Balloon tagging, though, would hardly do to track turtles over thousands of kilometers of ocean. In his book *The Sea Turtle: So Excellent a Fishe*, Carr told how he had dreamed of tracking his turtles with orbiting satellites, but dismissed the notion—at first—as an impractical fantasy:

> The ideal way to maintain contact with the island-seeking turtles cruising in the open sea would be by satellite. This thought occurred to me a long time ago, when Telstar was put into orbit; but I dismissed it as grandiose daydreaming, bound to lead only to discontent. Then, lo, a letter came asking me to submit a plan to use tracking facilities of one of the experimental satellites of the Apollo program my brother, Tom, who is a physicist and radio astronomer, and David Ehrenfeld and I quickly put together a proposal to be submitted to NASA for tracking turtles by earth-orbiting satellites.

The first attempt to use satellites to track sea turtles (and one of the first for any animal) was made in the early 1980s. Satellite tracking did not come into widespread use until the 1990s—too late for Archie Carr, who died in 1987. In the mid-1980s, Richard Byles began the first consistent satellite telemetry program, on loggerheads and Kemp's ridleys in Chesapeake Bay. George Balazs has been satellite-tracking green turtles on the French Frigate Shoals in the Leeward Islands of Hawaii since 1992. In 1994, Barbara Schroeder, Llewelyn Ehrhart, and George Balazs attached satellite transmitters to three greens, named Honu, Fairly, and Keya, nesting in the Archie Carr National Wildlife Refuge in eastern Florida.

Honu, Fairly, and Keya did not travel far

A female loggerhead at the Archie Carr National Wildlife Refuge in Florida is released after being fitted with a radio transmitter.

in the few months their transmitters operated. To the scientist's surprise, their feeding grounds seemed to be in Florida water. The tracking results helped to explain why no long-distance tag recoveries had yet been reported for green seaturtles tagged on Florida beaches. Satellite tracking has since revealed a great deal about turtle migrations. Over a 287-day period from August 1994 to May 1995, for example, Maurice Renaud and his colleagues tracked an adult female Kemp's ridley from her feeding grounds in Louisiana to the chief nesting colony of her species, at Rancho Nuevo in Tamaulipas, Mexico.

Satellite tracking, though, is expensive, and far from easy. Its cost—roughly US$2000 per transmitter, plus a further $2000 per turtle per year for satellite time—has limited the animals bearing transmitters to a select few. On the Turtle Islands in the Philippines, the Pawikan Conservation Project fitted 8071 turtles with flipper tags between 1984 and 1998, but has only been able to attach satellite transmitters to four. Jack Frazier and his co-workers fitted out the first two in October 1998.

The transmitters themselves, about the size of a portable tape player, are attached to the carapace with epoxy, or fiberglass cloth and resin (or, for leatherbacks, strapped on with harnesses or tethers). They must withstand the rigors of several months on a turtle's back. Their most fragile and exposed element, the antenna, is especially liable to be broken off as the turtle rubs itself against rocks, coral, or debris—or it may be knocked off by an amorous male attempting to mount.

The transmitter carries a microprocessor that continually gathers information (providing its batteries keep working). Depending on the sensors in the transmitter, its data can tell us not only how far a turtle swims, but how fast, and how often and how deeply it dives en route. The information is sent to satellites circling the earth in a polar orbit, gathering weather information for the U.S. National Oceanographic and Atmospheric Administration (NOAA). At least five carry instruments, manufactured by the French company ARGOS, capable of receiving the ultra-high-frequency (UHF) signals the transmitters send.

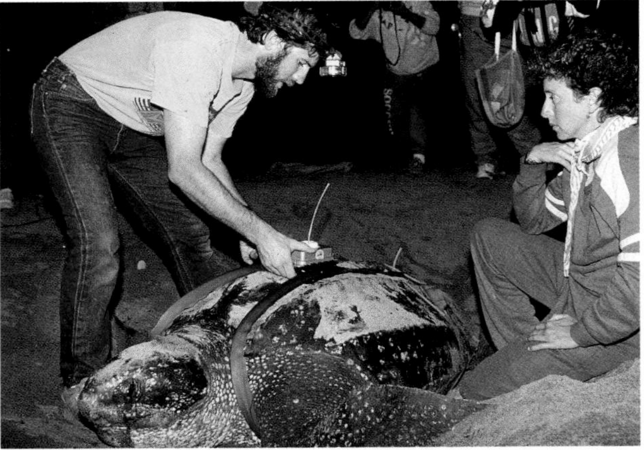

Scott Eckert and Laura Sarti attach a satellite transmitter to a leatherback on the Pacific coast of Mexico.

An ideal communication between turtle and satellite does not happen every day. The satellites must be "in view" above the horizon when the turtle is at the surface, the antenna is in the air, and the transmitter is sending a signal. Each satellite is likely to be in the proper position for only about two 12-minute passes per day, or less if the satellite passes low over the horizon. To get an accurate fix on the turtle's position, the satellite needs to receive at least four signals on each pass. It has to be in contact with the transmitter for 2–5 minutes overall. Despite these limitations, on a good track the signals come through often enough to tell us where the turtle is headed, and where it is spending its time.

Satellite data can tell us (as tagging data cannot) whether the turtles follow paths that allow them to take advantage of ocean currents or prevailing winds, or whether they stop to feed during long-distance migrations. The tale of a single turtle shows the power of satellite tracking. In 1982, flipper tag no. 6038 was attached to a male green sea turtle basking on the shore of French Frigate Shoals. The turtle returned in 1984, and in 1995 he was there again. The tag data told us only that he kept coming back to French Frigate; where he had been in the meantime no one could tell.

On June 7, 1995, George Balazs fitted him with a satellite transmitter. Twelve days later, 6038 set off for the open sea. For 30 days, he traveled southeast, swimming for 1200 km (746 mi), far out of sight of land, over water thousands of meters deep, till he reached the feeding and resting grounds of Kahului Bay on the northern coast of Maui. This time, biologists were able to follow him, by satellite and computer, all the way.

Satellite data also provide us with essential information for sea turtle conservation. By telling us exactly where the turtles go, the satellite data help scientists pinpoint potential danger zones. Richard Byles' Kemp's ridley study, for example, showed that the turtles do not migrate through the relatively safe open waters of the Gulf of Mexico, but along the coast in shallow water less than 46 m (150 ft) deep, where they may fall foul of fishing nets. Jeanne Mortimer's data on hawksbills in the Seychelles showed that adults breeding in the islands may spend their entire adult lives within the confines of the Seychelles Bank—providing a convincing argument to counter those Seychellois who objected to protecting turtles that they believed would only be slaughtered elsewhere (Chapter 10).

Tag data and satellite data can complement one another. In the North Pacific, loggerhead turtles nest in Japanese waters and feed in the ocean off Baja California in Mexico, thousands of kilometers away. Until 1996, when Wallace Nichols satellite-tagged a fully grown female Baja loggerhead named Adelita, nobody knew for certain if these were turtles from the same population. Nichols tracked Adelita for 11 months as she swam 10,000 km (6200 mi) across the Pacific to Japan. Nichols had no direct evidence, though, that turtles from Japan swam to Baja California (though it seemed obvious that they must do so, and genetic studies suggested as much).

The proof came from a flipper tag, recovered not from a turtle but from a Mexican fisherman who had kept it on his key chain for five years. The tag came from a turtle hatched on Yakushima Island, where a third of Japan's loggerheads nest. The hatchling had spent its first year at the Okinawa Aquarium, where it was tagged and released in 1988. Six years later, it fell into the hands of the fisherman on the other side of the Pacific.

In the Atlantic, Scott Eckert has begun satellite-tracking leatherbacks nesting in Trinidad. Two of his tracks show that the turtles are not as predictable as they might seem. Though they both started from the same point, one traveled north along the Gulf Stream to the North Atlantic, while the other headed almost due east, straight across the ocean, and spent the next few months wandering up and down the Atlantic coasts of Europe and North Africa.

Frank Paladino and his students have been satellite-tagging the leatherbacks that gather to feed on jellyfish in Monterey Bay, California. In 2000, Paladino tagged two females. Soon afterward, the turtles left the bay and set out across the Pacific, heading almost directly towards their probable nesting grounds in Papua New Guinea. They covered some 40 km (24 mi) a day. Because they made only shallow dives on the way, Paladino believes that they were not feeding, relying instead on the fat stored up from their gelatinous feast in Monterey Bay. We know that female greens eat little or nothing during the nesting period, and this may be true of leatherbacks as well.

Paladino's satellite tracks reinforce the conclusion, derived from mitochondrial DNA studies by Peter Dutton and his colleagues, that the leatherbacks off the west coast of North America come from the breeding population of the western Pacific, which nests in Malaysia, Indonesia, and Solomon Islands. The leatherbacks nesting in the eastern Pacific, on the shores of Mexico and Costa Rica, are genetically distinct, and range, instead, southward along the coast of South America.

We know that leatherbacks usually only come ashore to nest every two to three years. What are the chances that both of Paladino's turtles would set out for their nesting grounds soon after they were tagged? Was this simply a coincidence, or is Monterey Bay a staging ground where turtles pile on a final feast before crossing the Pacific to nest? Do they travel somewhere else to break their fast, only returning to Monterey Bay years later? With data from only two turtles, we don't know—yet.

### Homeward Bound

Now that we have some idea, based on tagging, genetic analysis, and satellite tracking, where sea turtles go on their breeding migrations and when, we can consider the more difficult questions of how and why. What sends sea turtles to one breeding area and not another? How do they find their way? And given that most sea turtles do not nest every year, what distinguishes a promising nesting year from a bad one?

No matter how much sea turtles from different rookeries may mix together on their feeding grounds, they go their separate ways, to their own natal beaches, when it comes time to nest. No leatherback tagged in South Africa, for example, has been found nesting anywhere else in the world. This attachment to the land of one's birth is termed *philopatry.*

Though we have known for some time that female turtles exhibit philopatry, Nancy Fitzsimmons and her colleagues have recently shown that, at least for Australian greens, the males, too, return to their native waters to mate. This pattern may not hold for all sea turtles. Male olive ridleys, in particular, may mate with any females they come across over a wide geographic area. But, for Australian greens at least, male philopatry may ensure, as we suggested early in this chapter, that their own breeding cycles coincide with those of the females. Even within Australia, some green sea turtle populations breed in summer and others in winter. Arriving among the right batch of females at the right time of year may be critical.

In the South Atlantic, one of the most important nesting areas for the green sea turtle is the remote volcanic island of Ascension (the site of the annual sea turtle mating bash in the comic strip "Sherman's Lagoon"). The turtles that nest on Ascension travel there from feeding grounds off the coast of Brazil, 2300 km (1430 mi) or more away to the west. What drives Brazilian turtles to Ascension, and how do they find this tiny spot, only 248 sq km (96 sq mi) in area?

Archie Carr and others suggested some years ago that sea turtles navigate across the open ocean by taking a heading from the direction of the rising sun. Once they get close to their nesting areas, the turtles would "home" in on chemical cues carried over the water from their final destination, much as migrating salmon find the rivers of their birth by the chemical "taste" of the water. The existence of these supposed chemical "signatures", though, has yet to be demonstrated.

C. W. Brown proposed, in 1990, that for Ascension at least this complex mechanism may not be necessary. Ascension lies in the middle of the South Atlantic Equatorial Countercurrent, a river of seawater that flows eastward through the surface layers of the ocean during the Austral summer. During the Austral winter, the surface waters between Ascension and Brazil are driven westward, this time by the southeast trades.

Turtles nest on Ascension from December through June, with a peak between late February and April. For adults headed for their nesting grounds in the Austral summer, the waters of the current provide a corridor through the ocean leading them to Ascension. Their eastward flow eases the turtles' swim, reducing the energetic cost of migration. In 51 to 57 days, turtles leaving the eastern bulge of Brazil could be at their island destination.

Since their eggs take roughly 60 days to incubate, most hatchlings emerge in the Austral autumn, just as the southeast trades intensify. The westward-flowing waters probably sweep the bulk of them to Brazil. Their parents, heading back to their feeding grounds at roughly the same time, of course swim on their own, but could certainly take

The breeding cycle of green sea turtles (*Chelonia mydas*) in the southwestern Pacific is affected by El Niño events.

advantage of the boost the trades provide.

Brown's mechanism requires no chemical cue. In fact, as he points out, an eastward-flowing current would sweep any chemicals from Ascension to the east, towards Africa, rather than westward, towards Brazil. The current provides both a direction for the turtles to swim, and a good reason—energy efficiency—for the turtles to follow it. This does not explain exactly how they find the tiny dot that is Ascension. To answer that, we may have to look at other navigational compasses turtles possess, including their geomagnetic sense.

While ocean currents may drive Atlantic greens to Ascension, ocean temperatures may be responsible for the breeding patterns of Indo-Pacific greens on the Great Barrier Reef of Australia. Colin Limpus and his colleagues have been studying nesting turtles there for over two decades. In 1974, Limpus arrived for his first season of turtle tagging on Heron Island, in the Capricorn Group near the southern end of the reef. He was prepared to tag no more than 200 to 600 turtles, the numbers an earlier researcher, Robert Bustard, had suggested were typical for a Heron Island nesting season. But, as he reports in a paper co-authored with Neville Nicholls:

> To our surprise, we tagged ~1200 nesting females in the 1974–1975 breeding season and the local residents spoke of the greatest number of nesting turtles in their memory. We returned the following summer equipped to tag thousands of nesting green turtles only to be met by a total of 21 females breeding on Heron Island for the entire summer. Understandably we were puzzled. Some locals suggested that "Limpus has scared the turtles away."

Heron Island did not see another peak breeding year for a decade, and then not again until 1996 and 1999. The clue tying these years together proved to be the fluctuation in ocean temperatures known as the El Niño Southern Oscillation (ENSO). Each peak season followed two years after a major El Niño event, and each crash, when breeding numbers plummeted, followed two years after its opposite, La Niña. Following the 1998 El Niño event in the Western Indian Ocean, Jeanne Mortimer reported similar cyclic fluctuations in the nesting density of green sea turtles.

What is going on here, and why the two-year delay? Individual female greens in eastern Australia breed only once every five or six years. It takes over a year for a green sea turtle to build up the resources she needs to travel to the breeding grounds and generate the yolk for her enormous store of eggs. Therefore the conditions for a successful nesting season—in particular, a good food supply—must be set well before a female

A pair of mating green sea turtles off Sipadan Island, Borneo, Malaysia, are followed by another male eager to get in on the action.

After a successful courtship, a pair of loggerhead turtles (*Caretta caretta*) mate in the waters off the Bahamas.

commits to migration and breeding.

The key, then, appears to be not how ENSO affects the turtles, but how it may affect their food supply. A bumper crop of turtle food plants supplies the resources that can go towards a successful breeding season in a later year. A bad year on the seagrass beds, or the algal mats, may deny the turtles the food they need to prepare for nesting two years later. So far, Limpus has not been able to tell exactly how ENSO affects turtle food supplies, but he has shown that failures of seagrass crops from other causes, like storms, floods, and algal blooms, have led directly to poor nesting years for the turtle populations of Moreton Bay and other sites on the Australian coast.

That it has taken over two decades for Limpus to establish the patterns we have seen so far, though, shows how enormous is the task of understanding the complex factors that affect sea turtles over their long lifetimes. If Limpus and Nicholls are right, and the relationship between the green sea turtle and

A male green sea turtle (*Chelonia mydas*) clings to his female as they mate near Sipadan Island, Borneo, Malaysia.

A pair of mating olive ridleys (*Lepidochelys olivacea*) off the Pacific coast of Costa Rica.

ENSO is tied to its herbivorous diet, the environmental factors that determine when other sea turtle species are ready to nest may be quite different. We do not even know, of course, whether ENSO has any effect on other greens outside the Western and Central Pacific and the Western Indian Ocean. As with so much else about sea turtle biology, we still have a lot to learn.

## Courtship at Sea

Sea turtle courtship and mating seems every bit as chaotic and brutal as it is for their cousins on land or in fresh water (see Chapter 7). Males bite the females on the head, neck, and flippers, and scratch the females' shells with their claws. A female, even in the act of mating, may be accompanied by several "attendant" males. Other males trying to horn in on the mating will often bite the male in position, sometimes causing serious injuries to his flippers or tail. In fact, one of the best indications we have that female sea turtles only mate before they lay their first clutch is that, over the rest of the nesting period, the open sores their mates inflict on them progressively heal.

Michael Frick and his co-workers were able to make one of the few sets of observations of sea turtle courtship while conducting aerial surveys of loggerheads off the coast of Georgia and northeastern Florida. By combining their observations on a number of turtle pairs, they were able to put together a sequence of courtship maneuvers. A male makes his first, slow approach to the female as she floats on the surface with her flippers and hindlimbs tucked against her body. He circles for two to three minutes, while she pivots about, facing him. The male then dives under her, and begins sniffing and nuzzling at the scent glands near her bridge (this may help him make sure that the turtle he is courting is actually a female). After a minute or so, his nuzzling gives way to vigorous biting at her hindlimbs. Sometimes, that is all that is necessary before she allows him to mount. If it is not, he resorts to gentler means, facing her and stroking her head and neck with the upper surfaces of his forelimbs. Three to five minutes of this may do the trick, but one male, even after all this effort, "was chased away by wild splashing and snapping from the female."

Julie Booth observed similar postures in

A mass of green sea turtle eggs, collected on Sangalakki Island off the coast of Kalimantan, Borneo, Indonesia.

courting green sea turtles, but argued that some of the positions taken up by the female are not so much part of a courtship ritual as attempts to keep males at bay. Tucking in her hindlimbs may prevent males from getting access to her cloacal opening. Female greens regularly take up this position just after egg-laying. A really unreceptive female adopts a "refusal" position, holding her body vertically in the water with her limbs spread wide. A male confronted with this posture usually gives up and swims off.

Mature males of cheloniid sea turtles develop, according to Thane Wibbels, David Owens, and David Russell, a soft, leathery spot in the center of their plastra that may provide them with improved traction as they cling to their females, or even provide them with an extra bit of tactile sensation as they mate. Jeffrey Miller describes the details of the actual mating process in *The Biology of Sea Turtles*:

> The male mounts the female quickly, usually at the surface with a lot of splashing, and hooks onto her carapace using the enlarged claws on his front flippers and the large claws on his hind flippers to hold himself in place. The male curls his long tail to bring their cloacae into contact. His penis is erected into her cloaca. The shape of the penis with the bifurcation of the sperm duct at the tip allows for the transfer of semen into each oviduct without passing through the environment of the female's cloaca. In captive situations the pair may remain coupled for 10 or more hours...

There is evidence that the longer the male remains in place, the better chance he has to fertilize the female. Annette Broderick and Brendan Godley recorded two male green sea turtles in Cyprus that were so persistent that each remained clinging to his female even after she dragged herself out of the water and crawled some 5–10 m (15–30 ft) up the beach.

During their receptive period, females may mate with several males, and males may mate with several females. The sperm the females collect will be stored in their oviducts, available to fertilize each clutch as it develops through the season. In both loggerheads and greens, scientists have found clutches whose eggs were fertilized by sperm from more than one male. Males may leave after the mating period, or hover off the nesting beaches throughout the season, making repeated (but usually unsuccessful) efforts to mount females after they have begun laying. Male olive ridleys off Orissa, India, concentrate in aggregations at sea; during mating, these aggregation areas may hold 26 mating pairs per square kilometer (or 68 per square mile).

### The Mating Cycle

The cycle of mating and nesting in sea turtles, as in other vertebrates, is regulated by the ebb and flow of hormones. It begins with a surge of testosterone, which may stimulate the onset of mating and nesting behavior. This happens in both males and females, but the male peak comes first, and is already on the decline before female levels reach their highest point. In the females, a rise in the hormone estradiol kicks the ovary into high gear. The follicles in the ovary begin to mature into yolk-filled eggs, a process called *vitellogenesis.* As the eggs mature, the calcium that will form the superstructure of their shells peaks in the female's blood plasma. Once the eggs are complete and ripe, another hormone, vasotocin, stimulates the female to deposit them in the nest.

Twenty-four to 48 hours after she returns to the sea, soon after the beginning of her internesting interval, a surge in progesterone and luteinizing hormone (LH) resets her ovarian clock for the next nesting cycle. How many times this happens, and how long the internesting interval lasts, depends on the species. It is usually anywhere from 10 to 18 days. Seawater temperature, which affects the rate at which the developing eggs mature in the female's body, may determine whether an individual turtle's internesting interval is at the longer or the shorter end of its species' range. The two ridleys have the longest and most irregular internesting intervals, 20 to 28 days for Kemp's ridley and 17 to 30 days for the olive ridley.

Kemp's ridleys nest no more than three times a season, each nesting supplied from ovaries successively depleted of their follicles until none are left by the end of the third nesting. Leatherbacks, on the other hand, may nest up to 10 times, ovulating every 9 or 10 days throughout the season—the shortest internesting interval for a sea turtle. They are able to lay so many clutches because at each nesting they deplete only 10 percent of the follicles in the ovary, compared to 50 percent for Kemp's ridley.

Multiple clutches may not be as universal among sea turtles as we once thought. William David Webster and Kelly Cook found that, of 127 nesting female loggerheads they tagged on Bald Head Island, North Carolina, over half—55.9 percent—laid only a single clutch per season. Possibly these were younger females; other, perhaps more experienced animals, nested anywhere from two to six times.

All sea turtles except the leatherback and the flatback commonly lay well over 100 eggs per clutch. The leatherback still averages over 80 m the Atlantic (fewer in the Eastern Pacific), though at 10 clutches per season that still amounts to an annual reproductive output, in a breeding year, of over 800 eggs, each one weighing 90 g (3.2 oz)—the largest of any sea turtle. Flatback clutches average just over 50 eggs each. Their eggs are almost as large as those of the much larger leatherback. This is probably related to the unusually large size of flatback hatchlings, which in its turn, as we have already suggested, may be an adaptive response to a high level of hatchling predation.

The huge clutches of eggs that sea turtles lay probably reflect the very small chances that any one of their offspring will survive. Archie Carr suggested an additional reason: since sea turtle nests are deep and hatchlings are small, below a certain clutch size there may simply not be enough turtles in a brood to make the group effort necessary to dig their way out of the egg chamber. Whatever their significance, large clutches require that each individual egg represent a proportionately small investment of energy by its parents.

A green sea turtle on Floreana Island in the Galápagos has abandoned its nesting attempt, leaving a "false crawl."

## A Place in the Sand

With her eggs fertilized, fully developed, and ready to lay, a female is finally ready to return to the land. She now faces a critical choice: where and when to emerge, dig her nest, and deposit her eggs. Daniel Wood and Karen Bjorndal have argued that when a female sea turtle selects a nest site, she is making an adaptive trade-off between the cost of her search (in terms of both the energy she may use up and the risk that she may be attacked by a predator while she is on the beach and relatively helpless) and the benefits of picking a site that is suitable for the successful incubation of her clutch.

The risk of predation on both adults and hatchlings probably drove the evolution of mass nesting in ridleys. On Playa Nancite, Costa Rica, American crocodiles (*Crocodylus acutus*) sometimes attack isolated olive ridleys on the beach; possibly the *arribada*, with its masses of turtles, may provide safety in numbers.

Though philopatry may guide a turtle back to the general area of its birth, it has little or nothing to do with where, in that area, she chooses to lay. Sea turtles often show *nest site fidelity*—a tendency to dig their successive nests, over the course of the season, in the same general area where they deposit their first clutch. Amy Chaves and her coworkers reported that leatherbacks nesting on Playa Langosta generally dug each nest within 100–300 m (330–990 ft) of their previous one.

It is not at all unusual, though, for a female to spread the nests she lays during a single breeding year across several beaches. Some females are very conservative, always return-

The tracks of a loggerhead sea turtle (*Caretta caretta*) over the beach end at the buried pit of its nest.

Female olive ridleys (*Lepidochelys olivacea*) taking part in an *arribada* at Ostional, Costa Rica, crawl ashore at sunset.

ing to the same short stretch of beach. Others in the same population may wander more widely. Over the long term, the conservative females maintain the existing rookery, while the more adventuous ones found new colonies—essential of the original beach is degraded or destroyed. For an individual female, spreading out her nest may well avoid the risk that a single, localized catastrophe could wipe out an entire season's output. Loggerheads in North Carolina, Japan, and South Africa all tend to space their successive nestings an average of roughly 4 km (2.5 mi) apart. In the Guianas, home of the largest leatherback nesting colonies left in the world, females may shift back and forth between nesting beaches in Surinam and French Guiana. In 2000, 18 percent of the

A green sea turtle (*Chelonia mydas*) deposits its eggs in its nest in the Turtle Islands National Park, Malaysia.

leatherbacks that dug 11,925 nests on three Surinam beaches had been tagged in the neighboring country.

A female sea turtle preparing to nest has a number of choices to make. Some she must make while still at sea: what beach she should select, and where on that beach she should emerge from the water. Once she hauls herself out on land, she faces a third decision: how far up the beach she should crawl before selecting the right spot to begin her excavations—or even if she should nest at all, or abandon the attempt and return to the sea, leaving behind only a curving track called, rather misleadingly, a "false crawl."

Exactly how she makes the first two decisions we do not know. Obviously a female cannot tell, while she is still in the water, whether any beach she picks will be ideal for digging—unless she has nested there before, or is emerging en masse, like the two ridleys, with a host of other turtles. She can decide, however, whether the beach is easy to reach. This is an important consideration for an animal so ungainly on land. Jeanne Mortimer found that green sea turtles prefer unlit beaches with wide sandy offshore approaches free of clutter.

For some populations at least, the slope of the beach seems to be a primary consideration. Wood and Bjorndal showed this for loggerheads nesting on the Archie Carr National Wildlife Refuge in Florida, as did E. Balasingam for leatherbacks in Malaysia. It is vital that the eggs in the eventual nest be above the high-tide line or the seawater table to avoid being flooded with salt water. The beach platform must be high enough in the first place, and the steeper the slope, the shorter the distance a female will have to crawl to reach a safe level above the sea and the easier will be her return—clearly a consideration for a 270 kg (600 lb) leatherback. The farther from the water's edge, though, the less likely a nest is to be destroyed by flooding or sea storms, so a wide beach may be a better choice than a narrow one, especially for smaller turtles such as the olive ridley.

A female olive ridley (*Lepidochelys olivacea*) prepares to cover her clutch of eggs after nesting at Ostional, Costa Rica.

A green sea turtle (*Chelonia mydas*) on Ascension Island, in the Atlantic, covers its nest with sweeps of its forelimbs.

The nature of the beach sand itself may be important. It must be easy to excavate, loose enough to allow air to get to the developing eggs, and yet firm enough for the nest to hold

A black vulture (*Coragyps atratus*) feeds on a hatchling olive ridley (*Lepidochelys olivacea*) at Ostional, Costa Rica.

its shape as the female digs. These characteristics are hard, if not impossible, to judge before the female actually crawls from the water, though Ahjond Garmestani and his co-workers found that loggerhead turtles in Florida's 10,000 Islands preferred to emerge on beaches with fewer shells. A shell-covered beach is probably more difficult to dig in, and may well be uncomfortable for a sea turtle to cross.

Beach topography may also be a consideration; according to Michael Salmon and his colleagues, loggerheads nesting on the oceanfront off Boca Raton, Florida, preferred to dig in front of tall objects, perhaps because they blocked light coming from the city beyond. Flatbacks at Fog Bay, Australia, nest at the base of rocky dunes; the dunes may provide some necessary shade for the sand over the nest. Nest temperature affects both survival and sex. Sea turtles exhibit Pattern Ib TSD (see Chapter 7). Their pivotal temperature is high, reaching 30.5–31°C (86.9–87.8°F) in olive ridleys.

Nonetheless, despite many studies of nest site selection among sea turtles, scientists have yet to come up with a list of specific features that always signify, to a turtle, the best available nesting site. Some things sea turtle eggs need to develop properly—high humidity and low salinity in the nest chamber, for example—may be out of the female's power to control. One researcher, perhaps in despair, suggested that once females pass the high-tide line, they simply crawl a random distance inland, stop, and dig where they are.

## Full Circle

At last, having grown, traveled the high seas, matured, navigated her way back to her native waters, mated, and selected her nesting beach, the female is ready to make what may be (except for those few green sea turtles that bask on Hawaiian beaches) her first journey on land since the night she hatched—or, if she is older, only one of many such journeys over a long reproductive lifetime.

She usually prefers to emerge either at, or just after, high tide. Not only does this shorten the distance she has to travel on land, it lowers the risk that an incoming tide will overtake her before she finishes her labors. Most sea turtles choose, too, to nest at night; daytime temperatures on a hot beach can be lethal. Almost 97 percent of the green sea turtles nesting at Ras Baridi, Saudi Arabia, emerged from the sea between 1900 and 0200 hours. The exceptions are the flatback, which regularly nests during the day, the two ridleys, whose mass *arribadas* may (and, in Kemp's ridley, usually do) take place during daylight hours and hawksbills in the Western Indian Ocean and, to a lesser extent, in the

A ghost crab (*Ocypode quadrata*) scavenges a dead loggerhead (*Caretta caretta*) hatchling at Juno Beach, Florida.

Persian Gulf.

After she ascends the beach—moving her forelimbs alternately if she is a loggerhead, hawksbill, or ridley, or "humping" along with her limbs moving together if she is a green or leatherback—flatbacks do both—she selects her digging spot. First, she prepares a body pit, sweeping debris out of the way with her flippers and excavating down to a level where the sand is firm enough to hold its shape as she digs the egg chamber. Like other turtles, she digs her nest with her hindlimbs, using them alternatively, like cupped hands, to scoop clumps of sand from the deepening pit and place them to one side and then the other. The shape and depth of her finished labor depends on the length and extent of her hindlimbs: shallow and rounded in hawksbills, deeper and more flask-shaped in greens. Her digging behavior is purely instinctive. A female with a missing hindlimb will still swing her stump in alternation with her functional flipper, as though it, too, were removing loads of sand.

The chamber finished, she releases her clutch of eggs into the hole, usually singly or in groups of two, three or four. As the female leatherback lays, the strange yolkless eggs—little sacks of albumin, surrounded by a shell, that are an unexplained feature of every leatherback clutch—emerge last from each of her two oviducts. Since the oviducts do not empty at the same time, though, they may not be the last eggs she actually lays. When her clutch has been laid, she scrapes the moist sand she has excavated back over her eggs, piling it in and compacting it in place with her hindlimbs. Then she crawls forward, throwing sand backward over the nest site with her front flippers, possibly to conceal the site, but more likely to add a layer of insulation on top of the nest. Only then does she return to the sea.

The eggs she leaves behind will incubate in their chamber for anywhere from 6 to 13 weeks, depending on the temperature, providing, of course, they are allowed to do so. Sea turtle eggs in an undisturbed nest have a naturally high success rate, typically 80 percent or more. Undisturbed nests, though, may be hard to come by. Their greatest enemy, on many sea turtle nesting beaches around the world, is our own species. In some places, poachers and collectors take every egg from every clutch (see Chapter 10). Even on protected beaches, though, nest failure may be very high.

Up to 60 percent of leatherback nests may be destroyed by tides, storms, and beach erosion alone, though this varies from place to place. The larvae of flies and beetles may infect sea turtle eggs, and ants may attack emerging hatchlings. On Sanibel Island, Florida, fire ants invading their nests are probably the worst enemy loggerhead hatchlings face. Feral dogs take large numbers of leatherback hatchlings in French Guiana, South Africa, and in Tortuguero National Park in Costa Rica. Half the green sea turtle nests at Tortuguero are destroyed by dogs, coatis (*Nasua narica*), black and turkey vultures (*Coragyps atratus* and *Cathartes aura*) and ghost crabs (*Ocypode quadratus*). In the Mediterranean, foxes are the chief threat to loggerheads nesting on beaches in Turkey.

At Ras Baridi in Saudi Arabia, Rueppell's fox (*Vulpes rueppellii*) digs up green sea turtle nests and hatchlings. According to Nicolas Pilcher and Mustafa al-Merghani:

> Foxes waited for the hatchlings to emerge, and then decapitated as many hatchlings as possible before they reached the sea. The foxes then returned to consume those they had killed. In this manner it was believed that they were able to consume more hatchlings than if they had consumed one at a time after emergence.

Gray foxes (*Urocyon cinereoargenteus*), feral pigs, and particularly raccoons (*Procyon lotor*) raid loggerhead nests on the Gulf Coast of Florida. In earlier days, black bears (*Ursus americanus*) visited the beaches around Naples, Florida, to feed on loggerhead eggs. Charles LeBuff, author of *The Loggerhead Turtle in the Eastern Gulf of Mexico*, was present when one of the last of them was shot, in the spring of 1958. Many other animals, including crows, herons, frigatebirds, and others that may be ill equipped to dig out eggs in their nests, will snap up hatchlings on their way to the sea.

Some eggs, though, do hatch, and some hatchlings do survive, and grow, and embark on the path sea turtles have followed for millions of years into the gyres of the sea. I have called this chapter, the last in this book to deal with turtle biology, "The Endless Journey." Of course, no journey, least of all that of a species, is truly endless; but whether the travels of sea turtles continue for millennia, or end in the next few decades, is up to us. In our last two chapters, we will turn from the biology of turtles to the most crucial of the issues that surround them: their conservation.

CHAPTER NINE

# *Peril on Land*

Seventeenth-century mariners provisioned their ships with giant tortoises, like these from Santa Cruz in the Galápagos.

IN THE YEAR 1684, four years after the last dodo is thought to have died on Mauritius, the explorer and sometime pirate William Dampier (1652–1715) visited the Galápagos Islands. His description of its tortoises is not only the first detailed account of them we have, but also gives us an idea of what was in store for them at our hands:

> The land-turtle are here so numerous, that five or six hundred men might subsist on them alone for several months, without any other sort of provision: They are extraordinarily large and fat, and so sweet, that no pullet eats more pleasantly. One of the largest of these creatures will weigh one hundred and fifty or two hundred weight, and some of them are two feet, or two feet six inches over the callapee or belly. I never saw any but at this place, that will weigh above thirty pounds weight. I have heard that at the island of St. Lawrence or Madagascar, and at the English forest, an island near it, called also Don Mascarenha or Bourbon, and now possessed by the French; there are very large ones, but whether so big, fat and sweet as these, I know not....

Freshwater turtles, mostly Indian flapshells (*Lissemys punctata*), on sale for food in Sylhet, Bangladesh.

The tortoises from the Mascarenes that Dampier refers to are now extinct; the Madagascar tortoises had already vanished by Dampier's time. The first Western settlers cut their island forests, introduced alien species and diseases, and killed them relentlessly for food. Mariners provisioned their ships with living tortoises to provide fresh meat for long journeys at sea. Sailors on six early voyages to the Mascarenes took almost 21,000 tortoises

Introduced red-eared sliders (*Trachemys scripta elegans*) are a potential threat to native turtles in many countries.

from the islands. We did not know, then, and probably would not have cared if we did know, that these were creatures of slow growth, long lifespan, and a limited ability to regenerate their numbers, ill equipped to handle the drastic short-term changes that we humans can inflict. They could not withstand us, or the pigs, rats, and other interlopers we brought with us, and they disappeared.

Today, we are doing to the whole planet what we once did to the islands of the giant tortoises. Our actions are taking a tremendous toll on turtles the world over. They are victims of almost the entire catalog of abuses we heap on our environment. We cut their forests, dam their rivers, drain their wetlands, and mine and develop their nesting beaches. By creating urban and suburban habitats for animals like raccoons and skunks in North America (not to mention domestic cats), we may stimulate huge increases in the populations of turtles' natural predators.

We compound the situation by introducing exotic alien species. Some exotics hunt turtles or destroy their nests, as foxes do in Australia (see Chapter 7). Some ruin their habitat. In the Alligator River region of northern Australia, introduced water buffalo (*Bubalus bubalis*) trample sandbanks that are nesting grounds for the pig-nose turtle (*Carettochelys insculpta*), eat the young waterside plants that turtles depend on for food during the dry season, and destroy the edges of billabongs where new food plants must grow. Other exotics compete for food or space. In the West Indies, introduced red-eared sliders (*Trachemys scripta elegans*) have hybridized with their native cousins, threatening the genetic integrity of the island species.

Pollution has been implicated in the spread of bacterial shell disease in freshwater turtles in the Rappahannock River, Virginia. Turtles accumulate dieldrin, PCBs, and other contaminants in their tissues, reducing their resistance to infections. Chemical pollution resulting from the Gulf War threatens the Euphrates softshell (*Rafetus euphraticus*), a threat compounded in Anatolia, Turkey, by, in the words of Ertan Taskavak and Mehmet K. Atatür, "the ongoing trend of dumping waste of every kind, domestic or otherwise, directly into the Euphrates and Tigris systems." Greenhouse gas pollution contributes to global climate change, which in turn could lead to destabilizing shifts in the sex ratios of turtles with temperature-dependent sex determination (see Chapter 7). Light pollution has been called the greatest threat to hatchling loggerhead sea turtles (*Caretta caretta*) on the Florida coast (see Chapter 8). It also misdirects hatchling painted terrapins (*Callagur borneoensis*) in Malaysia.

We hunt land and freshwater turtles, and their eggs, almost everywhere they occur. Those we do not eat directly we sell to seemingly insatiable food markets, or deliver into the wasteful and often illegal international traffic in reptilian pets. Our vehicles kill turtles on roads, sometimes in high enough numbers to threaten the survival of local populations. Our agricultural equipment plows up their nests, kills their hatchlings, and maims adults. Painted terrapins and other estuarine species collide with boats and drown in fishing gear.

Today, over 200 species and subspecies of turtles are included in IUCN's 2001 Red List of Threatened Species. The difference between William Dampier's contemporaries and ourselves, I hope, is that today we have a better understanding of the damage we do to wild species, and the potential harm that that damage can do to our own lives. But can we apply that understanding in time to save the world's turtles?

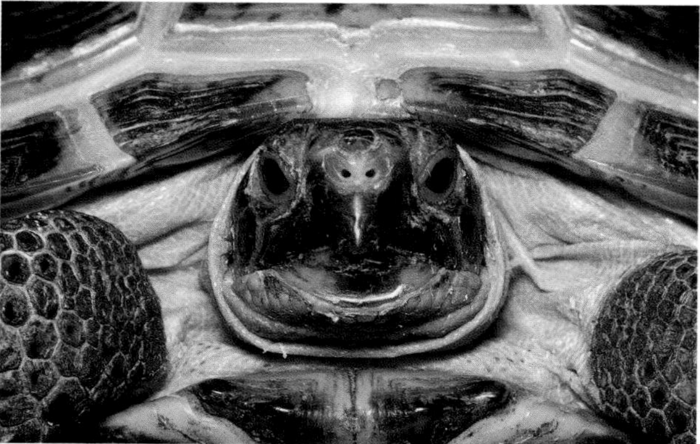

The Central Asian tortoise (*Testudo horsfieldii*) is losing habitat to irrigation projects in Turkmenistan.

### Habitat Loss, Pollution, and Disease

Turtle habitat is disappearing everywhere, whether through outright loss or through degradation so severe, as we convert it to our uses, that is no longer a suitable place for turtles to live. On the coastal plain of South Carolina, over 90 percent of bay wetlands, the favored habitat of the chicken turtle (*Deirochelys reticularia*), has been altered or eliminated. Southeastern longleaf pine forest, a favored habitat of the gopher tortoise (*Gopherus polyphemus*), has been reduced by 97 percent. River cooters (*Pseudemys concinna*) have become endangered in Illinois, according to Michael Dreslik, as swamps and oxbows have been drained, rivers channeled, and vegetation lost. Bog turtles (*Clemmys muhlenbergii*) disappear with their bogs, and wood turtles (*Clemmys insculpta*) with their woods.

The rising waters behind the Ataturk Dam in southeastern Anatolia have completely submerged the natural sandbanks where the Euphrates softshell once nested. Australian freshwater turtles face increasingly turbid rivers, banks devoid of waterside vegetation, the presence of introduced competitors such as European carp (*Cyprinus carpio*), and a list of other threats, current and potential, to the integrity of their habitat.

Chaco tortoises (*Geochelone chilensis*) have lost habitat to agriculture in Argentina, and Central Asian tortoises (*Testudo horsfieldii*) to irrigation in Turkmenistan. Drought and desertification, which have been linked both to climate change and to poor agricultural practices, may threaten the spurred tortoise (*Geochelone sulcata*) in countries like Mali along the southern edge of the Sahara. Agricultural practices, including the planting of large stands of single crops, plowing of virgin land, destruction of hedgerows, burning of stubble, spread of chemical fertilizers, and use of heavy agricultural machinery, have been implicated in the decline of spur-thighed tortoises (*Testudo graeca*) in the Balkans. The destruction of the native renosterveld in the Cape region of South Africa, for urban development and agriculture, is the chief threat facing the Endangered geometric tortoise (*Psammobates geometricus*) (see Chapter 4).

In southern Asia, loss of primary evergreen forest endangers Travancore tortoises (*Indotestudo travancoica*), Cochin forest cane turtles (*Geoemyda silvatica*), keeled box turtles (*Pyxidea mouhotii*), and spiny turtles (*Heosemys spinosa*). Sand mining threatens the nesting beaches of river terrapins (*Batagur baska*), painted terrapins, narrow-headed softshells (*Chitra* spp.), Asian giant softshells (*Pelochelys cantorii*), and a number of species of roofed turtle (*Kachuga* spp.).

In sections of the Perak river system, and in the Setiu River, home of the largest nesting population of painted terrapin in Malaysia, dams and watergates prevent both painted and river terrapins from migrating to productive areas upstream. On the Perak, unseasonal flooding of nesting banks, caused

The river terrapin (*Batagur baska*), a victim of habitat destruction and overhunting, is now Critically Endangered.

by the untimely opening of sluice gates and the deforestation of surrounding watersheds, has had "disastrous effects on the river terrapin population," according to Dionysius S. K. Sharma and his colleagues. Increasing contamination has polluted nesting beaches along the coast and degraded several important terrapin rivers, including the Perak and Setiu. In 1996 Sharma reported that a rubber processing factory on the Perak was releasing effluent directly into the river opposite painted terrapin nesting banks in the Bota Kanan Terrapin Sanctuary. Most of the terrapins had been driven away.

Wood turtles are highly vulnerable to habitat fragmentation as their forest habitat is cut into smaller and smaller pieces. The reasons why are not immediately obvious. Fragmentation creates more edge situations, and wood turtles actively seek out forest edges—so why do they not benefit as we chop up their territories? Raymond Saumure and J. Roger Bider suggested some reasons in a study of the effect of agricultural development on a population of wood turtles in southern Quebec.

In North America, the wood turtle (*Clemmys insculpta*) suffers from poaching and habitat fragmentation.

This sign north of Weaver, Minnesota, warns motorists to watch for Blanding's turtles (*Emydoidea blandingii*).

Turtles at forest edges risk injury or even mutilation beneath the wheels of our vehicles or the hooves of our cattle. Agricultural areas carry increased numbers of the wood turtles' most persistent predator, the raccoon (*Procyon lotor*), which hunts turtles along the woodland borders. Over 70 percent of the turtles sampled by Saumure and Bider were mutilated to some degree, either through accidents or encounters with predators; several had lost a limb or a tail. Though these sorts of injuries happened in forest areas, too, Saumure and Bider found that at their agricultural site where they encountered humans and their machinery, turtles had 2.7 times more mutilations than in their forest sites.

Turtles may suffer from multiple effects as we take over their habitats. Seekers for waterfront recreation on the eastern seaboard of the United States leave little safe room for the diamondback terrapin (*Malaclemys terrapin*). Increasing human use leads to bulkheading and other stabilizing techniques that make the shoreline better for people but eliminate terrapin habitat. Efforts to check local development in the name of terrapin conservation, however, can falter because, thanks to another effect of beachfront recreation, there may be no terrapins left to protect. More people

An epidemic of upper respiratory tract disease has infected populations of the desert tortoise (*Gopherus agassizii*).

along the coast has meant more use of inshore crab traps. Terrapins enter crab traps and drown; Willem Roosenburg found a single trap that had killed 49 terrapins. Fortunately, New Jersey and Maryland laws now require terrapin excluders—a sort of turtle escape hatch—on all recreational crab pots, and other states are considering introducing similar requirements.

Desert tortoises (*Gopherus agassizii*) in the American Southwest, and gopher tortoises in the Southeast, have become victims of a highly contagious upper respiratory tract disease (URTD). It is caused by *Mycoplasma agassizii*, one of a category of organisms that have been implicated in human diseases from pneumonia to Gulf War syndrome. The disease was first discovered in 1988, in the Desert Tortoise Natural Area (DTNA) of Kern County, California, when it was already well established. Its effect on the tortoises has been catastrophic. In a single 2.6 sq km (1.04 sq mi) plot in the DTNA, tortoise numbers fell from 204 counted in 1982 to only 13 ten years later.

The disease spreads easily among captive tortoises in the same pen, and the epidemic probably started in captivity. There may be more than 40,000 desert tortoises in captivity in Las Vegas Valley, Nevada, and over 200,000 in California. Owners of sick tortoises commonly release them into the desert, where they infect their wild cousins. Besides its direct effect on the animals, URTD hampers our efforts to conserve the desert tortoise. Relocating tortoises to areas where they have disappeared is an important part of managing species, but the stress of relocation may itself make the animals more susceptible to

An Australian Aboriginal hunter poses with his catch, a Northern long-necked turtle (*Chelodina rugosa*).

disease. Transferring animals in the early stages of URTD, before their symptoms become obvious, may inadvertently spread the disease to healthy wild populations.

**Turtles as Food**

Almost everywhere turtles occur, people eat them or gather their eggs. In the United States, diamondback terrapins, chicken turtles, common snapping turtles (*Chelydra serpentina*), and alligator snappers (*Macrochelys temmincki*) have been part of local cuisine for generations. Turtles can be easy to catch, easy to keep alive—at least, until they are ready for the pot—and are a ready source of protein. Their shells even provide built-in cooking vessels. Aboriginal women and children in northern Australia collect aestivating turtles by feeling for them in mud or shallow water, using their hands, their feet, or a stick. They dig pits where they store their catches, covered with layers of bark and earth, until the animals are ready to be eaten. The turtles, killed, gutted and washed, are roasted on hot coals in the shell, which will provide a serving plate when the cooking is done.

Not every turtle makes a worthwhile meal. Few people can get past the musky smell of helmeted terrapin (*Pelomedusa subrufa*). Though some inhabitants of the Petén region of Guatemala will cook white-lipped mud turtles (*Kinosternon leucostomum*) in their shells—while the turtle is still alive—recent immigrants to the area find the practice repugnant. Most other turtles, though, seem to be eaten by at least some people, and many are in high demand.

Today, however, in many parts of the world the combined effect of environmental stresses, the growth of human populations, the shift in settlement patterns, and the introduction of new technology has upset whatever balance existed between turtle hunters and their prey. The shift from traditional local use to commercial trade has increased the pressure on turtle populations as more and more people demand their meat and eggs. Tradition alone is no longer enough to protect turtles; without a scientific knowledge base and a management plan that is backed by government and supported by local peoples, animals that have provided food for generations may be eaten out of existence.

The remains of a Central American river turtle (*Dermatemys mawii*) have been unearthed from a Mayan burial ground in Uaxactún, in the Petén. The Mayans who brought it there—there is no suitable habitat for the species in the immediate area—undoubtedly valued its white flesh, and their descendants still prize it as the most delicious turtle meat in the Petén. The chief difference from Mayan times may be that today, *Dermatemys* is Endangered.

Though protected, at least on paper, by a number of local and international laws, Central American river turtles are under pressure throughout their range. They are harpooned, netted, and captured by divers. John Polisar reports that in Belize, although some hunters and villages are interested in conserving the river turtle:

> ...low income rural people who share riverine habitats with *Dermatemys* view it as a commodity roughly equivalent to a common catfish. Though disappointed if it disappears, their concern is solely economic. They are unaware of its endemism, its antiquity, or its ecological uniqueness.

In the Dominican Republic, the endemic slider *Trachemys decorata* fetches a high price as a gourmet food or a source of traditional medicine. Poachers concentrate on the larger females, especially as they come ashore to lay their eggs, and all known populations of the species have declined substantially in recent years.

Antandroy and Mahafaly peoples in the far southwest of Madagascar regard the radiated tortoise (*Geochelone radiata*) as *fady*, or taboo. Neighboring tribes, however, and less traditional Antandroy, consider it a delicacy. Other Malagasy peoples consume large specimens of the Malagasy big-headed turtle (*Erymnochelys madagascariensis*), and overexploitation for food by a rapidly growing population may be the main threat facing this unique species. In the large lakes and backwaters that were once its chief stronghold, populations of *Erymnochelys* have been hunted to near or total extinction. Today it probably thrives only in small, hard-to-reach lakes and swamps.

Nile softshells (*Trionyx triunguis*) are eaten in many parts of Africa, and giant softshells of the genus *Pelochelys* are avidly sought from India to New Guinea. Villagers along the Sepik River in Papua New Guinea use the bony carapaces of *Pelochelys* and the much commoner, and smaller, New Guinea snapping turtle (*Elseya novaeguinae*) to make elaborate ceremonial masks. The masks are subsequently sold to tourists, often at high prices.

Local peoples in both New Guinea and Australia prize the flesh of the pig-nose turtle. In New Guinea, where pig-nose eggs are a

In Merauke, Irian Jaya, Indonesia, a man carries a brace of Jardine River turtles (*Emydura subglobosa*) to market.

Aboriginal rock paintings with turtles, in the Nangaloar Caves, Kakadu National Park, Australia.

delicacy, the arrival of outboard motors, the decline of clan warfare, and the migration of hill peoples to coastal regions have turned what was once a strictly localized harvest into a commercial enterprise. The eggs are collected in huge and growing numbers for local markets. During a two-month survey in August–September 1998, investigators in West Irian, the Indonesian sector of New Guinea, recorded the collection of some 84,000 eggs. In one district alone, the regency of Merauke, collectors take an estimated 1.5–2 million eggs each year, despite the fact that this species is legally protected in Indonesia.

Overharvesting of eggs is the chief threat facing the Critically Endangered painted and river terrapins. In Peninsular Malaysia, terrapin eggs may fetch four to five times the price of chicken eggs, and in some areas they are valued as aphrodisiacs. Licensed egg collectors are allowed to take terrapin and sea turtle eggs from designated sections of beach, are supposed to observe a quota, and are required to turn over 70 percent of the eggs they collect to hatcheries run by the fisheries department. There is, however, often little enforcement. In 1993, only 30 to 40 percent of eggs harvested at Terangganu, where the population probably includes fewer than 200 nesting females, were sent to government hatcheries; this figure is probably typical. Despite numerous studies and calls for better enforcement and management, the painted terrapin remains one of the most endangered of Asian river turtles.

### Turtles and People in Amazonia

For centuries, riverside communities throughout the Amazon and Orinoco basins have depended on the eggs and meat of river turtles of the genus *Podocnemis.* As long ago as 1541, Father Gaspar de Carvajal, who crossed the Andes and journeyed down the Amazon with Francisco de Orellana, noted that turtles were the most common food item in villages along the river. He found over 1000 turtles in enclosures in a single village. By the middle of the 19th century, some 2 million turtles were being taken for food every year in the state of Amazonas alone. The great 19th-century naturalist Henry Walter Bates (1825–1892) found stocks of arrau (*Podocnemis expansa*) in every backyard pond. In his classic book *The Naturalist on the River Amazons* (1863) he described the many ways turtle meat was prepared:

> The entrails are chopped up and made into a delicious soup called sarapatel, which is generally boiled in the concave upper shell of the animal used as a kettle. The tender flesh of the breast is partially minced with farinha, and the breast shell then roasted over the fire, making a very pleasant dish. Steaks cut from the breast and cooked with the fat form another palatable dish. Large sausages are made of the thick-coated stomach, which is filled with minced meat and boiled. The quarters cooked in a kettle of Tucupi sauce form another variety of food. When surfeited with turtle in all other shapes, pieces of the lean part roasted on a spit and moistened only with vinegar make an agreeable change.

In Bates's time, arrau eggs were a major cash crop. With unusual insight for a naturalist of his day, he wondered how long the bonanza could continue:

> The destruction of turtle eggs every year by these proceedings [the immense egg

> harvest] is enormous. At least 6000 jars, holding each three gallons of the oil, are exported annually from the Upper Amazons and the Madeira to Para, where it is used for lighting, frying fish, and other purposes. It may be fairly estimated that 2000 more jars full are consumed by the inhabitants of the villages on the river. Now, it takes at least twelve basketsful of eggs, or about 6000 by the wasteful process followed, to make one jar of oil. The total number of eggs annually destroyed amounts, therefore, to 48,000,000. As each turtle lays about 120, it follows that the yearly offspring of 400,000 turtles is thus annihilated. . . . [The Indians] say that formerly the waters teemed as thickly with turtles as the air now does with mosquitoes. The universal opinion of the settlers on the Upper Amazons is that the turtle has very greatly decreased in numbers, and is still annually decreasing.

There were some limited attempts to control the harvest. State authorities nominated judges to oversee the collection of eggs on the nesting beaches; Bates reported that in the Tefé region the council appointed a yearly commandante to keep an eye on things, and sentries were posted on each beach to make sure that only authorized collectors were allowed to dig up the eggs. Nonetheless, by the 1890s even this system seems to have broken down.

Arrau populations continued to dwindle through much of the 20th century. In recent years, a number of South American governments have set up management plans for Amazonian river turtles. Bolivia, Brazil, Colombia, Ecuador, Peru, and Venezuela operate management projects for yellow-spotted river turtles (*Podocnemis unifilis*). The Brazilian government began managing the arrau in the mid-1970s (see Chapter 3), after warnings that river turtles might become extinct in 10 years at then-current levels of exploitation. In 1990 it set up a National Center for Amazonian Turtles (Centro Nacional dos Quelônios da Amazônia [CENAQUA]) to oversee its national management program. Ten percent of the hatchlings produced by the management program in each Amazonian state go to approved commercial turtle breeders who raise the animals for meat. According to a report by Jack Sites and his colleagues:

Researchers release hatchling yellow-spotted river turtles (*Podocnemis unifilis*) near Iquitos, Peru.

> . . . efforts from 1979–92 have resulted in the release of over 18,000,000 young turtles back into rivers after they hatched from nests on protected beaches. CENAQUA fields a crew of over 70 people during the nesting seasons, and teams monitor nesting activity on about

115 beaches in 15 regions on 12 rivers flowing through 9 Brazilian states.

Protecting nesting beaches alone, though, will not be enough to conserve Amazonian river turtles. Richard Vogt, who works in Brazil's Mamirauá Sustainable Development Reserve, believes that involving local villagers in the conservation of their turtles is absolutely essential if populations are to recover. In the Mamirauá Project, villagers themselves help monitor nesting populations and guard beaches from poachers.

Fundacion Natura has been doing similar work with five indigenous communities on the borders of Cahuinar' National Park in the Amazon region of Colombia. Claudia Nuñez and her colleagues counted about 5000 nesting arrau females there between 1994 and 1997. They enlisted residents to recover and incubate eggs in nests they considered to be at risk, and release the hatchlings in neighboring lakes.

Despite these and similar projects in other parts of the Amazon basin, river turtles continue to be exploited at unsustainable rates throughout most of their range. In some Peruvian nesting colonies of the six-tubercled Amazon river turtle (*Podocnemis sextuberculata*), collectors converging from local and surrounding areas take over 90 percent of the nests. On unprotected sand beaches in Colombia and Peru, the take reaches 100 percent. Even in the Mamirauá Reserve itself, poachers have eliminated the arrau and driven the yellow-spotted river turtle and the six-tubercled Amazon river turtle to historic low numbers.

**Can We Use Turtles Sustainably?**

Involving local communities in the conservation of the turtles they consume, as is being done in the Amazon, is both valuable and important. Does that mean that the key to the conservation of turtles, and everything else, is "sustainable use"—harvesting animals, for local use, commerce, or even international trade? Its proponents argue that sustainable use, by providing an economic incentive to conserve wild species, is the way forward for wildlife conservation in a more and more overpopulated and developed world. But sustainable use assumes that take levels will not rise so high that populations will be unable to make good their losses. For turtles, this resolves into a more basic question: is such use, above and beyond traditional subsistence levels, even possible?

Sustainable use assumes that wild populations produce surplus individuals. These individuals can be collected or consumed without affecting the underlying stability of the population. In other words, if you only harvest the surplus, the population that remains can replenish itself. Some herpetologists have argued, however, that whatever its merits as a conservation philosophy, adults turtles at least are not designed for sustainable use.

Turtles are long-lived animals, slow to grow and slow to mature. Breeding adults are not easily replaced. Because their reproductive potential is measured in years, if not decades, removing even one from a population may mean losing years of successful nesting. In many turtle populations, therefore, there may be no such thing as a surplus, at least not of adults. Turtles that lay large clutches of eggs and experience high hatchling mortality may well have surplus eggs or young—animals that will probably never reach maturity. For turtles that lay only one or a very few eggs at a time, such as the pancake tortoise (*Malacochersus tornieri*), even that surplus may not exist.

Decades of overexploitation have depleted populations of the alligator snapper (*Macrochelys temmincki*)

Traps like this one are traditionally set for turtles in the lower Mississippi River, U.S.A.

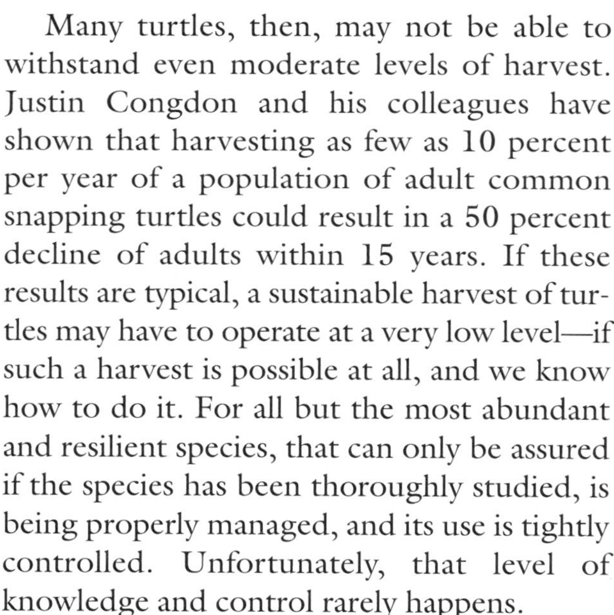

Many turtles, then, may not be able to withstand even moderate levels of harvest. Justin Congdon and his colleagues have shown that harvesting as few as 10 percent per year of a population of adult common snapping turtles could result in a 50 percent decline of adults within 15 years. If these results are typical, a sustainable harvest of turtles may have to operate at a very low level—if such a harvest is possible at all, and we know how to do it. For all but the most abundant and resilient species, that can only be assured if the species has been thoroughly studied, is being properly managed, and its use is tightly controlled. Unfortunately, that level of knowledge and control rarely happens.

For many turtles, the result of continued harvesting has been a steady, or even a drastic, decline. Commercial trappers stepped up their harvests of the alligator snapping turtle from the 1960s through the 1980s. Kevin Sloan and Jeffrey Lovich conducted a case study of exploitation of alligator snappers in Louisiana, where they have long been hunted for their meat. The Louisiana Department of Wildlife and Fisheries had estimated that at least 35 large-scale dealers in alligator snapper meat were operating in the state. Between 1984 and 1986, a single dealer in south-central Louisiana purchased 17,111 kg (37,737 lb) of alligator snapper meat, representing approximately 1223 individuals.

Larger turtles have a higher reproductive output, so a harvest that selects for them will have a serious effect on the health of the population. The dealers, however, preferred large animals because they had greater amounts of meat (meat with bones sold for up to US$13.20/kg [$6.00/lb] in parts of the state). The exploitation of alligator snappers, far from being sustainable, has led to a drastic decline throughout much of their range.

Does turtle farming, or ranching, provide a way to produce and sell large numbers of animals without depleting wild populations? A ranching project for the local race of slider (*Trachemys scripta emolli*) in the Caño Negro Nature Reserve in northern Costa Rica has the merits of involving local people. They collect eggs from the wild for the ranch, and in return are paid 50 percent of the funds generated by the sale of the turtles it raises. The project has reportedly led to the end of heavy seasonal hunting pressure on the wild turtles in the region.

Strictly speaking, a turtle farm is a "closed" operation that replenishes its stock entirely through captive breeding, while a ranch periodically brings in eggs or breeding adults from the wild. In practice, though, the terms "farm" and "ranch" are often used interchangeably. Most turtle farms rely heavily on adults collected from the wild to replenish their breeding stock. Though the hatchling red-eared sliders, once the standard dime-store turtle, have been banned from sale in the United States since 1975, huge numbers of the turtles are still farmed for export. In the 1980s, up to 100,000 red-eared sliders were being taken from the wild each year to supply farming operations in the United States, with a further 765,000 adults taken for direct export as food.

Thousands of juvenile turtles, like these red-footed tortoises (*Geochelone carbonaria*), end up in the pet trade.

Ironically, though 8 million farmed red-eared slider hatchlings are exported to some 60 countries every year, its population at home may be declining. Lisa Close and Richard Siegel compared harvested and undisturbed populations of red-eared sliders in southern Louisiana and western Mississippi, where the animals were captured for ranching operations, for food and for export, during 1990 and 1991. They found that "the most conspicuous difference among the populations was the virtual absence of turtles greater than 22 cm [8.7 in] carapace length from harvested sites; in essence, the entire upper end of the size distribution had been eliminated." Dealers the researchers interviewed agreed that the turtles were declining in both individual size and overall number. Mississippi has now banned the commercial sale of all native turtles except the common snapper.

Even if turtle farms are true "closed" breeding operations, they still raise concerns associated with all types of wildlife farming: the increased risk of disease, and the chance that farmed turtles could escape, as red-eared sliders have done, and establish themselves outside their normal range to the detriment of native turtle populations. Though farming shows some promise, the jury on its value to turtle conservation is still out.

### The Pet Trade

For many people, their first love for turtles came from owning a pet. Before the United States banned the domestic sale of turtle hatchlings in 1975—not as a conservation measure, but over fears that hatchlings spread diseases like salmonella—vast numbers of baby red-eared sliders became what Ted Williams, in a 1999 article in *Audubon Magazine*, called "disposable pets." Today, owning reptiles is a popular hobby in many countries; there may be somewhere between 2.5 million to 15 million pet turtles in the United States alone. The United States imports at least 30,000 turtles a year, the vast majority caught in the wild. Certainly many hobbyists cherish their animals, buy them only from reputable breeders, know what they are doing, and take an active interest in their conservation. Unfortunately, these reputable hobbyists are but a tiny part of a vast and wasteful international traffic in turtles.

The international trade in wild-caught pet turtles condemns many animals to slow and miserable deaths. Many die long before reaching a retail store, during capture or shipment. Some species suffer particularly high mortality in transport, reaching over 30 percent for Florida softshells (*Apalone ferox*) and map turtles (*Graptemys* spp.). Veterinarian Barbara Bonner, director of the Turtle Hospital of New England, told Ted Williams that:

> ...virtually all turtles from pet stores are desperately sick when purchased. "It takes about six months just to clean them up and rid them of parasites," she says. "Some are half the body weight they ought to be. It's estimated that 95 percent of the wild turtles that enter the pet trade are dead within a year. Pet stores don't make their money selling the turtle; they make their money selling the $250 setup that goes with it. So if your pet dies, it doesn't matter to them, because with that kind of investment, you're going to buy another.

The most unscrupulous practitioners of the trade defy national and international laws, often with impunity. Their eagerness to fill the demand for the latest rarity can have drastic effects. The Roti Island snake-necked turtle (*Chelodina mccordi*) was only described in 1994. It is confined to a single island in the waters south of Timor. There is nothing particularly extraordinary about it, as snake-necked turtles go, except its rarity and novelty, but that has been enough. Soon after its discovery, its price in the trade rose to US$2000. Within five years, collectors on Roti had trapped so many that the species could no longer be had. Apparently only a single dealer, who bought every specimen the collectors could find, profited from the destruction of the Roti Island snake-neck.

History may be repeating itself with Pritchard's snake-neck (*Chelodina pritchardi*). Another newly described species, it is confined to a tiny area east of Port Moresby, the capital of Papua New Guinea. It, too, has become an expensive novelty, and though it has legal protection is still being trapped and smuggled out of the country. Its fate is uncertain, but it is already considered to be Endangered, and may well follow the Roti Island snake-neck into near, if not total, oblivion.

Though rarities often fetch high prices when they first appear in international markets, their prices may soon fall. Traders and dealers often respond to the bonanza a rarity may bring by flooding the market. If breeders are successful with the new arrival, they may soon be able to undercut the price of imports with captive-bred animals. Unfortunately, this price fall does not necessarily take pressure off wild populations. As the price drops more and more, less affluent buyers can afford the animals, and demand may rise again. If the price drops far enough, cheap wild imports may undercut the prices offered by breeders, making it more economical to deal in wild-caught animals than to breed them in captivity. Michael Klemens and Don Moll described such a trade pattern for the pancake tortoise in the early 1990s, when legal exports of the species from the wild still occurred:

> This recent flooding of the U.S. market with pancake tortoises has saturated the market far beyond the limited number of serious collectors and institutions willing to pay hundreds of dollars for a single tortoise, causing a subsequent reduction in price. Under these conditions the demand for pancake tortoises will continue unabated as the price is now within the range of many casual pet owners. This devalued market will consume as many pancake tortoises as can be supplied, while providing no financial incentive to breed the species in captivity. A previous example of a "bottomless tortoise market" was the insatiable post–World War II demand for inexpensive Mediterranean tortoises (*Testudo graeca* and *T. hermanni*) which continued until 1988 throughout western Europe. Although many populations of *Testudo* were severely damaged by collecting activities, the wide geographic range of these species prevented their extinction. However, the pancake tortoise is much more vulnerable to this "bottomless market" type of exploitation because of its comparatively small geographic range, narrow habitat requirements, and localized populations.

The United States is one of the few countries to have issued meaningful penalties against turtle smugglers. A violation of the *Lacey Act*, which makes it illegal to import animals in violation of the laws of their home country, carries a maximum penalty of five years in prison and a $250,000 fine. In recent years, there have been a number of well-publicized Lacey Act indictments that involved turtle smuggling.

In 1998, Michael Van Nostrand, owner of Strictly Reptiles in Hollywood, Florida, pled guilty to charges of illegally importing reptiles including pig-nose turtles, chaco tortoises, red-footed tortoises (*Geochelone carbonaria*), and yellow-spotted river turtles. He was sentenced to eight months' imprisonment, eight months' home detention, and a fine of almost $250,000 to be paid to the Indonesian chapter of the World Wildlife Fund. In December 2000, Keng Liang "Anson" Wong, who spent nearly two years in a Mexican prison fighting extradition to the United States, pled guilty to 40 felony charges ranging from money laundering to smuggling the world's most endangered tortoise species, the angonoka (*Geochelone yniphora*). On June 7, 2001, Wong was sentenced to 71 months in prison—the maximum term under federal guidelines—and a fine of $U.S. 600,000.

Unfortunately, felony indictments and a scattered number of relatively stiff sentences do not seem able to stop the excesses of reptile smugglers. Even within the United States, poachers continue to strip entire watersheds of their populations of wood turtles. Part of the problem is the sheer size of the task of enforcement: even in the United States, which probably has more wildlife inspectors at its border crossings than any other country, no more than 10 percent of incoming shipments are inspected.

Imports do not have to be illegal to be appallingly wasteful. Ted Williams also interviewed Joe Ventura, a Fish and Wildlife inspector who works at the Port of Los Angeles:

> "We've seen turtles stacked on their sides like dinner plates so they couldn't extend their limbs," [Ventura] says. Turtles destined for the pet trade are often shipped in cardboard boxes that get crushed when cargo shifts, and it's not unusual to see these boxes soaked with blood. In one shipment from Tanzania, 511 pancake tortoises and 307 leopard tortoises had been packed on top of one another. Fifty animals were dead, 400 appeared near death, and almost all were grievously dehydrated. There was much blood, many broken carapaces, and dozens of missing legs. About 50 females carried broken eggs.

This should not happen. The International Air Transport Association (IATA) issues regulations for the main transport of live animals, and these regulations are a requirement for the parties to the CITES convention. That CITES cannot entirely stop the excesses that Williams describes is one of the frustrations for those of us who spend much of our lives dealing with, and trying to support, this powerful, often valuable, and yet at times painfully impotent international instrument.

| SPECIES | APPENDIX | DATE LISTED (on current Appendix) |
|---|---|---|
| Central American river turtle *Dermatemys mawii* | I | 1981 |
| River terrapin *Batagur baska* | I | 1975 |
| Painted terrapin *Callagur borneoensis* | II | 1997 |
| Wood turtle *Clemmys insculpta* | II | 1992 |
| Bog turtle *Clemmys muhlenbergi* | I | 1992 |
| Asian box turtles *Cuora* spp. | II | 2000 |
| Spotted pond turtle *Geoclemys hamiltonii* | I | 1975 |
| Indian roofed turtle *Kachuga tecta* | I | 1975 |
| Keeled hill turtle *Melanochelys tricarinata* | I | 1975 |
| Burmese eyed turtle *Morenia ocellata* | I | 1975 |
| American box turtles *Terrapene* spp. | II | 1995 |
| Coahuilan box turtle *Terrapene coahuila* | I | 1975 |
| Tortoises Testudinidae spp. | II | 1975 |
| Galapagos tortoise *Geochelone nigra* | I | 1975 |
| Radiated tortoise *Geochelone radiata* | I | 1975 |
| Angonoka *Geochelone yniphora* | I | 1975 |
| Bolson tortoise *Gopherus flavomarginatus* | I | 1975 |
| Geometric tortoise *Psammobates geometricus* | I | 1975 |
| Egyptian tortoise *Testudo kleinmanni* | I | 1995 |
| Hard-shelled sea turtles Cheloniidae spp. | I | 1981 (entire family) |
| Leatherback sea turtle *Dermochelys coriacea* | I | 1977 |
| Indian flapshell *Lissemys punctata* | I | 1995 |
| Cuatro Cienegas softshell *Apalone ater* | I | 1975 |
| Indian softshell *Asperidetes gangeticus* | I | 1975 |
| Indian peacock softshell *Asperidetes hurum* | I | 1975 |
| Black softshell *Asperidetes nigricans* | I | 1975 |
| Malagasy big-headed turtle *Erymnochelys madagascariensis* | II | 1975 |
| Big-headed Amazon River turtle *Peltocephalus dumeriliana* | II | 1975 |
| Amazon River turtles *Podocnemis* spp. | II | 1975 |
| Western swamp turtle *Pseudemydura umbrina* | I | 1975 |
| Ghana has listed the following species on Appendix III: Nile softshell (*Trionyx triunguis*); helmeted terrapin (*Pelomedusa subrufa*); Adanson's mud terrapin (*Pelusios adansonii*); West African mud terrapin (*Pelusios castaneus* ); African forest Terrapin (*Pelusios gabonensis*); African black terrapin (*Pelusios niger*) | III | |

Species of turtles listed on the Appendices of CITES (as of June 2001). The classification and arrangement follows that in the CITES Appendices, and differs slightly from that used elsewhere in this book.

## Turtles and CITES

The Convention on International Trade in Endangered Species of Wild Fauna and Flora (CITES) was signed in 1973, and came into force in 1975. Today, it has over 150 signatories, more than any other wildlife-related treaty. CITES is not an international endan-

Egyptian tortoises (*Testudo kleinmanni*) are still being sold illegally in Egypt, though the species is almost extinct there.

gered species list. Instead, it deals with a single issue: the need, as it states in its Preamble, to provide "international protection . . . of certain species of wild fauna and flora against overexploitation through international trade." How well it succeeds, though, depends on the will and resources of its signatory nations, or Parties, and both may fall short of the mark where turtles are concerned. The species that CITES protects, or seeks to protect, are listed on one of three Appendices.

Species listed on Appendix I are considered to be "threatened with extinction," and are, or may be, affected by international trade. Appendix I species may not be taken from the wild and traded internationally for "primarily commercial purposes." Appendix II species are "not necessarily threatened with extinction," and may be traded commercially, but their trade requires regulation to ensure that they do not become threatened. Species may also be listed on Appendix II as a control measure, because they look so much like another listed species that customs officials would be unlikely to be able to tell them apart. Only the Parties, voting as an assembly (the so-called Conference of the Parties) can add a species to Appendix I or Appendix II, remove it, or shift it from one Appendix to the other. However, any Party can add its own population of a species to Appendix III as a way of getting international help in controlling its trade.

Listing a species on a CITES Appendix can help bring it to the attention of conservationists, and can provide country governments with the framework to create and enforce national laws to control trade in turtles. Some have argued that listing species, especially those in demand for the pet trade, simply flags them for the attention of smugglers. There is little evidence that this has actually happened. CITES controls, though, can only work if there is the will, and the wherewithal, to enforce them—and smugglers have ways of getting around rules if the incentive is high enough.

The Egyptian tortoise (*Testudo kleinmanni*) was transferred from Appendix II to Appendix I—at the request of Egypt—at the 1994 CITES meeting in Fort Lauderdale, Florida (where, ironically, substantial numbers were being offered for sale in local shops). The transfer bans international commercial trade in the species. Smuggling, though, still goes on—after all, no law stops crime 100 percent—and Egyptian tortoises are still offered for sale in Egypt, although the species is all but extinct there. The tortoises on sale in Egypt probably come from neighboring Libya.

Meanwhile, other tortoises, also probably from Libya, are apparently being smuggled into Egypt, where over 900 of them have been exported since 1994 as "Egyptian Greek tortoises"; in other words, as spur-thighed tortoises from Egypt. The problem is, as far as we can tell, that there are no wild spur-thighed tortoises in Egypt. As James Buskirk con-

Louisiana has now banned exports of box turtles like this three-toed box turtle (*Terrapene carolina triunguis*).

cluded in his report on this situation: "A disastrous result [of these sorts of goings-on] is the unchecked exploitation of vulnerable chelonian populations 'laundered' for sale in the international pet trade by the complicity or ignorance of wildlife officials in countries that have no natural populations of these animals."

The increasingly rare and highly localized pancake tortoise (*Malacochersus tornieri*), a species listed on Appendix II of CITES, has been subjected to the same treatment. Pancake tortoises are known only from Kenya and Tanzania, where they suffer from habitat alteration in the form of shifting cultivation and poaching for the pet trade. In Kenya, almost the entire population lives outside of any protected area. Kenya bans their export, and Tanzanian exports are ostensibly restricted to animals raised in ranching facilities.

Pancake tortoises recently began appearing in international markets labeled as animals from Mozambique and Zambia, countries where they almost certainly do not occur. There are, of course, no Mozambican and Zambian laws preventing the export of animals they do not have. As a result, countries accepted the tortoises even though they were surely smuggled out of Kenya or Tanzania in the first place. At the 2000 CITES meeting in Nairobi, Kenya proposed that the whole species be transferred to Appendix I, but withdrew the proposal after Tanzania refused to support it. Tanzania did, however, agree not to allow the export of wild-caught tortoises (which have in any case been under a zero quota since 1992).

Once again, smuggling goes on. In 2001, the Uganda Wildlife Authority confiscated 209 pancake tortoises packed in an animal trafficker's hand luggage at Entebbe airport. Some of the animals were dead, or died from injuries shortly thereafter, but 190 survivors were returned to Kenya with the help of the International Fund for Animal Welfare. The Kenya Wildlife Service plans to return them to the wild, in Tsavo East National Park, where they can be monitored and protected.

For all that, CITES has helped to protect some turtles. It has played an important, and controversial, role in the conservation of the hawksbill sea turtle (*Eretmochelys imbricata*) (see Chapter 10). A listing of North American box turtles (*Terrapene* spp.) on Appendix II in 1994 (except for the Coahuilan box turtle [*Terrapene coahuila*], which was already on Appendix I) gave the U.S. federal government leverage to deal with exports of eastern box turtles (*Terrapene carolina*) from Louisiana, the only state with a legal export trade. Though Appendix II species may be traded, trade is only supposed to take place if the exporting country's Office of Scientific Authority (OSA), one of the administrative bodies each Party must set up, finds that the export will not harm the wild population—in technical terms, it must make a *non-detriment finding*.

Because there was no valid information on how Louisiana's box turtle trade, which was exporting thousands of animals annually, was affecting the animals, the OSA reduced the state's export quota to zero in 1996 and 1997. Louisiana filed, and lost, an appeal. A campaign launched by two Louisiana residents, Martha Ann Messinger and George Patton, finally convinced the state government to ban

An adult angonoka (*Geochelone yniphora*) at Project Angonoka in northwestern Madagascar.

After a theft in 1996, Project Angonoka installed an improved security fence to protect its tortoises.

the commercial harvest, sale, and trade of wild-caught box turtles in 1999.

The powers of CITES are not limited to trade bans. Through its Significant Trade Review, it has tried to persuade the government of Madagascar to bring the trade in its unique reptiles, including its tortoises, under control. Its actions have included a partial embargo on trade in Malagasy chameleons (Chameleonidae) and day geckos (*Phelsuma* spp.). One of the most important of its functions today is as a coordinator of cooperative international efforts to control unsustainable trade in broad categories of wild species. It is partly in that role that CITES has recently turned its attention to the most immediate, and probably the most serious, of all issues facing turtle conservation, the immense and devastating trade in the turtles of Southeast Asia.

**Madagascar: Tortoises on the Edge**

The unique fauna of Madagascar is one of the most beleaguered in the world, and its turtles, sadly, are no exception. Ron Nussbaum and Chris Raxworthy's commentary on the state of the radiated tortoise (*sokake* or *sokatra* in Malagasy) could well serve for most of them:

> Their habitat is being degraded and destroyed at an increasing rate, they are being harvested by the thousands every year, and local law-enforcement does little to mitigate the situation or stop such activities. The laws protecting sokatra are well-known to the Malagasy, but they have learned that these laws have been completely ignored.

Madagascar's capital, Antananarivo, lies in the deforested and degraded center of the island. It is far from the Malagasy coasts, where smugglers' boats put in to take on shipments of endemic, and supposedly protected, tortoises. The animals are frequently transshipped eastward to the island of Réunion. Pet tortoises are highly popular on Réunion, where they are often given as wedding presents or to promote good luck. The chief advantage of Réunion to a smuggler, though, is that it is politically a part of France. Once the tortoises reach the island, they are, for customs purposes, in Europe, and they may be shipped throughout the EU without a shred of customs documentation.

Poachers have threatened the integrity of one of the best-known tortoise conservation programs in Madagascar. By 1996, Project Angonoka, the captive breeding program for the Critically Endangered angonoka operated by the Durrell Wildlife Conservation Trust (formerly Jersey Wildlife Preservation Trust), had produced well over 150 hatchlings. Then, in May 1996, someone—it has never been established who—broke into the breeding compound at Ampijoroa and stole 73 hatchling tortoises and two adults.

Months later, young angonoka, surely from the breeding station, began appearing in European pet markets. In 1998, the Netherlands government confiscated 35 animals and sent them to the Bronx Zoo in New York for veterinary care. The Madagascar government filed a court action for their return—the tortoises were legally its property—and won. To avoid potential health risks, an arrangement was made to have the angonoka housed by a commercial dealer. Before the U.S. returned the tortoises, the animals were marked indelibly on their carapaces. That way, they could never be sold again.

Project Angonoka has recovered. When I visited the station in 1997, the project managers had already installed an improved secu-

Massive recent exports for the pet trade threaten the survival of the flat-tailed tortoise (*Pyxis planicauda*).

rity system, and at least 30 new baby tortoises had been hatched. Meanwhile, the project has conducted in-depth studies of the tiny wild population, and developed plans to protect its habitat and to help the people that live there to care for their local environment. It has established good relations with the people of Baly Bay, who, according to Lee Durrell, "are proud to be involved in the safeguarding of 'their' tortoise."

In 2000, the Madagascar government suddenly raised its CITES export quota for the rarest and most localized of its tortoises after the angonoka, the flat-tailed tortoise or kapidolo (*Pyxis planicauda*), from 25 to 800. At the same time, it raised the export quota for

In southern Madagascar, a Malagasy carries radiated tortoises (*Geochelone radiata*) home for a meal.

the flat-tailed's more widespread relative, the spider tortoise (*Pyxis arachnoides*), from 25 to 1000. Considering that the entire population of the flat-tailed tortoise must be small, there is no way that such a quota could be biologically justified. Nonetheless, hundreds of flat-tailed tortoises, certainly caught in the wild, began appearing in U.S. markets. According to U.S. government figures, imports for the year 2000 of flat-tailed tortoises, spider tortoises, and Malagasy big-headed turtles into the United States alone, all bearing export permits issued by the government of Madagascar, exceeded even the new global quotas by 179 percent. After protests from conservationists, Madagascar reset the quota to zero early in 2001, but the animals continue to appear, even though initial high prices have tumbled.

Though the flat-tailed tortoise is poorly known in the wild, there seems little doubt that, if its trade is not brought under real control, the species will suffer serious harm. Conservationists continue to work with the government of Madagascar to prevent this from happening. In May 2001, the IUCN Conservation Breeding Specialist Group facilitated a meeting of Malagasy and expatriate field workers in Madagascar to assess the status of many species, including the flat-tailed tortoise, and concluded that even without the trade it could become extinct within 30 years from habitat loss alone. A properly managed captive population may be essential for its survival, but massive uncontrolled exports are surely not, especially given the extremely low reproductive rate and longevity of the species.

The radiated tortoise may be in less danger than its rarer relatives, despite unrelenting pressure on its spiny forest habitat and the fading of the traditional *fadys* that have protected it in the heart of its range. In some areas, it is still a common animal. How long this will continue, though, is difficult to say. While the flat-tailed tortoise is hard to find, and the angonoka is in such a remote and tiny region that it is, perhaps, relatively safe from smugglers, radiated tortoises are easy to collect during the humid months when they are active. Ron Nussbaum and Christopher Raxworthy describe following a bus down one of Madagascar's ghastly country roads, watching it make repeated stops so passengers could pop out and pick up (quite illegally) radiated tortoises they had spotted on the way.

According to John Behler, it is relatively easy for collectors to sweep through a patch of tortoise habitat and pick up every single animal. This has happened even in national parks. Though collection for local use as food may be the biggest problem the species faces , the Malagasy government regularly confiscates radiated tortoises at Antananarivo airport. The species is on CITES Appendix I and is therefore barred from international trade. The incentive to smuggle radiated tortoises overseas is enormous; large pairs have changed hands for up to US$25,000, and end-market prices of US$5000 per animal are not uncommon.

Is it even possible to stop, or at least control, poaching and illegal trade in a country where the people are desperately poor and laws are not enforced? It is not as though there are no committed conservationists in Madagascar, including in the government, and the successes of Project Angonoka provide some hope that the conservation message is being heard in the animals' range. Nussbaum and Raxworthy suggest that the Appendix I listing of the radiated tortoise is counterproductive, driving up prices and condemning the tortoises shipped in defiance of it to transport under appalling and often lethal conditions. They suggest that

Though protected by law in Indonesia, the Malayan giant turtle (*Orlitia borneensis*) is exported in huge numbers.

Madagascar should supply the markets, instead, with captive-bred individuals, providing that a properly managed breeding operation could be set up. "Captive breeding" operations for reptiles in Madagascar, however, are almost entirely fronts for the laundering of wild-caught animals.

My own view is that it would be better to try, first, to make the listing work, if not in Madagascar itself, then in the market. Madagascar's poor may be driven to overcollecting tortoises (and the middlemen they sell them to surely have no interest in reducing their profits in the name of conservation). No one, though, is driven to buy a wild-caught Madagascar tortoise. It is not just Madagascar's poor that need to be educated, but the wealthy buyers in richer countries who take their animals from them.

**Crisis in Asia**

In July 1997, William McCord, a private collector, visited markets in Guangzhou and Shenzen, China, to videotape and photograph turtles. In his two days there, he saw, offered for sale, some 10,000 turtles of over 40 different species. Dealers told him that it would take only two to seven days to sell their entire stock. If the number of turtles McCord saw was typical, and turnover is really that high, the two markets must account for over one million turtles a year. There are many such markets in China.

Southeast Asia is being drained of its turtles at a rapid and growing rate. In December 1999, 45 delegates from 15 countries met in Phnom Penh, Cambodia, for a workshop organized by the Wildlife Conservation Society, TRAFFIC, and the World Wildlife Fund. They were there to assess the impact that rapidly expanding trade was having on the conservation of Asia's freshwater turtles and tortoises.

The picture that emerged from the workshop was worse than even many of its delegates had anticipated. The basic facts, as reported in the Executive Summary of the workshop proceedings, are startling enough:

> Based on the most recent available data, a minimum of 13,000 metric tons of live turtles are exported from South and Southeast Asia to East Asia each year. At least 5000 tons of wild-collected turtles are exported from Indonesia annually and 1500 tons from Bangladesh, as well as 4000 tons of farm-produced softshells from Thailand and 2500 tons of turtles from wild and farm sources in Malaysia. The actual amounts may be substantially higher and these numbers do not include amounts exported from Cambodia, Laos, Myanmar, or Vietnam.

The picture was the same in country after country. Bangladesh exported an estimated 10,000 metric tons (11,023 tons) per annum of Indian peacock softshells (*Aspideretes hurum*) between 1985 and 1992, before the

The once-common Chinese three-keeled pond turtle (*Chinemys reevesi*) has almost vanished from Chinese markets.

species was listed on Appendix I of CITES. Over 9 million live turtles of many species were imported to Hong Kong in 1998 alone—an increase of 28 times since 1992. The majority of them (81.5 percent) originated in Indonesia and Thailand. Most were probably destined for sale as food in Chinese markets. Though these figures include red-eared sliders and Chinese softshells (Pelodiscus sinensis) raised on turtle farms, they undoubtedly represent a huge number of wild-caught animals.

According to a case study presented to the workshop by Chris Shepherd of TRAFFIC Southeast Asia, two exporting companies in the provinces of North Sumatra and Riau, Indonesia, shipped more than 22.7 metric tons (25 tons) of wild-caught live turtles each week to China, Hong Kong and Singapore. Ten of Sumatra's 13 freshwater turtle and tortoise species were sold in the trade, including large quantities of Malayan giant turtle (*Orlitia borneensis*), a protected species. In fact, it was one of the two most heavily exported species, despite the fact that all trade in Malayan giant turtles is illegal in Indonesia.

The result of this unrelenting pressure is a grave threat to the world's largest and most diverse turtle fauna. Forty-five of the 89 species of turtles native to Asia are now believed to be either endangered or threatened, and probably none of the others are truly safe.

Many of the victims of the Asian turtle trade end up in the international pet market. Almost every species of turtle in Asia is available in the pet trade. Others are turned into curios, or are used for religious rituals in which celebrants release turtles into the wild, though often far from their natural habitats. The vast majority, though, are sold as food, or for use in traditional medicine. The bulk of them end up in markets in China.

Turtle imports into China have increased tenfold since 1977. This is partly because, in 1989, the Chinese government allowed its currency to be exchangeable with that of other countries, a change that led to a tremendous surge in imports. The increase is also happening partly because the fashion for turtle meat and other turtle products is growing in China, and is spreading to the north where turtles were, traditionally, not very popular.

The surge in imports is also happening because China has run out of wild turtles. Chinese turtle populations are severely depleted. Even the once-common Chinese three-keeled pond turtle (*Chinemys reevesi*), a well-known species from Japan, Korea, and China (including Hong Kong and Taiwan), has nearly disappeared from southern Chinese markets. Some species, including McCord's box turtle (*Cuora mccordi*) and Zhou's box turtle (*Cuora zhoui*) which herpetologists have never found outside of Chinese markets, may already be extinct in the wild (see Chapter 4). The same may be true for some

The three-striped box turtle (*Cuora trifasciata*) fetches high prices for its purported medicinal properties.

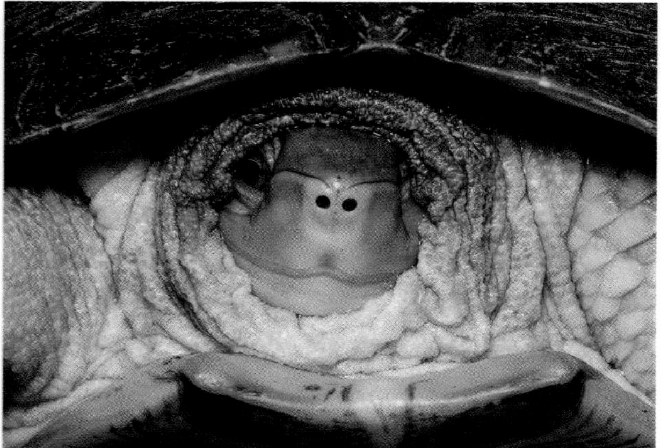

Today, turtle jelly in Hong Kong is often made with substitute species like the Asian yellow pond turtle (*Mauremys mutica*, above) or the Malayan flat-shelled turtle (*Notochelys platynota*, below).

other little-known and recently described turtles, such as *Ocadia glyphistoma*, though it is possible that these new "species" may be simply hybrids of better-known animals.

With China's own supply of turtles nearing exhaustion, the trade moved outward, first to the turtles of neighboring countries like Bangladesh, Myanmar, and Vietnam. As even these turtles became harder to find, the trade shifted onward to Indonesia, Papua New Guinea, and beyond. It has already had a drastic effect on distant turtle populations. The natural range of the Sulawesi forest turtle (*Leucocephalon yuwonoi*) is farther from China than that of almost any other Asian bataguird. Sulawesi forest turtles appeared in great numbers in Chinese markets for a short while after the species was described in 1995. It has now disappeared again, its numbers presumably so depleted that it is no longer worth collecting.

As the trade spread, turtle faunas of country after country collapsed in its wake. In Vietnam and Laos, turtles have become almost impossible to find, even for professional collectors. Ross Keister, who visited Vietnam in 1997 as a member of the Global Biodiversity Team for the United States Forest Service, reported that things had gotten so bad that "wherever we looked there was what we called a reverse pet shop. A storefront with signs advertising that they will buy any kind of turtle. We've been recently told that these stores are closing since there are no more turtles left to buy."

The demand for turtle meat in China is intimately tied to the demand for ground shell, bone, and other turtle products for traditional medicine. Historically, softshells have been considered to be a delicacy in China, while other species, the so-called "hard-

The Chinese softshell (*Pelodiscus sinensis*) is the most frequently farmed turtle in eastern Asia.

shelled turtles," were reserved for medicinal use. Today, however, both softshells and hardshells are eaten, and their meat is valued not only for its flavor but for its reputed medicinal properties.

Of all the turtles in Eastern markets, none carries a higher price than the three-striped box turtle (*Cuora trifasciata*). Even its common trade name, "golden coin turtle," conveys the image of a rare treasure. The reason is its medical reputation. Its meat and, particularly, its ground plastron, mixed with medicinal herbs and marketed as "turtle jelly," are believed to have cancer-reducing and detoxifying properties. Priced at only US$50–$100 in the mid-1980s, by the late 1990s a three-striped box turtle could fetch US$1000 or more in Hong Kong—if one could be found. Over much of its range, three-striped box turtles have become almost, if not entirely, impossible to locate in the wild. In Laos, where hunters avidly seek them to sell to Vietnamese traders, the numbers of turtles passing through the market towns of Ban Guner and Ban Maka-Nua fell from some 300 per year in the early 1990s to 10 or less by 1998.

Today in Hong Kong, turtle jelly is more likely to be made with substitutes like the Asian yellow pond turtle (*Mauremys mutica*) or the Malayan flat-shelled turtle (*Notochelys platynota*). Turtle plastra have been shown to contain traces of selenium, an element which apparently does have cancer-reducing properties. The amounts present are so slight, however, that it is highly unlikely that turtle jelly really has the properties touted for it.

Turtle farming has spread rapidly in East and Southeast Asia as demand continues to grow and wild populations vanish. The most commonly farmed species is the Chinese softshell, which has the advantage of being relatively prolific and quick to mature. There are reportedly over 100 farms rearing Chinese softshells in China alone. The output from softshell farms in Taiwan rose from 1.1 million hatchlings in 1994 to 32.7 million in 1997. In May 1997, Ron de Bruin and Harald Artner visited a large, newly developed softshell farm on the Chinese island of Hainan. At the time of their visit, it held some 8000 turtles. According to their report:

> The manager told us the farm was owned by a Hunan businessman, and that all *P. sinensis* in the ponds were wild-caught in Hunan Province, China. He also told us that the eggs were deposited at night in artificially created sandbanks, and that they were left to incubate at those sites. Immediately after hatching, the hatchlings were sold to businessmen in Hunan.

In January 2001, Haitao Shi and James Parham visited another Hainan turtle farm that, according to its owner, held more than

50 species of turtles and 50,000 individuals, including 30,000 Chinese softshells. His stock included at least 1000 three-striped box turtles, and between 7000 and 8000 Asian yellow pond turtles. Shi and Parham believe that the number of captive-reared turtles in Asia has been greatly underestimated.

In addition to turtles reared locally, huge numbers of red-eared sliders, both farmed and wild-caught, are imported into Asia every year. Red-eared slider imports into China reached 1,832,400 in 1997. At least 182,000 were imported into Taiwan for the pet trade alone between 1994 and 1998; Taiwan now bans the import of reptiles as pets.

Should turtle farming and ranching be promoted in Asia? In theory, farmed turtles should be able to relieve some of the pressure on wild populations. Unfortunately, there seems to be no clear evidence that farmed turtles really do cut into the demand for wild-caught animals. This may be because the demand is so enormous that all the farmed and imported turtles available, plus the wild turtles in the markets, cannot fill it. Farmed turtles may also, partly, be unable to replace wild-caught animals because wild turtles are widely believed to be of superior quality and potency, and therefore fetch a much higher price.

How, then, is the Asian turtle crisis to be solved? The delegates to the Cambodia Workshop called for stronger and better-enforced laws, guidelines to help authorities dealing with confiscated turtles, studies on the natural history of Asian turtles and the effect trade is having on them, expanded conservation breeding programs both in Asia and abroad, regulation and monitoring of turtle farming, a search for substitutes for turtle products in traditional medicine, and public education, particularly among local communities.

The Cambodia Workshop also called for increased CITES coverage, possibly including a blanket listing for all Asian turtles. In 2000, all species of Asian box turtle (*Cuora*) were added to CITES Appendix II. That still leaves three-quarters of Asian species unlisted; more listings may follow at the 2002 CITES Conference in Santiago, Chile.

Listing, though, means little without enforcement. In many Asian countries, CITES listings are flouted or ignored. Enforcement budgets are often so low that even the most willing customs officers cannot make a serious dent in the trade. The borders between China and many of its neighbors to the south are sieves as far as smuggling of endangered species is concerned, and it is highly unlikely that any national or international regulation will stop the trade by itself.

Some say that there is nothing we can do to save Asian turtles in the wild. In a review of the global decline of reptiles, J. Whitfield Gibbons and his colleagues concluded that "the only way to prevent the imminent extinction of a large number of the more than 80 species of turtles native to southern Asia will be to maintain populations in ex situ captive breeding and genetic reserve programs."

The need to develop coordinated, properly managed breeding programs for at least some Asian turtles seems unavoidable. A new coalition, the Turtle Survival Alliance, was founded at a conference at the Fort Worth Zoo in 2001, with the aim of bringing together biologists, conservationists, zoo managers, and private interests to, as Deborah Behler reported, "develop captive populations and management strategies for the most endangered Asian turtles in Europe, North America, and Australia, and link these programs with similar efforts in the turtles native Southeast Asia."

All Asian box turtles, including the golden-headed box turtle (*Cuora aurocapitata*, left) and the Malayan box turtle (*Cuora amboinensis*, right), have now been listed under CITES.

It would be wrong, though, to give up on the wild. If we only save turtles in captivity, we lose the roles they play in their native ecosystems. We also lose the chance to get the message out, if we can, that wild species in their natural environments are resources to be husbanded and protected, not mines to be pillaged. A turtle in a cage on a distant continent does little to convey that message.

Part of that message involves proving that turtles have a value to humanity unrelated to the pot or the pet shop. Freshwater turtles in Asia eat snails which are intermediate hosts of *Schistosoma*, the deadly human blood fluke. In 1983, as part of India's Ganga Action Plan to clean up the Ganges River, Indian softshells (*Aspideretes gangeticus*) and Indian peacock softshells (*Aspideretes hurum*) were recruited as a natural cleanup crew. Collectors gathered softshell eggs and transferred them to hatcheries at Lucknow and Varanasi. Here, the young were hatched and reared to a size that would protect them from many of the predators that attack hatchlings—a technique called *headstarting*. The headstarted turtles were released in large numbers into the Ganges, to feed on the carcasses of humans and livestock commonly dumped into the river.

A valuable start in carrying out the recommendations of the Cambodia Workshop has

already been made, in one of the countries worst hit by the Asian turtle crisis. In 1998, Fauna and Flora International helped set up a Turtle Conservation and Ecology Project in Cuc Phuong National Park, in the northern part of Vietnam. The objective of the project is "to initiate immediate and urgent action in response to the threat to Vietnam's native turtle populations resulting from the illegal wildlife trade." The project promotes public awareness and education, trains local authorities, performs research on captive breeding, translocates confiscated turtles into protected areas, carries out field surveys of Vietnamese turtle populations, and monitors the turtle trade in Vietnam. It has even produced a children's book, *The Adventures of Lucky Turtle*, that tells the story of "a forest turtle that is caught by hunters and sold to an evil trader, only to escape while being shipped to China with a truck full of other animals."

We are still far from a solution to the Asian turtle crisis. CITES Resolution Conf. 11.9, passed in 2000, calls on all parties to deal with the threats posed by the trade and freshwater turtles and tortoises. As this book goes to press, a new technical workshop is being planned under CITES auspices. A CITES Working Group, formed in December 2000 to deal with Asian turtle issues, will meet again in 2001, in Hanoi, Vietnam. As a member, I hope that being in the geographical heart of the crisis will help us all to understand, and to do something about, this desperate situation.

CHAPTER TEN

# *Peril at Sea*

HUMANS HAVE EXPLOITED sea turtles for millennia. Bones that may have come from green sea turtles (*Chelonia mydas*) have been found among the leavings of early humans in the Niah Caves of Borneo. Traditional sea turtle hunts and egg harvests are still an important part of the lives of indigenous peoples from Central America to New Guinea. Ritual and subsistence hunts may not have jeopardized turtle populations in ages past, but our relationship with sea turtles in recent centuries has been far from benign, and certainly far from sustainable.

In 1492, the *Santa Maria* carried Christopher Columbus into seas teeming with turtles. During the 17th and 18th centuries European sailors hunted the turtles relentlessly, for eggs, meat, shell, and the delicacies known as *calipash*, the gelatinous greenish cartilage lying in the carapace fontanelles that gives the green sea turtle its name, and *calipee*, its yellowish equivalent cut from between the bones of the plastron. Calipee and calipash are prime ingredients in turtle soup, and adult turtles were killed simply to extract a few handfuls of the stuff. Archie Carr, in *The Sea Turtle: So Excellent a Fishe*, described finding turtles that had not even been killed before the plastron was cut off for its calipee.

The turtles could not withstand such an intense commercial trade. Green sea turtle populations on Bermuda were so depleted by 1620 that the Bermuda Assembly passed laws to protect them—to no avail. Bermuda's nesting turtles were gone by the 18th century, making them probably the first well-documented rookery to be exterminated at our hands.

Rookeries in the Bahamas followed them, and then those on Caribbean coasts and islands. Since Columbus's time, green sea turtle populations in the Caribbean have probably declined by at least 99 percent. The immense nesting colonies in the Caymans collapsed in the 18th century. So did the fishery based on them. By 1802, Cayman islanders had to travel to Cuba, and then to Honduras and Nicaragua, to find turtles. Today, the only green sea turtles likely to be seen in the Caymans are the captive animals at the Cayman Turtle Farm, or the turtles the farm has released.

A similar story can be told for other sea turtle species, and other populations, in many parts of the world. Huge numbers of sea turtle eggs have been, and still are, taken on both sides of the Pacific. Greens and hawksbills

A poacher carries a sea turtle ashore on the Turks and Caicos Islands in the Caribbean.

Probably the only green sea turtles (*Chelonia mydas*) in the Cayman Islands today are at the Cayman Turtle Farm.

Olive ridleys (*Lepidochelys olivacea*), killed for the leather trade, in a slaughterhouse (now closed) in Oaxaca, Mexico.

(*Eretmochelys imbricata*) have been exploited along the coasts of East Africa for at least 2000 years; the tortoiseshell trade spread from the Red Sea, the source of tortoiseshell in antiquity, to Madagascar, where exports began as early as 1613. In the Seychelles, concern was expressed for the survival of turtle populations as early as the 18th century.

Between 1961 and 1972, the Philippines shipped tortoiseshell from approximately 45,000 hawksbills to Japan. At the same time, a large sea turtle leather industry sprang up in Mexico and Ecuador, which used skin from the animal's front flippers and the underside of the neck. The leather, mostly from olive ridleys (*Lepidochelys olivacea*) but also from green sea turtles, was exported to Japan, Europe, and the United States.

Much of the legal international trade in sea turtles and their products ended after 1981, when all of the members of the Cheloniidae were added to Appendix I of CITES (see Chapter 9). The leatherback had been on Appendix I since 1977. Trade continued, though, among non-Parties and CITES Parties, like Japan, that had entered *reservations* allowing them to be exempt from the treaty's provisions with respect to some species of sea turtles. Though Japan has now withdrawn its reservations, and few countries are left in the world that are not Parties to CITES, the killing and egg collecting still goes on in much of the tropics, and sea turtle populations continue to decline.

As the numbers of hunters increase, traditions regulating hunts grow weaker, and turtle habitat degrades, so that even traditional hunts are harming turtle populations today. In the Kai Islands of eastern Indonesia, leatherbacks (*Dermochelys coriacea*) were once hunted only for ritual and subsistence use, protected from commerce by a tradition called *adat.* Today, as the human population in the islands grows and changes, *adat* is being forgotten and leatherbacks are being killed for the market. Unfortunately, this shift in the hunt has happened at the same time as the nesting colonies of the Kai Island leatherbacks, 1000 km (620 mi) away in Irian Jaya, are suffering extensive losses from beach erosion, egg poaching, and nest raiding by feral pigs.

For many on the island of Bali in Indonesia, turtle meat is de rigueur for feasts and religious ceremonies. The largest turtle slaughterhouse in the world may be the Balinese market at Tanjung Benoa. Some 20,000 green sea turtles are killed in Bali each year, far exceeding legal quotas set by the government of Indonesia. Fishermen have depleted Bali's own turtle stocks, and now hunt farther afield within Indonesian waters, taking turtles that are mostly migrants from Australian stocks. In 1998, they took an estimated 5000 turtles in the Aru Islands, a harvest large enough in itself to cover the Balinese quota.

The Indonesian government is now trying to prosecute the most flagrant of the Balinese turtle dealers. In May 2001, the chief among them, Widji Zakaria or Wewe, was sentenced to a year in prison for illegal activities including poaching, transporting and selling green sea turtles. As of June 2001, though, he is still free.

### A Sea of Problems

Today sea turtles bear the brunt, not just of direct exploitation, but of a host of other pressures. Thousands drown accidentally every year, entangled or hooked in fishing gear. Turtle nesting grounds are degraded, or lost altogether, as we develop beaches and construct seawalls. Artificial lighting draws many hatchling turtles to their deaths (see Chapter 8). Leatherbacks and other sea turtles searching for jellyfish on the high seas choke on floating garbage, including plastic bags and deflated helium balloons that they apparently mistake for their prey. Pollution, and disturbance or damage by boat traffic, fouls feeding grounds on coral reefs or sea grass beds, and may spread disease. In such places as the Spratly Islands off Taiwan, and in the Philippines, reefs are further damaged by illegal fishing with dynamite and cyanide. Since all sea turtles exhibit temperature-dependent sex determination (TSD), global warming may perhaps pose the same threat to their reproduction as it does for TSD turtles on land or in fresh water (see Chapter 7).

The combined effect of all these threats, and the declines they have produced, have led the Marine Turtle Specialist Group of IUCN to classify six of the seven sea turtle species as

A sea turtle shrine in Bali, Indonesia, where thousands of green sea turtles (*Chelonia mydas*) are killed every year.

This green sea turtle (*Chelonia mydas*) has drowned in a "ghost" fishing net off the Cayman Islands in the Caribbean.

Endangered or, in the case of the leatherback, the hawksbill, Kemp's ridley (*Lepidochelys kempi*), and the Mediterranean population of the green sea turtle, Critically Endangered—IUCN's category of highest risk. Only the flatback (*Natator depressus*), with most of its population in protected waters off Australia, is listed in the lower category of Vulnerable.

These listings may seem to be overdramatic. After all, sea turtles still number in the thousands, despite appallingly heavy egg harvests and continuing environmental stresses. Such a view of their status, however, may be blind to the effect that sea turtle longevity has on our perceptions. As Karen Bjorndal has pointed out:

> ... delayed sexual maturity and the corresponding large number of immature age classes mask the effects of intense harvests, so that over-exploitation can be mistaken for sustainable utilization for years, with eventual disastrous results. Year after year for decades, every nesting female and every nest can be killed on a nesting beach, and still, against any reasonable expectation, hawksbills will continue to crawl out on the nesting beach. What appears to be astonishing resilience to total exploitation is in fact not resilience at all, but merely the harvesting of 20 to 40 years' worth of subadults as they become sexually mature and venture onto the nesting beach for the first time.

There are places in the world where sea turtles are holding their own, or even improving their status, though these are the exception, and in almost every case these exceptions result from years of careful and intensive conservation and management. Leatherback turtles in the Caribbean and South Africa, and Kemp's ridleys in the Gulf of Mexico, have proven that sea turtles do not have to be doomed if we take the proper actions to conserve them. Elsewhere, though, the story is often grim.

Colin Limpus, in a 1998 survey of turtle populations in the Indo-Pacific, reported declines in population after population, and in species after species. Even the relatively secure

A captive loggerhead sea turtle (*Caretta caretta*) snaps at a plastic bag, perhaps mistaking it for a jellyfish.

Destruction of coral reefs threatens the habitat of hawksbills (*Eretmochelys imbricata*, photographed in Malaysia).

A graduate student measures a flatback sea turtle (*Natator depressus*) on Curtis Island, Queensland, Australia.

Meat, eggs, and a flipper of a green sea turtle (*Chelonia mydas*) on sale on a street in Limón, Costa Rica.

flatback was threatened by near-total predation of its eggs by pigs on its major nesting beaches in the Torres Strait Islands between Australia and New Guinea, and by the death of hundreds of turtles annually in the northern prawn fishery. Limpus concluded:

> ... marine turtle populations in the Indo-Pacific region outside Australia are severely depleted and/or are subjected to overharvest and/or to excessive incidental mortality. Where census data exist, most populations show clear evidence of decline. At the present time, all available data indicate that the general conservation outlook for marine turtles in Southeast Asia is dismal.

In 2001, TRAFFIC North America, a joint project of the World Wildlife Fund and IUCN, released *Swimming Against the Tide*, a study of the exploitation, trade, and management of sea turtles in 11 nations and territories in the northern Caribbean. The study "confirmed that demand for turtle meat and eggs remained strong in the region, and the use of marine turtles continues in all areas surveyed, despite fully protective legislation in five of the 11 nations/territories reviewed." The harvests today seem to be driven largely by domestic demand for meat and eggs, though hawksbill shell products continue to be sold illegally to tourists in the Dominican Republic, Jamaica, and Mexico. The study concluded that "Today, marine turtles are swimming against a tide of deeply entrenched use patterns, insufficient law enforcement and awareness, and inadequate political commitment."

The problem, then, is clear; deciding what should be done about it is not.

The traditional approach to sea turtle conservation has been either to focus on protecting nesting beaches, or to collect and incubate eggs in turtle hatcheries and release the young into the sea either as soon as they hatch, or after they have grown past their initial and, presumably, most vulnerable, size—a controversial procedure called *headstarting*. These techniques are appealing. They provide the opportunity for hands-on work with turtles themselves, can generate a lot of favorable interest and publicity,

Hatchling green sea turtles (*Chelonia mydas*) raised in a hatchery in the Turtle Islands Park, Sabah, Borneo, Malaysia.

A ring of fence wire protects a sea turtle nest in the Turtle Islands Park, Sabah, Borneo, Malaysia.

and do not usually require one to go up against the fishing industry, beach-front developers, or other powerful interests.

Hatchery releases and beach protection have undoubtedly been of great importance to sea turtle conservation, but they have their drawbacks. Incubating eggs in hatcheries may produce large numbers of hatchlings, but this may not always translate into a real increase in the breeding population. Where hatchlings are held too long prior to release, or when large numbers of hatchlings are released at the same time and place (see Chapter 8), mortality may be particularly high. There have been controversies over such things as incubation temperature; is it better to pick a temperature that produces an equal mix of males and females, or should the sex ratio be shifted one way or the other? The fact that we cannot answer these questions satisfactorily suggests that we may be further, in our attempts at management, from understanding what sea turtles really need than we like to admit.

Using hatcheries to incubate eggs that would otherwise die on the beach, and releasing the young immediately upon hatching is still an important and valuable technique for sea turtle conservation. In recent years, however, headstarting has come under severe scrutiny. The most well-known headstarting program was initiated in 1978, in an attempt to establish a second nesting colony of Kemp's ridleys on the Texas coast. The hope was that turtles headstarted from Texas would return there to breed, instead of traveling to the main nesting beach in Mexico. The project became highly controversial, with questions raised as to whether headstarted turtles could survive in the wild after a year of being fed in captivity. In 1993, the project was abruptly terminated, and arguments about its merits still continue. Recently, some Kemp's ridleys have nested on South Padre Island in Texas, but whether these are the first fruits of the headstarting project remains unknown.

The protection of nesting beaches has been vital to the conservation of such populations as the leatherbacks of Tongaland in South Africa, the green sea turtles of Tortuguero National Park in Costa Rica, or the nesting loggerheads of Floida. Nonetheless, beach protection and hatchery releases can do little good by themselves if the chief problem is the killing of breeding adults, or large subadults, at sea. Since saving the life of a single breeding adult of a long-lived species like a sea turtle may have a far greater conservation impact than releasing hatchlings, biologists have recommended that the adults, not the young, should be the chief focus of sea turtle management.

Ideally, of course, conservation should focus on all of a sea turtle's life stages, from egg to breeding adult. If we are to have a real chance of success in sea turtle conservation, we will need the cooperation of the many countries through whose waters the animals pass in the course of their long lives. As Jack Frazier has pointed out, "These are shared resources, not owned by any one country, not the unique right of any single nation, nor the sole responsibility of any single state." No one program, however worthy, can save sea turtles by itself; sea turtle conservation must be an international concern and an international effort.

On the good side, sea turtles, more than any other reptile, have become international symbols of wildlife conservation, almost as well known, at least in the West, as such "charismatic megafauna" as the giant panda and the African elephant. That gives international initiatives to protect them a valuable, and perhaps vital, head start.

Volunteer Dayna Coutal counts eggs from a flatback (*Natator depressus*) nest on Curtis Island, Queensland, Australia.

### Tumors, Pollution, and Disease

In recent decades, growing numbers of sea turtles have been infected with a mysterious disease known as *fibropapillomatosis* (FP). Diseased turtles are disfigured with bulbous, fibrous tumors, or *fibropapillomas*, that may swell to 30 cm (12 in) across. Tumors usually grow on their skin, particularly around the eyes and on the soft skin on the neck and flippers, but roughly a quarter of the infected turtles have internal tumors as well, which grow in their lungs, heart, liver, kidneys, or gastrointestinal tract. Forty percent of infected green sea turtles studied by George Balazs and his co-workers on Oahu in the Hawaiian Islands had tumors in their mouths. Elsewhere, oral tumors have only shown up in a few turtles from Australia. FP most commonly infects green sea turtles, but it is also known in Kemp's and olive ridleys, flatbacks, loggerheads (*Caretta caretta*), and, occasionally, in hawksbills.

Though the tumors are not cancerous, turtles with the disease are often weak, emaciated, and anemic, with imbalanced body chemistry. Internal tumors can cause a variety of problems including bowel obstruction, kidney failure, and flotation difficulties. The disease is frequently fatal, particularly in juveniles, or at least it weakens its victims to such an extent that they die from other causes. Over 80 percent of the stranded turtles autopsied since 1996 at the National Wildlife Health Research Center in Hawaii

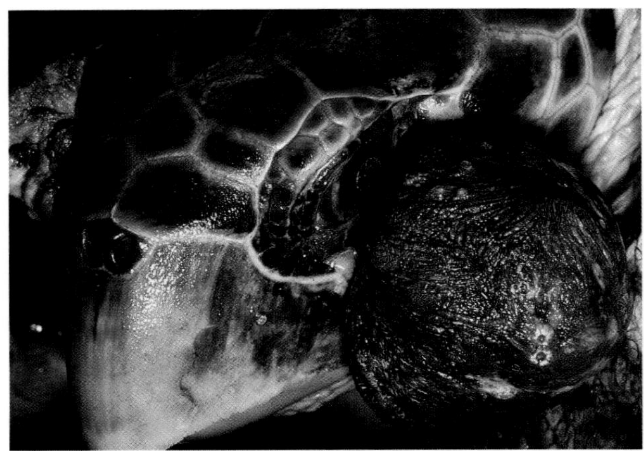

A hawksbill (*Eretmochelys imbricata*, possibly a hybrid) suffering from fibropapillomatosis (FP), at the Turtle Hospital, Marathon, Florida.

A green sea turtle (*Chelonia mydas*) with fibropapillomatosis (FP) on Maui, Hawaii.

have had FP.

FP has been known for decades; the first reported case was apparently a green sea turtle, brought to the Key West Aquarium in Florida in 1934, that developed tumors in 1936. Since the 1980s, the disease has appeared in other areas, and infected higher and higher proportions of sea turtles within infected populations. Much of the Caribbean is still almost free of the disease, though there are signs that it may be spreading there. Green sea turtles in Florida and Hawaii, however, have been particularly hard hit, with FP infecting one-half to three-quarters of the turtles at some sites. The infection rate is lower, but still significant—17 percent of greens and 8 percent of loggerheads—among the turtles of Moreton Bay in eastern Australia.

How serious is the disease? The confusing picture of FP infection led Peter Bennett and Ursula Kueper-Bennett, working with George Balazs, to begin following and videotaping Hawaiian green sea turtles on Maui. Of the 418 individual turtles the Bennetts identified in the years between 1988 and 2000, 245 showed evidence of the disease. In 21 of the 91 turtles they found in more than one summer up to 1998, their subsequent sightings showed that the tumors had regressed. That is partly, but not entirely, good news. Of the recovering animals, only one was a juvenile, and among adults, males seemed three times as likely to regress as females (though in more recent years more females may be regressing). If the pattern on Maui is typical, FP may still have a serious impact on the age and sex distribution of green sea turtles.

We still do not know exactly what effect FP is having on sea turtle populations. Nor do we have a clear understanding of its cause, or, perhaps more accurately, its causes. FP tumors appear to be associated with a number of viruses; more than 95 percent of the tumors studied in Florida carried a herpes virus. Turtles may be growing more vulnerable to the disease, though, because something is weakening their immune systems; Hawaiian turtles in advanced stages of FP have suppressed immune responses. On the other hand, laboratory research suggests that even turtles without suppressed immune systems can contract FP.

Outbreaks of FP may be linked to toxins produced by a number of different algae, whose growth in turn has been linked to marine pollution. High algal blooms tend to be commonest in enclosed bays that do not flush out easily, where pollution levels are often higher than in the open sea. In Australia, blooms of a blue-green alga called *Lyngbya* occasionally smother the sea grass beds in Moreton Bay. The blooms not only affect the turtles' food supply, but produce a toxin known to promote tumor growth. In Florida, blooms of the dinoflagellate *Prorocentrum*, a single-celled alga, produce biotoxins including okadaic acid, a chemical that produces tumors in mice. Green sea turtles pick up these poisons when they eat sea grass covered with the algal growth.

Recent research by Karen Holloway-Adkins and Llewellyn Ehrhart suggests that the incidence of FP in Florida is highest in places where the turtles are eating more *Prorocentrum*-covered sea grass. They found almost no FP at a Trident submarine base where only about 3 percent of the sea grass was covered with *Prorocentrum*, while at Mosquito Lagoon, where 79 percent of the sea grass is infested, the FP rate among green sea turtles is 72 percent—a figure that has grown from 29 percent in 1985.

There is much we do not know about FP. We do know, however, that it is spreading, that it is contagious, and that it is most severe in areas where humans pollute the sea. FP has consequences, too, for the already-contorversial practice of headstarting, in which large numbers of hatchlings are reared in captivity. If workers head starting facility are not extremely careful, they may be providing, by keeping unnaturally large numbers of turtles together, a golden opportunity for the disease to spread further.

### Hook and Longline

Fishing nets and lines have always caught more than their haulers were looking for. In recent decades, as the technology of the fishing industry has made it easier to spend longer and longer at sea, and to deploy ever more sophisticated and effective catching devices, bycatch—the unnecessary and wasteful destruction of animals the fishing industry does not even want—has become a growing conservation concern, and sea turtles have become some of its more high-profile victims.

After the international bans on driftnet fishing in the late 1980s, many fishing fleets turned to *longlining*. Longline fishing vessels set out great lengths of hook-studded monofilament line, baited for valuable catches like swordfish, tuna, and dolphin (the fish *Coryphaena hippurus*, also called mahimahi, not the mammal). A single high-seas longline may carry between 1000 and 1500 hooks, and trail out for over 60 km (37 mi). There may be thousands of them being drawn through the ocean at any given time,

This loggerhead sea turtle (*Caretta caretta*) was rescued before it could drown in a monofilament gill net.

particularly from the huge offshore fleets of Japan, Korea, and Taiwan.

Though longlines are not the indiscriminate "curtains of death" that driftnets were, they can be deadly all the same. Seabirds dive for the barbed baits and drown in vast numbers. Longlines have been blamed for spectacular plunges in the numbers of albatrosses, greatest of the pelagic seabirds, in the Southern Ocean.

Less well known, but equally devastating, are the longline catches of sea turtles. Off the Pacific coast of Costa Rica, coastal fishing vessels operating within its Exclusive Economic Zone (EEZ) tow 24–32 km (15–20 mi) lines bearing 400 to 700 hooks each. The ships are not fishing for sea turtles, but the turtles, nonetheless, have become the second-largest component of their catch. Between August 1999 and January 2000, Randall Arauz monitored nine such longline trips for the Sea Turtle Restoration Project. The boats caught 253 turtles, almost all of them olive ridleys.

If Arauz's figures are typical, the inroads of longline fleets off the Pacific coast of Latin America must be substantial indeed, even if some of the hooked turtles survive. Vessels from Costa Rica, Ecuador, and Chile hunt mahimahi and swordfish. The longlines they set for mahimahi lie near the surface, where a hooked turtle can snatch a breath and may survive its ordeal. The lines for swordfish, however, are set deeper, and a turtle caught in them will almost certainly drown.

The eastern Pacific is now being invaded by high-seas fleets from Taiwan, longlining for shark fins, and from Spain, in pursuit of tuna. What they may be doing to sea turtles there no one knows, though tuna swim deeper than either mahimahi or swordfish, below the levels turtles usually reach.

A similar story is unfolding in the Atlantic, around the Azores. Here, loggerheads fall victim to swordfish longliners. Alan Bolten has estimated that every time a longline goes over the side turtles are snagged at a rate of 1.7 for every thousand hooks—a very substantial number, considering the intensity of longline fishing in the area.

Researchers, often working with the fishing fleets, are trying to find ways to minimize the threats longlining poses for turtles. Off the Azores, Bolten has been experimenting with different types of hook to see if changing the gear the fishing vessels use can reduce turtle mortality. In Hawaii, researchers have been testing captive turtles to see if the baits longliners use can be made less attractive to them. Attempts to reduce seabird mortality by using baits that are dyed blue, presumably making them harder for the birds to see in the water, have met with some success, but the turtles seem to learn to eat them after only a few days' experience.

Perhaps the most critical longlining victim is the Pacific leatherback. Longliners may catch few of these giants—Alan Bolten's studies in the Atlantic found that of 237 turtles caught in 93 longline sets off the Azores, only four were leatherbacks—but leatherback numbers in the Pacific are already so low that even a small catch could deliver the coup de grâce to the whole Pacific population.

James Spotila and his colleagues have been sounding the alarm over the fate of the Pacific leatherback since the mid-1990s. Between 1982 and 1986, the world population of adult female leatherbacks fell from an estimated 115,000 to 34,500. The Atlantic population is still fairly healthy, but in the Pacific, the picture is not so much one of decline as of collapse. The famous nesting beaches of Terangganu in Malaysia, once used by thou-

A leatherback (*Dermochelys coriacea*) from the greatly depleted Pacific population swims in the seas off the west coast of Mexico.

sands of leatherbacks, were almost deserted by 1994. At Playa Grande in Costa Rica, Spotila's team counted 1367 nesting females in 1998–89, but only 117 a decade later. At the once-huge colony on the Pacific coast of Mexico, numbers fell from 70,000 in 1982 to fewer than 250 in 1998–99.

In a paper published in the journal *Nature* in June 2000, Spotila and four other well-known sea turtle biologists placed the blame directly on the fishing industry. They noted that, by a conservative estimate, "longline and gill-net fisheries killed at least 1500 female leatherbacks per year in the Pacific during the 1990s. These included Asian trawl, longline and drift-net, Central and South American longline and gill-net, and Hawaiian longline fisheries."

Whatever the cause, the *Nature* paper paints a dire picture: "The total adult and subadult population is about 2955 females—compared with over 91,000 adults in 1980. We conclude that leatherbacks are on the verge of extinction in the Pacific." If that is to be avoided, the annual mortality rate must fall from its current 23 percent per year to 1 percent—or less than 50 adult females per year in the whole of the Pacific.

In 2000, the threat to the leatherbacks was serious enough to convince a Hawaiian court to close a vast stretch of the central Pacific—as large as the North Atlantic—to unrestricted longlining. In part of this region, the number of lines that can be set out is restricted; in another, longlining is banned altogether. Over the rest of the area, which includes the seas around the main Hawaiian Islands, longlining is only permitted for tuna. Sea turtle biologists and environmentalists applauded the decision (though longliners, as might be expected, bitterly attacked it), but they were concerned that it did not go far enough. Larry Crowder wrote in *Nature*:

> Closures in U.S. fisheries alone will not resolve the problem—leatherbacks in both the North and South Pacific are killed by fishing vessels from several countries.
>
> The leatherbacks' nesting habitat must be protected throughout the Pacific. Direct harvest of eggs or turtles must be banned everywhere. Wherever fishing occurs, the bycatch of leatherbacks must be reduced to as close as possible to zero. Once countries such as the United States have minimized their own bycatch, they should encourage other nations to adopt protection measures. The U.S. government and environmental organizations need to support research and management programs in less developed countries, with environmental organizations educating the public and encouraging consumers to avoid products resulting from longline fishing.

An international treaty now in the final stages of negotiation, the Convention on

Conservation and Management of Highly Migratory Fish Stocks, has provisions to reduce bycatch. Sea turtle biologists like Crowder have insisted that the treaty contain explicit language on turtles.

In May 2001, charging that longliners were simply relocating from Hawaiian waters to the seas off California to avoid the ban, the Turtle Island Restoration Network and the Center for Biological Diversity sued the National Marine Fisheries Service (NMFS), to force it to consider the impact of the fishery on the Pacific leatherback population under the U.S. *Endangered Species Act.*

A Kemp's ridley (*Lepidochelys kempii*) entangled in a monofilament gill net.

## TED Wars

No issue better demonstrates the sociological, economic, legal, and political ramifications of sea turtle conservation than the controversy over Turtle Excluder Devices (TEDs). What started out as a straightforward technological fix, intended to keep sea turtles from drowning in shrimp trawls, became a political nightmare of lawsuits, regulatory battles, protests, massive demonstrations at sea, and, eventually, international legal battles before the tribunals of the World Trade Organization (WTO).

The issue of turtle mortality in shrimp trawls first came to public notice during the 1960s and 1970s, a time when conservation awareness was finally beginning to move beyond the narrow circles of scientists and wildlife managers. By the 1980s, the U.S. government had identified the drownings as the major factor causing the deaths of loggerheads and Kemp's ridleys, in particular, along its Atlantic and Gulf coasts.

In 1980, the U.S. National Marine Fisheries Service (NMFS) presented their solution: the Turtle Excluder Device or TED. This was a metal cage, or, in later versions, a grid, installed in the trawl. It allowed shrimp to pass through and be caught, but forced turtles out to safety through a trapdoor at the top or bottom. Trials by NMFS showed that TEDs reduced sea turtle mortality by 97 percent. NMFS later approved a number of modified TED designs developed by the industry itself. At first, the industry agreed to adopt TEDs voluntarily. Supporters of TEDs pointed out that shrimpers stood to benefit from the devices. By excluding not only turtles but also unwanted larger fish, TEDs would make the trawls easier to pull, require less fuel to operate, and reduce damage to shrimp that would otherwise be crushed by the larger animals in the net.

Nonetheless, shrimpers, by and large, saw TEDs not as a boon but as a threat, and were reluctant to cooperate with the government. Finally, in 1987, threatened with a lawsuit by the Center for Marine Conservation, Greenpeace, and the Environmental Defense Fund under the provisions of the *Endangered Species Act*, NMFS brought in regulations requiring the seasonal use of TEDs on shrimp trawls in offshore waters from North Carolina to Texas.

The industry, which had never before faced across-the-board federal regulation, reacted in fury. Louisiana challenged the order in court and passed a law prohibiting the enforcement of the federal regulations in the state. Louisiana governor Edwin W. Edwards told thousands of cheering shrimpers, "Perhaps some species were just meant to disappear. If it comes to a question of whether it's shrimpers or the turtles . . . bye-bye turtles." In July 1989, as the regulations were about to be reimposed following a court order, hundreds of shrimp vessels staged a 36-hour blockade of Gulf ports in Texas and Louisiana. Despite the protests, in 1991 further regulations made TED use mandatory year-round in most parts of the southeastern United States.

The two sides, shrimpers and conservationists, remained far apart. The shrimpers saw TEDs as the ruin of their industry. They claimed that the heavy metal grates could be dangerous when the net was hauled back on board. They argued that TEDs allowed many shrimp to escape, and could become tangled with bottom debris, causing fishing gear to be ruined or lost. TED supporters argued that there was little evidence supporting these claims. A 1992 report from the Center for Marine Conservation, the World Wildlife Fund, and the National Wildlife Federation pointed out that shrimp catches actually went up in the first two years following the imposition of TEDs, claims for gear loss and damage went down, and there was not a single report of a TED-related injury.

The shrimpers denied that they were a major cause of turtle deaths. In 1987, however, NMFS estimated that 47,973 turtles were captured annually in commercial shrimp trawls, of which 11,179 drowned. In 1990, the National Academy of Sciences estimated that up to 55,000 turtles drowned per year in U.S. coastal waters in trawls not equipped with TEDs.

This shrimp trawl net carries a "Georgia Jumper," a variety of Turtle Excluder Device (TED).

These estimates did not include trapped turtles that were still alive when the trawl was hauled in, and were released back into the sea. It appears that many of these animals, particularly juveniles, died shortly after being released, probably from shock as a consequence of being deprived of oxygen while they were still in the net.

Shrimpers have been blamed for increases in the number of dead turtles washed up on beaches in the southeastern Atlantic states and the Gulf of Mexico, under the assumption that most of these animals were victims of the trawls. In 1991, Charles Caillouet and his colleagues published a study showing that dead turtles washed up on the beaches of southwestern Louisiana and Texas in significantly higher numbers where shrimping boats were most active offshore.

By 1989, shrimpers had moved the TED issue onto the world stage. Faced with mandatory TED use themselves, they persuaded Congress that other countries exporting shrimp to the United States should be forced to play by the same rules. In November 1989, Congress passed a law, which became s. 609 of the 1989 U.S. *Endangered Species Act*, banning imports of shrimp unless the exporting countries met certain conditions relating to sea turtle conservation. Any country seeking to export trawler-harvested shrimp to the United States had to be *certified*; in effect, it had to prove that its sea turtle protection measures were up to U.S. standards.

Originally, the law was applied only to 14 countries in Central and South America, giving them three years to phase in the use of TEDs in shrimp fisheries. By and large, the countries cooperated, assisted by training missions from NMFS. Only French Guyana, and, for a time, Surinam, failed to achieve certification. In 1996, however, Earth Island Institute challenged the interpretation of the law in court, arguing that it should apply to all nations whose shrimp fisheries could harm sea turtles. They finally succeeded in the U.S. Court of International Trade in 1996, a decision that expanded the number of countries affected by the U.S. law from 14 to about 70.

NMFS and the U.S. State Department immediately began training missions in Asia and Africa. Some Asian countries, including Thailand, instituted laws of their own requiring TED use. A number of Asian countries, however, saw the U.S. law as both an egregious case of eco-imperialism and an attempt to discriminate unfairly against their shrimp fisheries.

In 1997, Thailand, Malaysia, India, and Pakistan lodged formal complaints against the United States with the World Trade Organization. Thailand, which already used TEDs and had been certified by the U.S. in 1996, claimed to be joining the action as a matter of principle. In March 1998, the WTO ruled against the United States. Pressed by environmental groups, the U.S. appealed, but on October 12, 1998, the Appellate Body of the WTO ruled against it again. One of the Appellate Body's chief objections was that the U.S. law "*requires* other WTO Members to adopt a regulatory program that is not merely *comparable*, but rather *essentially the same*, as that applied to the United States shrimp trawl vessels." In other words, the law was objectionable because it required exporting countries to use TEDs, not some other technology that worked as well (for example, the "beam trawls" used by some Chinese shrimpers).

The Appellate Body tried hard to avoid the appearance that it was opposed either to environmental protection or to sea turtles:

> We have *not* decided that the protection and preservation of the environment is of no significance to the Members of the WTO. Clearly, it is. We have *not* decided that the sovereign nations that are Members of the WTO cannot adopt effective measures to protect endangered species, such as sea turtles. Clearly, they can and should.

This sort of language allowed both the United States and the complainant countries to claim a measure of victory. Nonetheless, environmental organizations, particularly in the West, excoriated the WTO, accusing it of placing trade concerns above the needs of the environment. The decision made sea turtles flagship symbols, not just of the environment, but of the victims of globalization. Some of the protesters agitating against the WTO during the "Battle for Seattle" in 1999 showed up at the barricades dressed in sea turtle costumes.

Meanwhile, TEDs, the cause of all the concern, continue to be adopted. Evidence that TEDs really do make a difference has

Demonstrators in sea turtle costumes protest against the World Trade Organization in Seattle, November 1999.

continued to grow. Maurice Renaud and his colleagues reported in 1997 that the number of sea turtles taken in shrimp nets around the southeastern United States had fallen by an order of magnitude since TED use became mandatory. In 1999, after federal and state enforcement officers in the state of Georgia began boarding shrimp trawlers to check compliance with TED regulations, the numbers of sea turtles stranding on Georgia beaches dropped precipitously. The adoption of TEDs has probably been a major contributor to the changing fortunes of Kemp's ridley, once the most endangered of all sea turtles. Since 1986, nesting Kemp's ridley numbers have increased tenfold.

The United States has adopted a compromise policy to meet WTO requirements (to the dismay of some environmental groups). In June 2001, a WTO dispute settlement panel, rejecting a further complaint from Malaysia, found that U.S. policy was no longer discriminatory. As of April 2000, 41 countries have become eligible to export shrimp to the U.S. On April 11, 2001, Orissa's Chief Minister, Naveen Patnaik, announced that TED use would become mandatory for all mechanized fishing boats operating off the Orissa coast in India, where 80 percent of India's sea turtles and perhaps half of the world's olive ridleys nest. As important as they are, though, TEDs are not a complete solution to the problems of trawling; heavy trawls dragging along the bottom may still damage turtle habitat, and in some sensitive areas may need to be excluded altogether. That, however, is another battle.

**Turtle Treaties**

During the 1990s, fear of fisheries sanctions, like the U.S. TED law, helped drive the development of international agreements for the cooperative management and conservation of sea turtles. In the Old World, these agreements have been concluded under the auspices of the Convention on the Conservation of Migratory Species of Wild Animals (CMS), often called the Bonn Convention.

The CMS applies to sea turtles in two ways. All species, except the flatback, are listed on Appendix I (not to be confused with Appendix I of CITES) and are supposed to receive strict protection on a country-by-country basis from the 73 (as of March 1, 2001) CMS Parties. This protection can include local projects approved by the CMS Scientific Council, funded through the treaty itself. Sea turtles are also listed on the much larger Appendix II, in order to stimulate countries to conclude specialized "Agreements" (in effect, mini-treaties created under CMS auspices) for their conservation.

Twenty-five African countries are now Parties to a CMS-brokered Memorandum of Understanding Concerning Conservation Measures for Marine Turtles of the Atlantic Coast of Africa, signed in 1999. A country does not have to be a member of the CMS to

Green sea turtles (*Chelonia mydas*) mate in the Turtle Islands Heritage Protected Area, Borneo, Malaysia.

participate in an Agreement; in July 2000, 38 countries bordering the Indian Ocean, including a number of non-CMS countries in Asia, concluded a Memorandum of Understanding on the Conservation and Management of Marine Turtles and their Habitats of the Indian Ocean and South-East Asia (IO-SEA), under CMS auspices, to come into effect on 1 September 2001. According to the press release CMS issued at the time:

> The Memorandum of Understanding puts in place a framework through which States of the region—as well as other concerned States—can work together to conserve and replenish depleted marine turtle populations for which they share responsibility. It acknowledges a wide range of threats to marine turtles, including habitat destruction, direct harvesting and trade, fisheries by-catch, pollution and other man-induced sources of mortality. The Memorandum recognizes the need to address these problems in the context of the socio-economic development of the States concerned, and to take account of other relevant instruments and organisations.

In a bilateral agreement outside the CMS, the governments of Malaysia and the Philippines have established the world's first trans-frontier protected area for sea turtles. The Turtle Islands lie partly in Philippine waters and partly in the Malaysian province of Sabah. They are important nesting grounds for green and hawksbill sea turtles. Satellite data show that the turtles that nest there do not leave the area immediately, but may spend up to three weeks near the islands before heading for more distant feeding grounds. In 1996, the governments of the two countries signed a Memorandum of Agreement establishing a jointly managed Turtle Islands Heritage Protected Area (TIHPA).

Although by and large the Americas have stayed out of the CMS, a number of Latin American nations have been cooperating on sea turtle conservation for some time. The Wider Caribbean Sea Turtle Network (WIDECAST), a coalition of national coordinators and nongovernmental organizations from 30 wider Caribbean nations and territories, has held annual regional meetings since 1984, promoting coordinated management action among governments in the area.

Two recently concluded treaties, both strongly supported by WIDECAST, give American sea turtles potentially powerful legal protection. The Protocol Concerning Specially Protected Areas and Wildlife (or the SPAW Protocol) entered into legally binding force on May 25, 2000, slightly more than 10 years after it was first signed in Jamaica. The protocol is a legally binding addendum to the Convention for the Protection and Development of the Marine Environment of the Wider Caribbean Region (or the Cartagena Convention).

Gathering sea turtle eggs for a hatchery, in the Turtle Islands Park, Sabah, Borneo, Malaysia.

A handful of hawksbill (*Eretmochelys imbricata*) eggs, in the Bay Islands, Honduras.

The SPAW Protocol is designed to protect both species and their habitats. Member countries are required to take some very strict measures to protect endangered species (including sea turtles), in particular the prohibition of "taking, possession or killing" and "commercial trade in such species, their eggs, parts or products." How this will work out in practice remains to be seen; the signatories to the protocol include Cuba, which allows a legal harvest of hawksbills.

SPAW applies only to the wider Caribbean region, and is not, strictly speaking, a sea turtle convention. The first independent treaty exclusively devoted to sea turtles is the Inter-American Convention for the Protection and Conservation of Sea Turtles (IAC). The IAC, which has the potential to apply throughout the Americas, entered into force on May 2, 2001, after a negotiation and ratification process that began in September 1994. The IAC has won the support not only of WIDECAST, but of fisheries organizations including the Latin American Association for Fisheries Development (OLDEPESCA), which has been involved in the IAC process from the first, though much of OLDEPESCA's support was because they saw the IAC as a preferable alternative to U.S. embargoes on Latin American shrimp exports. OLDEPESCA's support initially put off some sea turtle conservationists, who saw the initiative as a disguised attempt to support the shrimp industry.

By 1996, though, the conservationists were won over, and the treaty as finally drafted is, if not perfect, clearly a step forward for sea turtles. So far, Venezuela, Peru, Brazil, Costa Rica, Mexico, Ecuador, The Netherlands (for the Netherlands Antilles), Honduras, and the United States have taken the final official step required to bind them to its terms by depositing their Instrument of Ratification in Caracas, Venezuela. Belize, Nicaragua, and Uruguay have signed the treaty but have yet to ratify it.

Parties to the IAC are required to take measures "for the protection, conservation and recovery of sea turtle populations and their habitats," including "the prohibition of the intentional capture, retention or killing of, and domestic trade in, sea turtles, their eggs, parts or products." Other required measures include protection of sea turtle habitats in nesting areas, promotion of research and environmental education, and the reduction, "to the greatest extent practicable, of the incidental capture, retention, harm or mortality of sea turtles in the course of fishing activities" through regulation and the use of appropriate gear, specifically including TEDs. The prohibition against killing sea turtles is not quite as draconian as the one in the SPAW Protocol, because it allows for exemptions "to satisfy economic subsistence needs of traditional communities."

There are a number of other important provisions in the IAC, including the establishment of a Consultative Committee and a Scientific Committee, and a requirement for the Parties to meet regularly to consider how well the treaty is working. These may seem less interesting than blanket bans on killing, but it is provisions like these that can allow the IAC to grow keep in touch with ground-level realities, respond to change, and improve. Now that it is in force, its supporters can only hope that it will.

### The Battle for the Hawksbill

During the late 1990s, while SPAW, the IAC and the CMS agreements signaled new levels of international cooperation for sea turtle conservation, CITES became the arena for a bitter and divisive battle over a single species, the hawksbill.

The hawksbill has been listed on Appendix I of CITES since 1977, making all commercial trade in its shell illegal under the treaty (see Chapter 9). When Japan joined CITES in 1980, it took advantage of a provision of the treaty allowing new members to declare themselves exempt from the consequences of certain CITES listings. Japan entered these declarations, called *reservations*, against a number of species, including the hawksbill. Other CITES Parties remained bound by the listing—except for France, which held a reservation on behalf of its overseas territories until 1984—and therefore could not export tortoiseshell to Japan. Non-Parties, however, were perfectly free to ship Japan all the *bekko* (the Japanese name for tortoiseshell) they liked.

In the 1980s and early 1990s, the Japanese Bekko Association, the professional organization of Japanese tortoiseshell workers, became the world's largest importer of hawksbill shell. Among its chief suppliers were Cuba and Mexico, which did not join CITES until 1990 and 1991, respectively. Cuba entered a reservation against the hawksbill listing when it joined, as had St. Vincent and the Grenadines when they joined in 1989, but Mexico did not. That meant that Mexico could no longer trade with Japan, but Cuba could continue to export tortoiseshell as long as Japan maintained its reservation.

In the early 1990s, though, the United States certified Japan under the Pelly Amendment, a law that allows it to bring sanctions against any country that compromises the effectiveness of a wildlife conservation treaty. Faced with potential penalties, Japan withdrew its reservation against the hawksbill listing in 1994. Overnight, the bekko trade became illegal. Even though Cuba still held a reservation against the hawksbill, it could no longer export tortoiseshell to Japan.

Japan and Cuba were not prepared to let the matter rest. Since a reservation, once dropped, cannot be taken up again, their only option was to get the hawksbill transferred from CITES Appendix I, which bars legal trade, to Appendix II, which allows it. The Marine Turtle Specialist Group of IUCN, however, had recently classified the hawksbill as Critically Endangered throughout its range, a step that made transfer to Appendix II for the entire species a political impossibility.

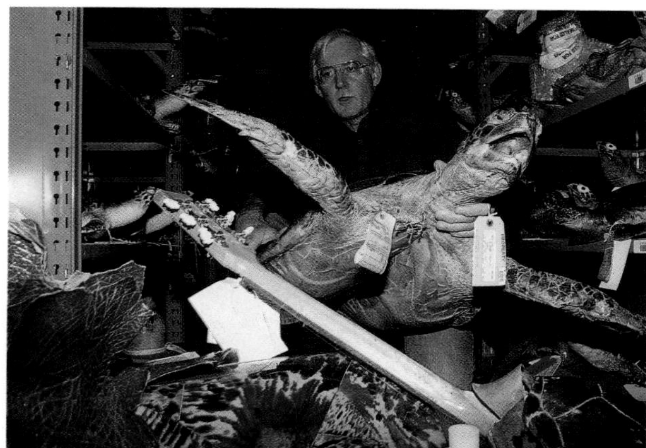

Ken Goddard displays confiscated sea turtle products at the U.S. wildlife forensics lab in Ashland, Oregon.

Some of Cuba's supporters, notably Nicholas Mrosovsky of the University of Toronto, have attacked the IUCN listing, suggesting that it is politically motivated. They argue that it is ridiculous to place an animal ranging throughout the warmer waters of the world, with a population in the tens of thousands, in the same category of risk as such rare and localized species as the Sulawesi Forest turtle (*Leucocephalon yuwonoi*) or the black softshell (*Aspideretes nigricans*).

However, IUCN categories are not just based on range and population, but on rate of decline. That rate is measured not in years but in generations, because generation length is crucial to a species' ability to replace its numbers. For a sea turtle, decades of decline may happen during a single generation. In a 1998 review, the Marine Turtle Specialist Group reaffirmed that the evidence of decline in global hawksbill populations supported Critically Endangered status for the entire species.

At the 1997 CITES conference in Zimbabwe, Cuba—strongly supported by Japan—introduced a proposal to transfer its population alone to Appendix II. Shell was not to be sold under the terms of the transfer unless it came either from government-registered stockpiles, or from a traditional harvest or experimental ranching program. Only a single trading partner—obviously Japan—was to be allowed, providing that it had agreed not to re-export the shell. This arrangement was in order to prevent laundering of tortoiseshell in other countries.

The proposal was defeated. In 2000, Cuba tried again, submitting two proposals: one, cosponsored by Dominica, that was essentially the same as the 1997 proposal, and a second that would have restricted exports to stockpiled shells only. The proponents and their supporters hoped that if the first proposal still seemed unpalatable, the second might succeed. Nonetheless, after intensive lobbying on both sides, the proposals failed to meet the required two-thirds majority needed for approval. As of this writing, it is highly likely that another Cuban proposal to trade in tortoiseshell will be on the table before the CITES Parties at their next meeting in 2002.

It is only fair to the reader to admit my own biases on this contentious issue. I represented the International Wildlife Coalition at both the 1997 and 2000 CITES meetings, where I fought the Cuban proposals. I was concerned that much of the support for Cuba came from countries that support Japan on issues like commercial whaling, and that the Cuban plan to sell its tortoiseshell stocks directly to Japan was practically identical to proposals to sell southern African stocks of ivory—proposals that succeeded at the 1997 CITES meeting. I believed, and still do, that sea turtles were being used as pawns in a wider battle that had little to do with their own conservation.

This is not to say that Cuba's argument was totally without merit, or that Cuba was merely acting as a front for Japanese interests. The Cubans vigorously defended their hawks-

A Caribbean hawksbill (*Eretmochelys imbricata*) off Little Cayman, the subject of ongoing CITES controversy.

bill program. They argued that Cuba's domestic harvest provided a valuable economic incentive for the conservation of its turtles, and created revenue that funds scientific research necessary for their conservation. Supported by a few sea turtle biologists like Nicholas Mrosovsky, they asked the CITES community to at least give them a chance to make an expanded commercial harvest work in the interest of conservation.

The debate, however, turned on more than the merits of Cuba's own program. There were certainly questions raised, particularly by scientists from the Marine Turtle Specialist Group, about the hawksbill population figures provided by Cuban wildlife managers. They questioned whether we know enough about the biology of hawksbills in Cuban waters to conduct an expanded commercial harvest safely and responsibly. Most of the countries that found the proposals unpalatable, however, were probably more concerned about the effect that granting them would have on hawksbill conservation around the world.

Neighboring countries like the Bahamas and Mexico argued that the Cuban "population" actually included turtles from elsewhere in the Caribbean. Mexico pointed to the improvement in the status of its own hawksbill population after Cuba stopped exporting tortoiseshell to Japan, suggesting that Cuba had actually been harvesting Mexican turtles. Hawksbill range states outside the Caribbean were concerned that reopening legal trade, even in the limited terms proposed by Cuba, would provide laundering opportunities that could stimulate tortoiseshell poaching in their countries. Many sea turtle biologists agreed. According to Colin Limpus, active harvest of hawksbills continued in Indonesia as of 1996, with tortoiseshell jewelry and ornaments still on sale in shops. Limpus concluded that hawksbills could only recover in the long term if Japan abandoned its attempts to reopen legal international trade.

In 1999, Karen Bjorndal wrote:

> It is critical to understand, however, that the motivation to continue the total harvest of hawksbills remains strong. Hawksbills and their eggs continue to be taken as a source of food. More importantly, however, hawksbill scutes can be stored easily for long periods of time with no degradation. The scutes are too valuable a commodity for fishermen to stop catching hawksbills because of what may be perceived as just a pause in international trade. Every case of illegal trade and every request to re-open any form of legal international trade encourages fishermen to continue to stockpile scutes, in the belief that eventually they will be rewarded when markets re-open or when opportunities for illegal trade arise. By continuing to vacillate in the commitment to end

international trade in hawksbill products, we prevent the cessation of trade from having its full impact on the conservation of hawksbills.

**Must Sea Turtles Pay Their Way?**

The central point of Cuba's argument at CITES—that wildlife in developing countries will not survive unless it has an economic value—has not been widely accepted by sea turtle biologists, at least as far as a commercial harvest is concerned. In the minds of some, this has acted to the detriment of sea turtle conservation. Supporters of the Cayman Turtle Farm, a commercial operation that has failed to gain international permission to market its products, called a book on its history *Last Chance Lost.* The problem, though, is the same as it is for land and freshwater turtles (see Chapter 9): the characteristics of turtle life history, and the demography of turtle exploitation, may make a sustainable harvest impossible.

One modern sea turtle harvest might reasonably claim to be sustainable, though the proof has yet to be gathered. This is the controversial egg collecting program at Ostional, on the Pacific coast of Costa Rica. Ostional is a major *arribada* site for the olive ridley, with a long history of human exploitation. During the 1980s, scientists noticed that egg mortality during the Ostional *arribadas* was unusually high. So densely packed were the nests on the beach that successive waves of turtles crawling ashore often destroyed the eggs laid by their predecessors. The rotting eggs, in their turn, provided sources for infections that could affect other, healthy clutches.

Under these circumstances, there seemed to be no harm in allowing villagers to collect eggs laid during the first nights of an *arribada*; such eggs were highly unlikely to survive anyway. A legal harvest might under-

Visitors to the Cayman Turtle Farm examine a tank of green sea turtles (*Chelonia mydas*).

cut egg poaching on other beaches, because the eggs from Ostional could be sold at a lower price. The money earned from legal egg sales could also provide a valuable source of income for villagers who had little opportunity to make money in other ways.

In 1987, the Costa Rican government legalized egg collecting on the beach, provided that it took place only during the first 24 (later 36) hours of the *arribada*. The harvest was to be controlled by a locally established cooperative, the Asociación de Desarollo Integral de Ostional (ADIO). Eighty percent of the revenue from the harvest originally went to the cooperative. This was later reduced to 60 percent, with the rest going to the government.

Thirteen years on, the project does appear to have reduced egg poaching. It has certainly put money in the pockets of Ostional villagers. Revenues from the harvest have been used, among other things, to improve the local school, build a community center, and bring electricity to the village. However, a key question for conservationists—whether the harvest is having a negative effect on the local ridley population—remains unanswered.

There admittedly seems to be no evidence that the project is causing harm. Even if such evidence were to appear, the villagers have become so dependent upon the project that it might be difficult to make substantial adjustments to it. In recent years, the Ostional project has become mired in controversy, not so much because of its role in turtle conservation but because of questions concerning its management. There have been accusations of corruption. Randall Arauz, who works with the project, admits the problems. He believes, though, that it is still important to work with the villagers to make the project as successful as possible.

The peculiar circumstances at Ostional, where the eggs that are harvested would otherwise be destroyed in the course of the *arribada*, may make similar projects difficult to establish elsewhere. This brings us back to the problem posed by the Cuban hawksbill harvest. Except in a few special cases, it is probably impossible to exploit sea turtles without causing their decline. As Colin Limpus has written, "The basic problem facing those who want to retain traditional uses of turtles by coastal peoples is that there are increasing numbers of people and a diminishing turtle resource. To continue managing turtles as we have in recent decades is a recipe for their eventual demise."

If the only way to convince peoples in developing countries to conserve their sea turtles is by providing them with an economic incentive, can sea turtles be saved at all?

A partial answer is that it is quite possible to make money from sea turtles without killing them or collecting their eggs. Sea turtle nesting beaches have proven to be highly popular tourist attractions in a number of different countries.

Ecotourism, though, cannot be the answer for every nesting beach. Even in the poorest countries, people may need to cherish their sea turtles, or learn to cherish them, without expecting any economic return. In several places around the world, local organizations,

Villagers from Ostional, Costa Rica, collect olive ridley (*Lepidochelys olivacea*) eggs during an *arribada*.

Bags of turtle eggs, legally collected at Ostional, Costa Rica, are loaded onto a truck for delivery to market.

Sea turtles as a tourist attraction: beachgoers watch a loggerhead (*Caretta caretta*) cover her nest on Juno Beach, Florida.

village cooperatives, and dedicated individuals are proving that this can, indeed, happen.

In Tongaland, South Africa, according to a 1996 report by George Hughes of the Natal Parks Board:

> ...a lucrative tourist industry is developing around the nesting [leatherback] turtles, thereby creating employment and revenue-earning opportunities for the local community... The demand for these tours—both on foot and by vehcle—is growing exponentially and they are emormously valuable for environmental education as well as for gaining support for sea turtle conservation. As a result of their very existence, the leatherbacks are now provding benefits in a far more sustainable manner than when conservation authorities first stopped the exploitation of eggs in 1963.

## Hope

In 1992, a village schoolboy in the south Indian state of Kerala saw a story in the newspaper about the plight of the olive ridley. Poachers were stealing the turtles and eggs to sell in local markets as a curative for hemorrhoids. The boy started his own community organization to protect the turtles. Though his group soon had help from the government forest department, the stimulus for its efforts to protect the nesting beaches came entirely from the community. Youths in the group tried to fence off nest sites, and even built their own hatchery out of local materials. They soon faced a bigger threat than poachers. Illegal sand mining on nearby river estuaries caused the nesting beaches to erode, wearing them away until, in 2001, erosion even claimed the hatchery.

Even if their efforts have been in vain—and they are not giving up—the villagers who supported the group gained something. They have been recognized by the machinery of the state, have won its respect, and have gained attention that is helping them to improve such things as the water lines supplying their village. By protecting the turtles, they may have achieved a level of empowerment that will help them better their own lives. There are more reasons for local communities to protect their turtles than money.

On the Caribbean coast of Costa Rica, the Caribbean Conservation Foundation's (CCF) Sea Turtle Migration-Tracking Education Program is designed to involve the people around Tortuguero National Park in conservation. Its satellite-tagging operations have become a village event. In 2000, the program tagged six green turtles and two hawksbills. The local schools held contests to name the turtles; Elder Esposito, a student in the third grade, named one of the greens *Mariposita del Mar* (Little Butterfly of the Sea). Tourists in the nearby ecolodges came to watch the festivities, and local and international press arrived to broadcast the details.

After the tagged turtles set off, their

A young girl helps release hatchling turtles on the coast of El Salvador in October 1997.

travels were plotted on the Internet (at *http://www.cccturtle.org/sat20.htm*). Most of the green turtles headed for the Miskito Coast off Nicaragua, but Zenit (Zenith), a turtle tagged on July 19, 2000, swam as far as Belize in little over a month—and tracking its progress became a joint activity for schools in both Belize and Tortuguero.

On the Pearl Cays off the Caribbean coast of Nicaragua, the Wildlife Conservation Society (WCS) is also combining satellite tracking with community outreach. WCS researcher Cynthia Lageux and her colleagues released three satellite-tagged hawksbill turtles there on August 2 and 3, 2000, with a crowd of some 40 Nicaraguans looking on. The data from the transmitters will feed into a wider program studying the movements of hawksbills throughout the Caribbean, and at the same time nine local communities have agreed to protect hawksbill nests on the Pearl Cays.

In the Indian Ocean, satellite-tracking data from the Seychelles has helped conservationists there make a political point. As Jeanne Mortimer explained in 1999:

> In the past, some Seychellois complained that it was unfair and futile to expect the people of Seychelles to protect turtles that would only be slaughtered when they migrated from Seychelles to the national waters of other countries. Data from the tracking study, however, indicate that adult hawksbills nesting in Seychelles may remain within the territorial waters of Seychelles throughout their adult lives. As such, they are a resource belonging to the people of Seychelles, whose responsibility it is to ensure their long-term survival.

The government of the Seychelles banned trade in hawksbill shell products in 1994, and purchased and locked away 2.5 tons (2.27 metric tons) of raw shell. In November 1998, it took the extraordinary step of burning the entire stockpile—as part of a Miss World Pageant being held in the islands! By doing so, the government sent a message that it considers live hawksbills more valuable to the islands, as a tourist attraction, then dead ones as items of trade, and that turtle poaching would no longer be tolerated.

In recent decades, poachers robbed almost every single nest along the south coast of Sri Lanka, where the greatest concentrations of the island's turtles lay their eggs. In 1993, a Turtle Conservation Project began in Sri Lanka, focusing on Rekawa, a small village on the South Coast whose beach is a major rookery for five species of sea turtle: the loggerhead, olive ridley, hawksbill, green, and leatherback. Its aims were to gather information about the nesting sea turtles of the area and, probably more importantly, to carry the message of sea turtle conservation to the villagers. The project brought more than education to Rekawa—it brought help. It established nighttime "Turtle Watches" for paying visitors, with part of the fees going to a loan scheme for

Tourists watch a nesting leatherback (*Dermochelys coriacea*) on the protected beach at Matura, Trinidad.

members of the community. It has been providing English classes since 1994, and in 1998 it established a rural medical clinic.

The project has already begun to work, though it continues to depend on external funding. Following a series of workshops in local schools, over 450 pupils and teachers volunteered to help with beach surveys. Today, 17 former poachers are working with the project, and with government officials, to gather biological data and to protect turtle nests on the beaches. Project director Thushan Kapurusinghe told an international conference in 1999 that, where once not a single turtle egg survived, 98,198 hatchlings reached the sea under the watchful eyes of the project and its volunteers.

Back in the Caribbean, one of the most successful sea turtle conservation efforts began as a village initiative on the East Coast of Trinidad. Nature Seekers Inc., which started out in 1990 as a tour guide service for visitors to the leatherback turtle beach near the village of Matura, has become a model for local conservation organizations. It has won seven ecotourism and conservation awards, including enrollment in the Global 500 Award Roll of Honor in 1993.

Nature Seekers was born when the government closed off the nesting beach near Matura. The villagers, who had once depended on the turtles for meat, negotiated with the government for access to the beach. The government granted their request, but only on condition that they become protectors of the turtles. What started as a necessity soon became a matter of pride. Nature Seekers has converted poachers, educated other villagers, developed a locally run tourist industry and craft market, and, in short, has made conservation a way of life in Matura. Today, it runs regular beach patrols, and, as a member of WIDECAST, plays a part in the broader management of Caribbean sea turtles. Nature Seekers even sent a representative, Solomon Aquilera, to the 2000 CITES meeting, where he was an eloquent opponent of Cuba's hawksbill proposals.

Local organizations, indeed, have developed not only pride in their turtles, but some degree of political clout in their protection. When one of the most important loggerhead and green sea turtle nesting beaches in Mexico, at Xcacel in Quintana Roo, was sold to hotel developers in 1997, members of the local community joined international organizations and sea turtle biologists to fight the Mexican government. In April 2001, the Mexican Environment Ministry succumbed to the pressure, annulled the hotel project, and agreed to a public review.

Stories like these may not be the general rule, and perhaps it is too early to say if they represent a trend that may turn the tide that runs against sea turtles. What we can say, though, is that enthusiasm for sea turtles, and an interest in their conservation, crosses cultures and bridges economic gulfs. It is felt in the developed countries of the north, on the shores of the Indian Ocean, and on islands in the Caribbean. Surely we can derive some hope from that.

# *Bibliography*

For reasons of space, the full bibliography for *Survivors in Armor* could not be included in the book itself. For those interested, I have posted the full list of references I consulted while writing this book on the internet at http://members.home.net/ornstn/Bibliography.htm. Those unable to access this site can write me at 1825 Shady Creek Court, Mississauga, Ontario, Canada L5L 3W2, or send an email to ornstn@home.com.


Alderton, David. *Turtles & Tortoises of the World.* New York: Facts on File, 1997.

Bartlett, Richard D., and Patricia Bartlett. *A Field Guide to Florida Reptiles and Amphibians.* Houston: Gulf Publishing, 1999.

Bjorndal, Karen A., ed. *Biology and Conservation of Sea Turtles.* rev.ed.. Washington, D.C.: Smithsonian Institution Press, 1995.

Boycott, Richard C., and Ortwin Bourquin. *The South African Tortoise Book: A Guide to South African Tortoises, Terrapins and Turtles.* 2nd ed. Hilton, South Africa: O. Borquin, 2000.

Branch, Bill. *Field Guide to Snakes and Other Reptiles of Southern Africa.* 3rd rev. ed. Sanibel Island, FL.: Ralph Curtis Books, 1998.

Campbell, Jonathan A. *Amphibians and Reptiles of Northern Guatemala, the Yucatán, and Belize.* Animal Natural History Series ; Vol. 4. Norman, OK: University Of Oklahoma Press, 1998.

Cann, John. *Australian Freshwater Turtles.* Singapore: Beaumont, 1998.

Carr, Archie Fairly. *Handbook of Turtles: The Turtles of the United States, Canada, andaja California.* Comstock Classic Handbooks. Ithaca: Comstock Pub. Associates, 1995.

—. *So Excellent a Fishe: A Natural History of Sea Turtles.* New York: Scribner's, 1984.

*Cites Identification Guide — Turtles and Tortoises.* Ottawa: Minister of Supply and Services, 1999.

Conant, Roger, and Joseph T. Collins. *A Field Guide to Reptiles and Amphibians: Eastern and Central North America.* The Peterson Field Guide Series: 12. 3rd ed. Boston: Houghton Mifflin, 1991.

Ernst, Carl H., and Roger W. Barbour. *Turtles of the World.* Washington, D.C.: Smithsonian Institution Press, 1989.

Ernst, Carl H., Jeffrey E. Lovich, and Roger W. Barbour. *Turtles of the United States and Canada.* Washington, D.C.: Smithsonian Institution Press, 1994 (2000).

Froom, Barbara. *The Turtles of Canada.* Toronto: McClelland and Stewart, 1976.

Gibbons, J. Whitfield. *Life History and Ecology of the Slider Turtle.* Washington, D.C.: Smithsonian Institution Press, 1990 (2000).

Glaw, Frank, and Miguel Vences. *A Fieldguide to the Amphibians and Reptiles of Madagascar.* 2nd ed. Bonn: Zoologisches Forschunginstitut und Museum Alexander Koenig, 1994.

Harding, James H. *Amphibians and Reptiles of the Great Lakes Region*. Great Lakes Environment. Ann Arbor: University of Michigan Press, 1997.

Harless, Marion, and Henry Morlock. *Turtles: Perspectives and Research*. Malabar, FL: Krieger Publishing, 1989.

Klemens, Michael W., ed. *Turtle Conservation*. Washington, D.C.: Smithsonian Institution Press, 2000.

LeBuff, Charles R. *The Loggerhead Turtle in the Eastern Gulf of Mexico*. Sanibel, FL.: Caretta Research, 1990.

Lim Boo Liat, and Indraneil Das. *Turtles of Borneo and Peninsular Malaysia*. Kota Kinabalu, Borneo: Natural History Publications (Borneo), 1999 (2000).

Lindsay, Charles. *Turtle Islands: Balinese Ritual and the Green Turtle*. New York: Takarajima Books, 1995.

Lutz, P.L., and J.A. Musick, eds. *The Biology of Sea Turtles*. Boca Raton, FL: CRC Press, 1996.

McNamee, Gregory, and Luis Alberto Urrea. *A World of Turtles: A Literary Celebration*. Boulder, CO: Johnson Books, 1997.

Mitchell, Joseph C. *The Reptiles of Virginia*. Washington, D.C.: Smithsonian Institution Press, 1994.

National Research Council, ed. *Decline of the Sea Turtles: Causes and Prevention*. Washington, D.C.: National Academy Press, 1990.

O'Keefe, M. Timothy. *Sea Turtles: The Watcher's Guide*. Lakeland, FL: Larsen's Outdoor Pub., 1995.

Pilcher, Nicolas, and Ghazally Ismail, eds. *Sea Turtles of the Indo-Pacific: Research, Management and Conservation*. London: ASEAN Academic Press, 2000.

Pough, F. Harvey, et al. *Herpetology*. 2nd ed. Upper Saddle River, NJ: Prentice Hall, 2001.

Powell, Robert, Joseph T. Collins, and Errol D. Hooper. *A Key to Amphibians and Reptiles of the Continental United States and Canada*. Lawrence, KS: University Press of Kansas, 1998.

Pritchard, Peter C. H. *Encyclopedia of Turtles*. Neptune, NJ: TFH Publications, 1979.

Pritchard, Peter C. H., and P. Trebbau. *The Turtles of Venezuela*. SSAR Contrib. Herpetol. 2: 1984.

Ripple, Jeff. *Sea Turtles*. Stillwater, MN: Voyageur Press, 1996.

Rudloe, Jack. *Search for the Great Turtle Mother*. 1st ed. Sarasota, FL.: Pineapple Press, 1995.

—. *Time of the Turtle*. New York: E. P. Dutton, 1989.

Van Abbema, J., ed. *Proceedings: Conservation, Restoration, and Management of Tortoises and Turtles—An International Conference*. State University of New York, Purchase: New York Turtle and Tortoise Society, 1997.

Walls, Jerry G. *Tortoises: Natural History, Care and Breeding in Captivity*. Neptune City, NJ: TFH Publications, n.d.

Zug, George R. *Herpetology: An Introductory Biology of Amphibians and Reptiles*. 2nd ed. San Diego: Academic Press, 2001.

# Index

# *Photo Credits*

**Banfi/Innerspace Visions**, 228
**R. E. Barber**, 86 (top), 115 (bottom), 238, 246, 248 (left), 268 (left) 272 (right) 283 (left);
**Peter Bennett and Ursula Keeper-Bennett**, 274 (right);
**Canadian Press Archives**, 281, 290
**John Cann**, 1, 4 (upper right), 6, 13 (bottom), 18, 43, 48, 51 (top), 52, 53, 54, 55, 56 (top), 81 (top), 117, 129 (bottom), 131, 139, 141, 143, 147, 150, 152 (left), 159 (upper right), 172, 175, 179 (top left and bottom), 185, 187, 190 (left and right), 191, 198 (left) 199, 200, 202, 243, 244, 249, 277, 282
**Indraneil Das**, 11 (left), 15 (bottom), 16 (left), 64, 65, 70, 71, 72, 73, 89 (left), 92, 93 (lower right), 95, 104, 108, 126, 142, 152 (right), 155 (left), 158 (right), 159 (left), 176 (upper right), 178 (bottom), 236, 240, 253, 259, 262, 265, 269
**Ivor Fulcher**, 283 (right); **François Gohier**, 106, 115 (top), 120 (bottom left), 121 (left) 144, 173
**Ryan Hagerty/US Fish and Wildlife Service**, 221
**Anne Heimann**,116, 176 (top left), 270 (left), 271 (right), 275, 278, 279
**Paul Humann**, 287
**Frank Ippolito**, 35, 36 (right) 39, 44, 45 (bottom)
**Marilyn Kazmers/Innerspace Visions**, 226 (bottom)
**Maris and Marilyn Kazmers/SharkSong**, 135 (bottom), 213, 217 (left), 231 (left), 289
**W. B. Love**, 10, 13(top), 51 (bottom), 56 (bottom), 57, 59 (bottom), 61, 75, 76, 81 (bottom), 83, 90 (right), 91, 93 (left, top and bottom), 97, 100, 101, 102, 105, 120 (bottom right), 124 (left), 160, 170, 176 (bottom), 212 (right), 241 (left), 257, 261, 273
**John Matthews**, 86 (bottom), 119, 179 (bottom right)
**Colin MacRae**, 25, 27 (left), 36 (left)
**Daniel McCulloch/Innerspace Visions**, 266
**Michael Patrick O'Neill/Innerspace Visions**, xi
**Ronald Orenstein**, 9, 16 (right), 99, 255
**Doug Perrine/seapics.com**, x, 2 (left), 67, 68, 114 (top), 205, 208, 209, 210, 215, 227 (right), 231 (upper right), 232 (upper right), 233, 270 (right), 271 (left), 274 (left), 288
**Doug Perrine/Innerspace Visions**, ii-iii, xii, 21, 69, 112, 118, 128, 134, 135 (top), 204, 211, 212 (left), 214, 216, 217 (upper and lower right), 219, 222, 226 (top), 227 (left), 231 (right), 232 (bottom), 234
**Peter C. H. Pritchard/Innerspace Visions**, 219 (bottom), 268 (right)
**Jeffrey Rich Nature Photography**, 80 82 (bottom), 164, 285
**Eda Rogers**, 111, 120 (top, right), 124 (right), 178 (top), 179 (upper right); James P. Rowan, 23, 58, 84, 109, 168, 237, 272 (left)
**Andre Seale/Innerspace Visions**, 207 (bottom)
**Kay Shaw**, 230
**A. B. Sheldon**, 3, 4, 5 (left and right), 11 (right), 14 (top), 17, 19 (left and right), 20, 79, 82 (top), 87, 88, 90 (left), 94, 117 (bottom), 121 (right), 122, 127, 129 (top), 132, 140, 144, 149, 151, 153, 157, 158 (left), 162, 163, 166, 167, 169, 177, 183 (left, top and bottom), 186, 194, 197, 201, 202, 239, 241 (right), 242, 248, 254, 260, 261 (upper right)
**Dennis Sheridan**, 2 (right), 59 (top), 89 (right), 103, 107, 155, 159 (lower right), 198 (right) 261 (bottom right), 264
**Rob and Ann Simpson**, 14 (bottom), 15 (top), 77, 114 (bottom), 125, 136, 138, 146, 291
**Gary M. Stolz**, 8
**Masa Uchioda/Innerspace Visions**, 225, 286
**Ingrid Visser/Innerspace Visions**, 207 (top)

Illustrations:

**Michael de Braga**, 29 (top)
**D. W. Miller**, 24
**David Peters**, 29 (bottom), 32 (right), 33, 45 (top), 46
**Colin South**, 183 (right, top and bottom)
**Dan Varner**, 32 (left), 40, 41

All maps created by **Lightfoot Art and Design**
All diagrams by **Jean Peters**